Studies in Big Data

Volume 69

Series Editor

Janusz Kacprzyk, Polish Academy of Sciences, Warsaw, Poland

The series "Studies in Big Data" (SBD) publishes new developments and advances in the various areas of Big Data- quickly and with a high quality. The intent is to cover the theory, research, development, and applications of Big Data, as embedded in the fields of engineering, computer science, physics, economics and life sciences. The books of the series refer to the analysis and understanding of large, complex, and/or distributed data sets generated from recent digital sources coming from sensors or other physical instruments as well as simulations, crowd sourcing, social networks or other internet transactions, such as emails or video click streams and other. The series contains monographs, lecture notes and edited volumes in Big Data spanning the areas of computational intelligence including neural networks, evolutionary computation, soft computing, fuzzy systems, as well as artificial intelligence, data mining, modern statistics and Operations research, as well as self-organizing systems. Of particular value to both the contributors and the readership are the short publication timeframe and the world-wide distribution, which enable both wide and rapid dissemination of research output.

** Indexing: The books of this series are submitted to ISI Web of Science, DBLP, Ulrichs, MathSciNet, Current Mathematical Publications, Mathematical Reviews, Zentralblatt Math: MetaPress and Springerlink.

More information about this series at http://www.springer.com/series/11970

Martin Holeňa · Petr Pulc · Martin Kopp

Classification Methods for Internet Applications

 Springer

Martin Holeňa
Institute of Computer Science
Czech Academy of Sciences
Prague, Czech Republic

Petr Pulc
Czech Technical University
Prague, Czech Republic

Martin Kopp
Czech Technical University
Prague, Czech Republic

ISSN 2197-6503   ISSN 2197-6511   (electronic)
Studies in Big Data
ISBN 978-3-030-36964-4   ISBN 978-3-030-36962-0   (eBook)
https://doi.org/10.1007/978-3-030-36962-0

This Springer imprint is published by the registered company Springer Nature Switzerland AG
The registered company address is: Gewerbestrasse 11, 6330 Cham, Switzerland

Preface

This book originated from an elective course called *Internet and Classification Methods* for master and doctoral students, which has been taught since the academic year 2013/14 at the Charles University and the Czech Technical University in Prague. The course is intended for students of the study branches Computer Science (Charles University) and Information Technology (Czech Technical University) and its main purpose is to make the students aware of the fact that a key functionality of several very important Internet applications is actually the functionality of a classifier. That functionality is explained in sufficient detail to remove any magic from it and to allow competent assessment and competent tuning of such applications with respect to that functionality. We expect, and the first years of teaching the course confirm it, that this topic is particularly interesting for those who would like to develop or to improve such applications.

The Internet applications we consider are:

1. *Spam filtering,*
2. *Recommender systems,*
3. *Example-based web search,*
4. *Sentiment analysis,*
5. *Malware detection,*
6. *Network intrusion detection.*

We consider them very broadly, including topics that are even loosely related to them, as long as the classification functionality is relevant enough. For instance, we consider also example-based search within pictures and other non-textual modalities of data because it is used in many recommender systems.

The above six kinds of applications are introduced in Chap. 1 of the book. However, they are not described thoroughly, as the students attending the course have other, specialized courses to this end. Similarly, the readers of the book are expected to be familiar with such applications already from elsewhere. We provide them with references to relevant specialized monographs or textbooks. On the other hand, we deeply discuss the classification methods involved, which are the focus of the remaining five chapters of the book, though it is also there illustrated on examples concerning the considered kinds of Internet applications. From the point of view of computer scientists, the classification is treated rather on a graduate than on an undergraduate level. And although the book does not use the mathematical style of definitions, theorems and proofs, all discussed concepts are introduced with full formal rigour, allowing interested readers to understand their explanations also in purely statistical or mathematical books or papers.

In Chap. 2, concepts pertaining to classification, in general, are discussed. In particular, classifier performance measures, linear separability, classifier learning (both supervised and semi-supervised) and feature selection. The chapter also addresses the difference between classification and two other statistical approaches encountered in the considered Internet applications, namely, clustering and regression. The introduced concepts are illustrated on examples from spam filtering, recommender systems and malware detection.

Chapter 3 gives a survey of traditional classification methods that have not been developed with the specific objectives of high predictive accuracy, nor comprehensibility. The chapter covers, in particular, k nearest neighbours classification, Bayesian classifiers, the logit method, linear and quadratic discriminant analysis, and two kinds of classifiers belonging to artificial neural networks. The methods are illustrated on examples of all considered kinds of Internet applications.

In Chap. 4, support vector machines (SVM) are introduced, a kind of classifiers developed specifically to achieve high predictive accuracy. First, the basic variant for binary classification into linearly separable classes is presented, which is then followed by extensions to non-linear classification, multiple classes and noise-tolerant classification. SVM are illustrated on examples from spam filtering, recommender systems and malware detection. In connection with SVM, the method of active learning is explained and illustrated on an example of SVM active learning in recommender systems.

The topic of Chap. 5 is classifier comprehensibility. Comprehensibility is related to the possibility to explain classification result with logical rules. Basic properties of such rules in Boolean logic and the main fuzzy logics are recalled. This chapter also addresses the possibility to obtain sets of classification rules by means of genetic algorithms, the generalization of classification rules to observational rules and finally the most common kind of classifiers producing classification rules, namely, classification trees. Both classification rules, in general, and obtaining rules from classification trees are illustrated on examples from spam filtering, recommender systems and malware detection.

Chapter 6 of the book deals with connecting classifiers into a team. It explains the concepts of aggregation function and confidence, the difference between general teams and ensembles and main methods of team construction. Finally, random forests are introduced, which are the most frequently encountered kind of classifier teams. The concepts and methods addressed in this chapter are illustrated on examples from spam filtering, recommender systems, search in multimedia data and malware detection.

In spite of its focus on the six kinds of Internet applications in which classification represents a key functionality, the book attempts to present the plethora of available classification methods and their variants in general: not only those that have already been used in the considered kinds of applications, but also those that have the potential to be used in them in the future. We hope that in this way, the influence of the fast development in the area of Internet applications, which can sometimes cause a state-of-the-art approach to be surpassed by completely different approaches within a short time, to be at least to some extent eliminated.

Prague, Czech Republic   Martin Holeňa
                                                                                      Petr Pulc
                                                                                   Martin Kopp

Acknowledgement Writing this book was supported by the Czech Science Foundation grant no. 18-18080S. The authors are very grateful to Jiří Tumpach for his substantial help with proofreading.

Contents

Chapter 1
Important Internet Applications of Classification

1.1 Spam Filtering

The word *spam*, earlier in internet history introduced as a quotation from Monty Python's Flying Circus [1], is today a generic term covering unsolicited or unwanted messages. The derived term *electronic spam* covers more specifically our scope of view—email messages sent usually without any personalization to huge list of recipients without explicit agreement between the sender and the recipients.

Spam is legislatively regulated in many countries, but different requirements have to be met due to different laws. For example, the explicit user agreement upon sending (opt-in) is very important in Europe and Canada, because the local legislation requires it [2, 3]. However, unsubscription instructions (opt-out) have to be included in nearly every message in the USA, required by the CAN-SPAM Act [4].

To some extent, the path of electronic spam can be compared to leaflets distributed by a nearby shop. Once the leaflets are printed out, they are handed to a distribution company and delivered to a list of addresses. But if you have some kind of "No Junk Mail" sticker on your postbox and the distributor has no bad intention, the leaflet will be returned to the shop.

The difference is that sending a huge amount of emails to the distribution network is nearly for free and the simple sticker is replaced by a rather complicated set of rules and evaluations. But also this replacement works mainly if the senders of electronic spam do not have bad intentions. As a matter of fact, they usually have.

From the beginnings of spam filtering, spammers and bulk email senders are trying really hard to get around that protection. To the more recipients they get and the more people will not recognize their spam at first glance, the better. It may have been safe to detect spam by its language some time ago if different from your native one, but this does not count today. Some attempts of automatic translations are still hilarious, but what makes smile freeze is that they are getting better. The personalization of spam is another great problem that we will have to face in the near future.

According to the Symantec Internet Security Threat Report from July 2019 [5], the global spam rate is 55%, in some countries up to 67%. And still, an average

© Springer Nature Switzerland AG 2020
M. Holeňa et al., *Classification Methods for Internet Applications*,
Studies in Big Data 69, https://doi.org/10.1007/978-3-030-36962-0_1

project leader spends half or more of the shift skimming through emails. Can you then imagine life without spam filtering?

1.1.1   The Road of Spam

Actually, sending a spam costs something: a computer running time and a decent internet connection. Because of that, spam rarely origins from the machine of a spammer. More often, using viruses or other known vulnerabilities of computers, the same message is sent from a set of infected computers, sparing the costs and hiding the IP address of the spammer. As a bonus, the spammer can possibly get access to contact list of the infected user, if the mail client is infected. This means more possible targets that are verified to exist.

Even if spammers will use only their computers, it can be very well assumed that the IP address was covered by some VPN and the like, and is thus untraceable.

After the message enters the Internet, the main goal of all systems it interacts with is to deliver it as fast as possible to the recipient. Mail transfer agents relay the message to a final destination server and a mail delivery agent stores the message into the recipient's account. If users access their email box through a web interface, the email processing ends here. However mail user agents (or mail clients) can be used to download messages to the user's machine.

This provides three possible places, where the filtering process can be employed.

Mail Transfer Agents, referred to as SMTP servers because SMTP is the protocol used, would be ideal blockers for spam messages if the message could be simply deleted. This would not introduce unwanted traffic in the first place. But sadly, this cannot be done. It may happen that the message is incorrectly considered as a spam and then an important message could be lost for good.

An alternative is to only mark message as a spam and forward it anyway. But if this server is not yours, will you trust it? On the other hand, message can bounce off multiple mail transfer agents and if the passed messages are not directly connected to you and your users, why would you spend processing time on spam filtering? Therefore this approach makes really sense only in a business environment, on the border of an enterprise network.

Mail User Agents (Mail Clients) are on the other side of spectrum. You, as a user of your software, can select which spam filters will be applied. And possibly learn and tweak them as much as you want. The most personalized and computer-intensive filtering can be applied here. Also, if you are keen on privacy, this is the only way how to possibly ensure that you are the only one analysing your messages.

But your client will usually have only a small set of pre-programmed patterns. No clue about current trends or other user spam message information. And you will possibly need to filter your messages on all devices, for example on your mobile devices.

Mail Delivery Agent therefore seems to be the best place for spam filtering nowadays. The spam filters can be learned on directly accessible datasets of multiple users, and it will not be a duty of the users to have installed and up-to-date spam filter any more. On the other hand, users usually cannot modify the filter in other way than correcting its false decisions.

1.1.2 Collaborative Approach

To overcome the problem of small datasets, the collaborative filtering approach has been developed. If a user marks an email as spam, a signature of it is recorded and stored in shared knowledge base for further message filtering. This way, even administrators of smaller Mail Delivery Agents or individual users can filter spam efficiently.

However this makes the strong assumption that all users will consider the same messages as spam. Therefore, [6] introduces the idea of personalised, yet collaborative spam filtering called CASSANDRA (Collaborative Anti-Spam System Allowing Node-Decentralised Research Algorithm).

Another collaborative approach uses social networks build on top of email addresses [7]. It is assumed that in ordinary communication, a standard user will communicate a lot with a narrow group of users and much less with people outside that group. However, spammers will simply send the same messages to all users every time.

1.1.3 Spam Filters

An email message consists of following parts, differently related to spam filtering:

Header contains information about the sender, the path across mail transfer agents to the recipient and other fields.

Body contains the actual text of the message, typically either plain text or HTML formatted.

Attachments are actually stored in the body of the email message as other sections of the multipart document. For the sake of simplicity, however, we will consider non-HTML attachments a separate category.

When considering what parts of email messages to analyse in a spam filter, we have basically three possibilities, shown in Fig. 1.1. Either we use properties of the whole message, or separately the header and/or the body. The contents of each can be then either treated as a set of tokens (usually as space-delimited words), or some semantics can be assigned to each part.

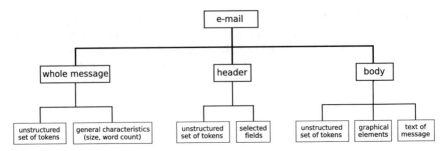

Fig. 1.1 Possible message parts for spam filter analysis

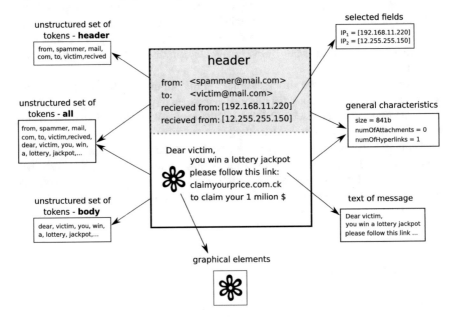

Fig. 1.2 Example of message data that can be used by a spam filter

Thus we can base our analysis on general properties of the message (eg. size, number of attachments), individual header fields (e.g. IP addresses, To and From fields) or properties of message body (e.g. distribution of words). An example of such data is shown in Fig. 1.2.

A spam filter can thus make use both of content-based and of non-content features of the message.

1.1.3.1   Content-Based Features

Possibly the most obvious way of detecting spam is by scanning message body (or text-based attachments) for certain words or patterns. With some extensions, this approach can detect also messages with obfuscated text, text hard to tokenize, mis-spelled words or synonyms. Some spamming techniques however involve appending spam message to longer regular text and thus distorting statistical properties of such a document.

1.1.3.2 Non-content Features

In combination with content-based features, many other message properties can be considered in message classification.

For example, message sender, origin IP or the path through the Internet can be considered and matched against lists kept in the system—blacklist, whitelist or greylist.

If an IP address is blacklisted, all traffic is marked as spam or blocked. Whitelist is the mere opposite, these entries are considered as trusted. These lists are usually distributed alongside with the spam filter to eliminate a learning period. Greylisting usually works as a temporary gate—the first mail is intentionally blocked to check if another try for delivery will be made. This basically filters well designed SMTP servers from the forged ones.

Also a discontinuity in SMTP path or evidence of changed header is considered as a security violation and makes email more likely to be marked as spam. Generally, a presence of some fields in the message header can lead to easy spam recognition, as they can differ a lot between spam and legitimate email.

Other non-content based features can be extracted from attachments. The presence of certain file types, scripts, masked content, or their size and properties can be considered in spam filtering.

1.1.4 Image Spam

On the beginning of the millennium, spam started to be delivered also in the form of one or more images, either attached to or inserted directly in an otherwise empty message. This was a short-time victory above spam filters, as they were not pre-pared to classify images or text contained in them. Another problem was that spam containing images is usually much larger and thus uses even more resources.

In the spirit of "Use a picture. It's worth a thousand words.", image can contain a lot of information. Image processing thus returns usually huge amounts of low-level information and is quite expensive. The extraction of high-level information suitable for spam filtering is still a matter of ongoing research. However, image spam usually contains some form of textual information, more or less cleverly hidden inside the image.

Therefore, an optical character recognition (OCR) capability has been added to spam filters and the recognized text is now analysed alongside with the text from email body. But OCR will not always cut it. Such a text can be distorted, covered in noise or can use other intricate ways to make it readable by people, but difficult for a computer recognition procedure. Pretty much the same way as CAPTCHA (Completely Automated Public Turing test to tell Computers and Humans Apart) is used by computer to check whether a human user is interacting with the software.

OCR needs a lot of resources. If we want to extract distorted or handwritten text dissimilar to known set of computer fonts, the problem can be up to unsolvable. Therefore, instead of recognising text letter by letter, some other high-level information can be extracted if we have a clue that some text is present in the image. For example, text position, orientation, font size, whether any known text obfuscation has been used, or the proportion of image taken by text.

But there is also a possibility, that a spam image will not contain any text. And it does not make sense to run OCR on images without text as it will return gibberish. But we should be able to classify such images as well. In these cases, we can make use of image saturation, colour and gradient distribution and heterogeneity, file format, size, aspect ratio, compression, etc. In [8], the image spam has been detected by means of image dimensions, aspect ratio, file type, size and compression.

However, spammers are always one step ahead and use for example randomized image cutting to confuse automated recognition.

1.1.5   Related Email Threats

Closely related to spam is also phishing, as its main medium of delivery is also email. The name originated as homonym of fishing, because both use a bait to capture the attention of the prey. In the case of phishing, this means to capture users' secure or personal information. To quote the Kaspersky Lab's quarterly spam report for the third quarter 2018 [9]: "In Q3 2018, the Anti-Phishing system prevented 137 382 124 attempts to direct users to spam websites 12.1% of all Kaspersky Lab users were subject to attack."

Basically, this security thread exploits client trust in presented message, which looks usually very official and trustworthy, however uses links to forged websites that gather user names, passwords, numbers of important document (passport, ID, driving licence, credit card) and other confidential information from users. Such information can be then used for user identity theft. Separate security application or specialized spam filter can recognize these false links and at least warn the user, or disable those links completely.

1.1.6   Spam Filtering as a Classification Task

From this brief overview of spam filters, it should be clear that their main functionality is discrimination, based on various features of email messages, between different email categories, typically between spam and regular mail (by analogy to spam sometimes called "ham"). Hence, the filter must be able to assign each feasible combination of values of the considered features to one of those categories. Mathematically, that functionality is accomplished by a mapping of the set of feasible combinations of values of features into the set of email categories. The construction and subsequent use of such a mapping is the usual meaning of the concept of classification, and the mapping itself is then called classifier.

1.2 Recommender Systems

Already in the past, people tended to ask for recommendations. What to buy, where to go, what to do, etc. Either to gather information about yet unknown field or to confirm it. In any case, many people don't want to base their selection only on their own opinion. And so it is convenient to ask for a recommendation.

The same reason, however in much larger scale and with influence to possibly many people, can be found also in large corporations. If the board is unable or not willing to make a decision on their own, a consultation agency is invited—to make a recommendation.

The real urge for developing a system that would recommend next action or alternate product emerged at the end of 20th century, when many markets got much better connected and much more goods were available. Customers started to be faced with so many options that the process of choosing was difficult or even impossible. Psychology recognises such state as a "choice overload" or simply "overchoice" [10].

Although the introduction of the Internet, and successively internet search, online shopping and price comparison sites, have somewhat helped the user to find a best deal for a specific product, other web portals enabled everyone to buy or sell anything, at any time and from any place. Therefore, the initial selection of goods or services is in fact increasingly complicated.

To help in such situations, recommender systems have been proposed—software tools that recommend certain choices based upon the collected information.

The first studied case of feature-based recommender systems revolved around cinema [11]. It was assumed that people with similar taste would like to see similar movies. In the simplest case, if there is a person who had watched the same movies as you did, as well as some additional ones, there is a possibility that you would like to see these additional movies as well.

Soon after, advertising companies started to be more focused to smaller groups of people with higher probability of purchase (or more broadly, key conversion[1]). Selection of advertisements on the Internet, selected upon the same principle of collaborative filtering (see Sect. 1.2.3), have been even patented in year 1999 [12].

---

[1]Conversion in general is usually defined as any action that has been taken by the customer based on a given offer. For example submitting a form, scanning a coupon or subscribing to a newsletter. Key conversion presents the ultimate of these actions leading to fulfilment of designed goal. For example purchase, order or visit.

On the Internet, offers and advertisements are currently almost personalised to reflect needs and wishes of the potential customer and subsequently increase the probability of conversion. And recommendation systems are tightly connected to such customisation.

1.2.1   Purpose of Recommender Systems

Recommender systems are commonly used examples of artificial intelligence; for common Internet users, they might be also the most visible. In this section, we will mention only few examples how recommender systems can be used.

E-commerce

Because a confused or puzzled visitor will almost certainly *not* place an order, the major purpose of recommender systems in e-commerce is to propose the visitor the precise goods and services they are looking for as soon as possible.

To achieve such goal, both the auction and direct sale portals (Amazon, eBay, AliExpress, or any other e-shop) use a similar tactic as regular stores. They place the most wanted, greatly discounted or otherwise attractive goods and services into the stores window—the entry page—as "Top Picks" or "Hot" product categories. This way, both the looks of displayed goods and introduction of only a limited number of items at a time creates craving to go shopping.

However, regular shops can only assume what is likely to be sold in general and propose a single set of items. E-shops have, on the other hand, a virtually unlimited possibility to customise the entry page for each incoming customer. Existing customers can be presented with goods and services that can complement their previous purchases or items they might like. But even newcomers may reveal through their browser an estimated location, preferred languages and a referral URL from where the visitor came, and thus providing some information for a customisation.

More recommendations can be given on the product detail page, which usually contains a list of related products. Both to provide possible alternatives or products complementary to the currently shown.

Content consumption

Media consumption sites (Google News, YouTube, Reddit, and local news, amongst many others) recommend not only the most current content, but also content popular in the long term. To do so, they have to analyse the content for similarities, as well as monitor the behaviour of users. Either to recommend further reading on a current topic, or to recommend further reading for specific or unspecified user in general.

Multimedia is also commonly used for pure entertainment (for example via Netflix, YouTube, Vimeo, Pandora Radio, Last.fm, Google Play, Audible, Books at Amazon). Because these services usually need a sign-up or even subscription, basic user profile information can be gathered during registration. During use of the service, the basic user profile can be also complemented by information about previous consumption, reviews, likes, comments or social interactions. Based on collected data, the system then recommends what to watch/read/listen to as the next.

Social networks

Direct recommendations can be also given on social networks (Facebook, Twitter, LinkedIn, eHarmony and if we are vague enough also Digg, Del.icio.us, YouTube and many others) as hints whom to follow/befriend/meet/employ/subscribe to or generally spent time with. However, as social networks work with specific data, we will discuss them separately in Sect. 1.2.5.

Advertising

A bit more hidden recommender systems are used in advertising. To attract new people to the store, merchants use advertisements to highlight their brand and subsequently promote sales. However, the strategy needs to be properly selected. Very wide advertising to many people during a relatively short period may boost sales for a short period of time, however usually has no long-term effects. Focusing advertisements to much smaller interest groups, which the Internet enables with very small granularity, is assumed to be much more efficient.

One possible way to limit the amount of addressed users is to set-up display of the ad only if certain keywords appear along. This approach is still used (Infolinks, Kontera, Etarget), but as it depends on shown text, it may not completely correspond to the individual user.

Advertisement services (such as Google AdWords) therefore like to track the behaviour of the user. They can do so on all sites with their advertisements. Larger companies that usually posses also a search engine or a social platform then base the provided advertisement also on the behaviour of the user on their site. And such companies also may sell their collected information to others. So, if you have searched recently for any given product or service, expect to get a lot of advertisements connected to such items in next few weeks.

Recommenders of proposed actions

A different recommender system may be hidden in generic problem solving. When the user is faced with a problem to solve, a plethora of solutions may emerge. Each with their possible disadvantages. Similar to the e-commerce use, the user may be paralysed with the overwhelmingly many possibilities, worried to not chose a wrong one. Such recommendation systems then tries to objectively weight all possibilities and propose the best one for a current scenario.

One of such problems may be an investment decision. As future outcomes are not known, rigorous analysis of any possible case is impossible and estimation based only on models may be insensitive to current events. Recommendation based on statistics and behaviour of other investors may be however a valid reason to choose certain strategy.

1.2.2 Construction of a Recommender System

The recommendation systems generally consist of three basic parts: (1) data collection, (2) generation of recommendations and (3) presentation of the results. For example, the construction of YouTube recommender is briefly described in [13].

The data collection part is responsible for gathering large amounts of data inserted into the system or generated by users' interaction with the system. For example, the new content, ratings, comments, likes, other social interactions, visits of individual web pages, time spent on these pages, viewed content, mouse movement and many other events can be collected. Even a moderately detailed capture may, therefore, result in an excessive amounts of data. As there is no possibility to process all events from all users, the resulting capture may be quite noisy, with false, outdated, incomplete or completely missing information. To be able to work with the logged data, this stage also needs to generate a smaller amount of signals, ideally few numerical values directly representing key properties of an inserted object or an engagement of the user.

Based on these collected data, recommendations can be generated either in advance or on request. Computation of recommendations in advance may use available processing time in moments of lower load and much more complicated data filters, which will be described in following subsections. On the other hand, such recommendations are never up-to-date for frequently changing or used content, and take-up significant storage space. On-request deduction of the recommendation from on-line data needs to use parallel processing and map-reduce engines with very simple operations, which reduces the overall possibilities of such recommendation system.

The resulting recommendations are then presented to the user. Because the filters usually rely on some similarity measure (typically distance-based or correlation), we are able to rank the individual recommendations according to that similarity, and show them as an ordered list. This part of recommendation system can also gather a direct feedback to proposed recommendations. For example, if the user often selects not the first proposed option, their profile in the system may be changed or even a retraining of internal parameters or models may be executed to enhance the future recommendations.

The iterations of the recommender system can be also made very short, so that it resembles a conversation. More on the conversational and critique-based recommender systems can be found in [14].

1.2.3 Content Based and Collaborative Recommenders

There are two distinct ways how to deduce a recommendation. The first approach is based on filtering the objects themselves. Processing their data or metadata reveals a similarity between individual items and thus the closest matches to another item or user profile can be returned as recommendations. The second approach, called

Fig. 1.3   Detail page of the e-book *Pride and Prejudice* on Amazon book store, as of June 2016

collaborative filtering, uses data collected from the behaviour of many users in connection with the considered items. In this way, the items are recommended to the current user based mainly on their actions in the past. Both of these approaches are commonly combined into a hybrid recommender.

Based on sources of the data, other classes of recommender techniques have been proposed [15–17]—such as demographic, utility-based or knowledge-based. However, as we will later discuss only the consequences for classification, the classes mentioned in previous paragraph will suffice.

A use of both of these basic approaches can be shown on an example of online bookstore. Amazon, for example, shows two lists of recommended books right on the beginning of product detail page (see Fig. 1.3). The upper list displays books that were purchased alongside the current one. In case of Pride and Prejudice by Jane Austen (a free e-book), this list consists of other novels by Jane Austen and some other, possibly content-related, but mainly free or cheap books that suggest a bundle purchase. As such, this is an example of collaborative filtering. The second list on the bottom shows a list of sponsored products (thus technically advertisements) that are related to the current item content- or metadata-wise. For example, the first book—Find You in Paris—claims to be inspired by the Pride and Prejudice.

However, Amazon does not stop here. Under reviews, there is a list of items actually purchased after viewing this one, recommending possible alternatives of the product. And at the very bottom of the page is the list of personalised recommendations based on viewing history of the current user. The first one is yet another example of collaborative recommendation, the later is an instance of hybrid recommender, as it combines filtering based on content of the books with usually a very little information about the behaviour of the current user.

1.2.3.1   Content Filtering

Recommendations based on content uses the information contained in or associated with the individual items. In case of text documents, usually the presence of certain keywords or frequency of terms is used to create a description vector. Such information can be also enriched by already extracted information or other structured data.

In case of books, we have not only the full text and structured information about the author, but thanks to semantic web, we are also able to gather a lot of related information from many domains. The concept of semantic web is very simple: accessible data is transformed according to rules (ontologies) [18] into triplets of two objects and the relation between them. Transition across just one possible triplet from DBpedia as

$$\text{Book('Pride and Prejudice')} \rightarrow \texttt{Film}$$
$$\text{Book('Pride and Prejudice')}.\texttt{FilmVersion}= \texttt{Film}(\text{'Pride and Prejudice and}$$
$$\text{Zombies')}$$

will connect the book to a movie. Other ontologies, describing for example the Internet Movie Database (IMDb), can then connect the book to individual screenings of the movie. The recommender system can therefore propose also visiting a cinema.

Simple recommendations based on user-provided data can be also achieved by content filtering, as user profiles can be considered individual items in the object space. Such profiles, however, have to contain a lot of explicitly filled user data to work properly. This may not be a major issue for dating sites or job-seeking portals, where the user is highly motivated to have the profile as complete as possible. However, in case of books, movies or other goods recommendation, filling up of such profile would be highly impractical.

A great advantage of the content-based recommendation is, that the underlying data are not as dynamic. Books are a good example of immutable data: once they are printed and published, the content of the book will not change.

Extracted data and even recommendations can be therefore computed in advance, stored and indexed. The process of recommendation is then just a matter of search in a database.

The dependency on the available data is, however, also a great disadvantage. In some cases—as with images, audio or video—features naturally representing the content that can be directly used for deduction of distance or correlation may be hard to gather from raw data automatically. In such cases, only user generated metadata are usually available and we cannot expect to have high-quality metadata for all items.

1.2.3.2 Collaborative Filtering

The collaborative recommenders use the input data from many users (such as purchase history, reviews and store navigation history) to create a custom profile for each identified user and later filter and recommend options based on distances between such profiles. Users can usually enhance the creation of their profile as well, by providing structured information about themselves on their own will, or such information can be gathered automatically from social networks or other third parties. An agreement to use personal information is a common part of Terms and Conditions for the service (e-shop, forum or other), to which the user needs to agree upon registration.

However, sites may monitor users behaviour also without registration. Cookies, a tiny fragments of data stored in web browsers, enable a tracking of user. Once a user enters a site, a unique session identifier is generated and sent to the browser to remember. On each other visited page, this identifier is sent back to the site. By design, cookies shouldn't be shared across multiple domains. However, as advertisements from one agency, or the Facebook "Like" button are hosted always from the same domain, they are able to follow a single web browser, and thus possibly one user, across many web sites.

If we stick to the Amazon example, the profile of each existing user contains the list of books ever purchased, visited or added to wish list. The site also collects the ratings and reviews of the individual books. This way, each user can be represented as a point in a highly-dimensional space, where each dimension represents some relation between the book and a user (whether the user ever visited the detail page of the book, whether he added it into a wish list, whether he bought it, with how many stars he rated it, whether he wrote a review of it, etc.). As only small fraction of users will interact with the same books, collected information is very sparse. Transformation of the data can be then used to reduce dimensionality and allow an easier assessment of similarity between users.

With such a prepared space, a recommender system usually proceeds in three steps: (1) finding users with similar behaviour patterns to the current user (neighbours), (2) computing a similarity between neighbours and the current user and (3) generating the recommendation. The second step is crucial for quality of the recommendation, as the computed similarity is used for weighting recommendations from the individual neighbours. We will discuss two distinct approaches:

Rating-based approach, proposed for example in [19], requires less processing and provides good results when sufficient data is available. On the other hand, each behaviour pattern needs to be transformed into a single number. The most natural source of information for this approach is therefore the numerical rating of products by individual users. The Pearson correlation coefficient (defined in Sect. 2.5) is then used to deduce similarity between such users.

Preference-based recommender systems are based on the creation of user preference models, where known user preferences are extended to all items in the system. Such model can be therefore used for users with a very small set of preferences (rated products) or very little overlaps with other users. Preferences are usually represented as all possible ordered lists of all products. Similarity between the complete preferences are then, according to [20], computed by averaging Spearman's ρ (also defined in Sect. 2.5) across all lists in the model.

With such information, behaviour patterns from the neighbours are weighted and recommended to the current user. In case of book ratings, for example, the ratings of neighbours are weighted by their correlation with current user. The resulting list of weighted ratings is sorted and first few, not yet purchased books are displayed as recommendations. This principle is, however, much more general and can be applied in many scenarios.

A great advantage of this approach is that collecting information about user activities and interactions is possible for any content. Even in case of pictures or very short video content with little to none textual information (Vines, 9gag), new content can be recommended based on user interactions as simple as visiting the item.

To propose better and more robust recommendations, the recommender system needs to gather a lot of information, possibly rich in content or semantics, such as ratings with numerical value. However, all new items and users start with an empty set. This major issue, known as "cold start", makes the collaborative recommender systems hardly suitable for the recommendation of new items, or to recommend anything to a new user.

The problem of cold start can be somewhat mitigated. When registering to a social network, at least the full name, birth date and gender is required to sign up, and the user is then asked to connect with at least few friends. When signing up to a professional network or job-seeking portal, even more information is required and user is challenged by some "profile completeness" gauge to fill as many fields as possible.

E-shops can then introduce "wish-lists" or possibility to "favourite" the products. This way, the system may possibly get some input information from the user even before the first real transaction.

But even after a significant time, not very much of usable information may be collected from users, as no-one will review all items and some items will get no reviews at all. And the more items and more users are in the system, the worse the problem is—basic collaborative recommendations have a poor scalability.

We are also building on top of an assumption, that users can be divided into distinct groups, and members of such groups have similar behaviour patterns. Some of the users may, however, be "grey sheep" with constantly changing membership in such groups, or even "black sheep" that does not belong to any. Recommendations for such users are intrinsically hard or impossible to deduce.

Collection of information from user behaviour can be also attacked by "shilling". This attack is based on overloading the recommender system by a lot of (positive or negative) actions with certain items. Usually, the recommender system has no (or very limited) way of distinguishing honest actions of real users from false or even automatically generated ratings, reviews, traffic, etc. The system then favours (or denies, respectively) some of the items in unfair manner. Such attacks and the development of defence against them are matter of ongoing research [21, 22].

1.2.3.3 Hybrid Filtering

The two approaches of collaborative and content based systems may be also combined into a hybrid recommender, that uses all available information. In reality, many of the existing recommender systems belong to this group. Personalised advertisements are, for example, based on the currently viewed page, but also on the information about lately visited pages, last search queries, similarities to behaviour of other users and either given or deduced user profile with age, gender, location and other personal data.

The aim of hybrid filtering is to overcome possible disadvantages by appropriate combination of methods. Either by running a separate recommender for each type of filtering and combining their results, or by combining both recommendation system types in one algorithm internally. The latter approach includes also recommender systems that are based mainly on one of both described types, but use features of the other type to overcome some deficiencies [23].

Another list of recommender hybridisation methods proposed in [24] includes:

- weighted combination of scores from several recommenders,
- switching between recommender techniques as needed,
- displaying results of multiple recommenders at once,
- combining of gathered features in one large recommender,
- refining results from one recommender by the other,
- using outputs from one recommender as the input to another,
- creation of a model by one recommender and passing it to the other.

1.2.4 Conversational Recommender Systems

A different kind of recommender systems is based on the principle of conversation, where the user queries the system with a vague definition of products or services, but is able to specify the requirements better throughout time, based on the presented choices. Conversational systems keep the whole context of the communication and earlier requirements do not need to be repeated.

Such systems also allow to use the context of the currently proposed item. The user is therefore able to express which aspects of the current offer suits him but which aspects should be changed in what manner for the next round of recommendation, such as: "Like this, but cheaper and bigger."

The recommendation therefore does not need to be perfect, as the user will redefine the query on every turn. And as users like to browse through limited and diverse catalogues, also the recommender system utilised in such conversation needs to be aware of the diversity of items it proposes [14].

1.2.5   Social Networks and Recommendations

Recommenders on social networks have, in addition to previously mentioned features, a very different source of background information—the actual graph of, sometimes even annotated, relations between individual users. A recommender based on information from social networks can therefore use a direct distance between friends and colleagues. Groups formed on the social network then usually join together people with similar interests or goals.

Profiles of the users contain a lot of personal data, and are frequently updated. The users do so not only to be discoverable, but because they want to share personal news with friends. This creates an excellent opportunity for recommendation systems to use such profiles for content-based recommendations, such as finding people living nearby, or classmates from same high-school.

The recommendations based on a social graph are therefore relatively easy to gather. If many of our friends have one common friend that is not in our friend list, it may be safely assumed that we would like to be friends of such a person as well. The same assumption can be applied to events or participation in groups. However, the main problems are caused with the high amounts of data being constantly pushed into the social network and dangerously high "WOW" factor caused by very high popularity of very few items in a short time. For answers to such challenges, new approaches needed to be developed [25].

With the help of both profile and social information, advertisements can be very precisely targeted to narrow groups of users. However, if the shown content is similar to a banner ad, it also has a similar effect. As social networks are primarily based on sharing of content and serve as a communication tool between individual people, simple announcements from advertising agencies with no engagement potential are unsuccessful. Even if the posts try to pretend content- and language-wise to be a post from a friend.

On the other hand, social networks and social media seem to be successful in case of brand self-propagation amongst its followers. As this model of one-to-company communication on a human level strongly resembles a casual talk between peers, people feel more engaged and tend to spread the (possibly commercial) announcement to other friends of their own by directly sharing the story, or by performing any other action later processed by the recommender.

1.2.6   Recommender Systems and Mobile Devices

Better affordability, higher processing power and increased use of mobile devices had several major consequences for recommender systems. As discussed in [26], users of mobile devices have two exclusive features: they are ubiquitous (they can be accessed easily by notifications or other messages in real time) and localizable (their physical position can be deduced and used as a valuable information source).

With the omnipresence of data connection through Wi-Fi or cellular networks, exchange of information required for recommendations can be done at any time. For example, the social networks for drivers are able to recommend a different road in case of traffic accident or any other congestion.

The location information can be used also directly for the recommendation of nearby places to visit. This has a major consequence for both travelling and local businesses, as the users can be presented with possibilities that not only suit their needs and preferences, but more importantly, are in their reach.

1.2.7   Security and Legal Requirements

As we have already mentioned, recommender systems need a lot of data to work. And some of the information provided by the users or automatically collected from their behaviour can be considered as private. Also, not only marketers and entrepreneurs with intent of selling goods may be interested in such data. Just few pieces of collected information can be possibly enough for blackmailing, fraud, theft or other criminal act. Although the local e-shops have to be registered as personal data processors according to local law for a longer time, some international portals were still not following the Data Protection Act in year 2016 [27].

One of the underlying technologies that can reveal a digital trace of a user, is the previously mentioned cookies retention. European Union have passed a directive 2009/136/EC that regulates the use of cookies for identification purposes and describes all required steps to undertake [28]. Since the directive has been passed, many of the web services (and foremost the ones based in the EU or oriented towards the EU) needed to add a cookie consent form.

1.2.8   Recommendation as a Classification Task

As we have discussed in this section, recommendation of the best product, service, book, movie or any other item is a classification task. Its aim is to find and propose the best items or solutions for given user and context. In other words, to map items to two disjoint classes (to recommend, not to recommend) possibly with some estimation of the degree of membership to the "recommend" class (with implied complementary degree of membership to the "not recommend class") through a preference scale.

Such recommendations can be based on properties of the objects (content-based) and/or behaviour of other users in context of the recommended item (collaborative). Content-based recommendation uses features created from the object properties that are typically used to measure similarity between objects. User behaviour patterns (ratings of the individual books, purchase history) are transformed into a space of user profiles and the similarity between such profiles is used as a similarity between the users.

1.3 Sentiment Analysis

As we have discussed in the previous subsection, orientation in the vast amounts of review data is complicated. And we are constantly being overwhelmed with them. However, while recommendation systems tend to work with objective or measurable data, such as star ratings and simple properties of the items, people like to decide upon subjective information as well. And sometimes the subjective information even overrides some of the objective ones.

For example, imagine there is a tie between two similar products with the same or very close technical specification and star rating. One product is bit more expensive, yet text reviews on aesthetics of the product are much more positive—a feature without a field for numerical rating. Surveys have shown [29, 30] that in such case, even the more expensive product is bought more often.

1.3.1 Opinion Mining—Subjectivity, Affect and Sentiment Analysis

Similarly to recommendations, the opinions have been shared by a word-of-mouth. The introduction of user interactivity on the Internet brought the opinionated text from printed media to mailing lists, discussion forums and other group communication tools. Open text field have been also added to many of the review forms to complement a grade-like evaluation.

According to the two previously mentioned surveys amongst more than 2000 adults in America, 81% of Internet users have done on-line research on a product at least once and 20% do so every day. Such research usually involves not only the comparison of technical aspects and price, but most importantly a research of opinions of other users. This is even more true in case of holidays, hotels, restaurants and other services with influence of up to 87%.

Collecting opinion is however not limited to goods and services. Opinion is also being searched in case of important decisions or acts, for example in politics. Survey held by Rainie and Horrigan [31] during the 2006 elections with over 60 million Americans revealed that 34% of these people use Internet to search for opinions outside their community and 29% even look on-line for an opposite opinion.

As users share their opinions, not only other users, but also entrepreneurs, brand holders and politicians are interested what is a general opinion on them or their goods, services and laws respectively. In such a case, the gathering of the opinions is not trivial, as it has to combine opinions from many possible sources of information including many different styles of evaluation and rating.

And as we express ourselves on-line more and more, also the area for subjective text had increased significantly. Major change have been brought by the blogging services, where virtually anyone can write long articles of opinions without much of a moderation. Media houses have also introduced commentary sections for verified users. Publishing an opinion is therefore much easier, not only for the "amateur journalists", but for everyone.

Although longer sections of text can be perceived as more trustworthy, shorter text on social networks is published immediately, shared much faster and may have much bigger impact. Also, the possibility to response swiftly to a current event enables the user to express current emotions. Social media is therefore a great place to gather immediate opinions about current affairs, freshly revealed products and many more.

So far, we have been talking mostly about opinions, however once we discover the subjective (opinionated) parts of the text and gather affection of the author to an emotion connected with discussed product, service, person or other item, sentiment can be revealed.

The intermediate steps—subjectivity recognition and affect analysis—are two very important fields that help to increase precision of the sentiment analysis systems. Subjectivity recognition denotes approaches to decide in general upon objectivity or subjectivity of a given word, sentence or document. And in case of subjective text (private state, opinion or attitude), also on the polarity of the text (typically positive or negative). In case of sufficient information from lexical databases, sentiment analysis does not even need any further training by human annotation, as concluded in [32].

Affect analysis deals with the detection of human emotions and attitudes, possible output is therefore more detailed. Such analysis systems usually use human-curated lexicons, where each word bearing an affect-related meaning or connotation is mapped to a set of affect categories describing emotions, human traits, states, etc. As many words can be ambiguous or express multiple affects, words can be mapped to multiple categories. To capture the correlation of the affect word and its category, the lexicon contains also an information on centrality (to what extent is the word related to the affect). To enable further processing, words are also marked with strength of the affect in their respective category. A list of "atomic" affects and a method of visualising newspaper article affect is presented in [33].

Despite all that, sentiment analysis can be theoretically built on a simple measure of entity co-occurrences with positive and negative sentiment words in the same sentence [34]. Even though such system is quite simple, it can discover for example that F-1 driver Fernando Alonso is connected with strong positive sentiment and convicted war criminal Slobodan Milošević with strong negative sentiment.

When the questions can be redefined into "in favour" and "against", sentiment analysis systems can give even more answers. As an example, the Brexit referendum results have been accurately estimated (with very little difference between real vote percentage and percentage of positive sentiment) from social media by multiple sentiment analysis systems hours before the official result announcement. For example by SENSEI [35] with their press release [36] and Twitris [37] as presented by TechCrunch [38].

Yet, on the same example we can illustrate the dynamics of social media and that it may be sometimes misleading for any primary conclusions.

To sum up this introduction, sentiment analysis can use a simple method relying only on co-occurrences of certain words, but also a more complex approach that discovers the main idea of an opinion and possibly assigns an emotion. The sentiment of the user is then gathered for each product, service, person or other item. Opinion mining and sentiment analysis are therefore clearly related fields—some authors even consider sentiment analysis a synonym to opinion mining.

Other closely related systems are recommender systems that we discussed in the previous section. The main idea of connecting recommender systems with sentiment analysis is that what we perceive as recommendations may be given in various forms (including star-like, grade-like or even open text). In such challenging cases, sentiment analysis is able to create a unified set of emotion values, on top of which a simpler recommendation system can be built.

1.3.2   Sentiment Analysis Systems

Sentiment analysis is one of very interesting fields of natural language processing. Therefore, with advances in part-of-speech tagging, named entity extraction and automatic text summarization, also the sentiment analysis started to be used in automated systems.

The main purpose of a sentiment analysis system is to monitor a given or discovered set of text documents, gather subjective text and return a polarity or emotion of such text. Gathered data can be then used for the deduction and monitoring of sentiment concerning items discussed in the text.

In a simplified case, sentiment analysis can be considered as a mapping of an input text into few classes (such as *positive*, *negative* and *neutral*) in a unified output template (containing, for example, the *holder* of the opinion, the *type* of emotion and its *strength*) as proposed in [39].

1.3.2.1   Words as Sentiment Bearers

The use of whole text documents or whole sentences for training of sentiment analysis systems does not make much of sense because such system would require vast amounts of annotated text data and the resulting mapping from the large space of sentences to the possible sentiment class would still entail a high probability of no match.

The natural landscape and some of the scenes are **overwhelming** and *spectacular**! The camera-work is so immersive, you believe are a part of **Hugh Glass**[E]' journey through the wilderness and back to civilization. Also with *great* performances not only by DiCaprio, but also Hardy, as the **unsympathetic** fellow fur **trapper** leaving Glass behind.

Story-wise, it is a bit thin for a 156 min picture. Glass' quest for **vengeance** is sometimes lost* as he utters **few words** about his drive and is being more or less, chased himself. The story arc of the Indians quest for their daughter felt a bit out of place and strange*. We also get to see the fur **trappers** p.o.v. that left Glass behind and the Captain way *ahead* of them. Which in my opinion takes a **little bit** of the magic of Glass' total **perilous** journey.

All my stars goes to the *beauty*, production value and performances alone! Regardless, this is one of those overlong movies one like, but would not sit out for another viewing!

Fig. 1.4 An example result of Semantria sentiment analysis (Lexalytics). Recognised phrases and entities with sentiment score are emphasized on scale from **negative** to *positive*. Star (*) denotes intensification by the previous word, [E] denotes a detected entity. The used text is a review of the movie *The Revenant* by *BoxOfficeKid* from Norway on the *Internet Movie Database* (IMDb). The star rating associated with this review is 7 out of 10.

Therefore, a massive simplification has to be made and a sentiment is usually recognised on the presence of individual words (unigrams) in the text, or at most commonly occurring short n-grams of words (bigrams or trigrams). This uses an assumption that each of such "phrases" contained in the text contributes in some positive or negative extent to the target sentiment.

Semantria, a sentiment analysis tool from Lexalytics, is an example of a system based on the recognition of such phrases and labelling them. Generally, this system detects the sentiment of topics, entities and phrase segments. To capture at least partially the context of an individual phrase, Semantria recognizes the presence of intensifiers (words that make emotion stronger or weaker) and negations.

An example result from processing of a movie review is shown in Fig. 1.4. The system uses its own proprietary lexicons and models, however, user can modify sentiment of individual segments if needed. The resulting sentiment score of -0.174 is a simple average of the individual sentiment scores on the scale $(-2, 2)$. Comparing to the original star rating provided with the review (7 stars of 10), the rating calculated by the system equals approximately to 4.5 stars.

The main reason for the detection of slightly negative sentiment may be, however, hidden in text itself. For example, the review favours only certain aspects of the movie and in the end the reviewer even states, that he would not like to see the movie again.

This example also shows many caveats connected to automated sentiment analysis. The unigrams and short n-grams are unable to capture their full context. Example of such flaw can be seen clearly on the detection of the word "overwhelming" by Semantria as negative. Although being overwhelmed by many things can be considered negative, the author of the review was most certainly overwhelmed by beauty of the scenes in the movie, especially if we consider the conjunction with the word "spectacular." The longer the n-gram is, the more precisely is the context captured, yet the size of lexicon increases dramatically and the extraction of sentiment becomes over-fitted.

And still, the context of longer text sections does not always capture enough information. For example "unsympathetic fellow fur trapper" describes just the quality
of a figure in the movie, yet not the quality of the movie itself. A keyword-based
system is hardly able to recognise and capture such nuance, which however leads to
false sentiment rating.

Another very serious issue on the level of individual words is caused by homo-
graphs (words that have the same spelling, yet different pronunciation and meaning)
or homonyms (same spelling and pronunciation, yet different meaning). Without a
context, it is virtually impossible to tell a difference between "lead" [li:d] (being first,
at the beginning and therefore positive) and "lead"[lɛd] (poisonous metal, associated
with negative sentiment). Or "tip" [tɪp] with same pronunciation, yet several positive
(advice, sharp point) and negative (to knock over, to drink, to dump waste) meanings.

To overcome some of the linguistics-related problems and also speed-up the creation of positive and negative keyword lexicons, advanced linguistics resources can
be used. Such resources enable to discover synonyms, deduce the meaning of homo-
graphs and homonyms [40] from a context and by applying local templates. One
of the most widely used resource for English language is WordNet [41]—sets of
synonyms (synsets) connected one to another trough linguistic relations (such as
hyper/hyponymy for nouns, troponymy for verbs, antonymy for adjectives, etc.) or
general similarity and derivation of words. WordNet can be also used as a lemmatizer
and Parts-of-speech detector to enhance the extraction of meaning even more—and
subsequently the extraction of sentiment. To capture all of this, a broader context
containing more words is, however, needed.

Another project, based on WordNet and much closer to the requirements of
sentiment analysis, is SentiWordNet [42]. This uses a basis of 7 "paradigmatically positive" *(good, nice, excellent, positive, fortunate, correct and superior)* and
"paradigmatically negative" *(bad, nasty, poor, negative, unfortunate, wrong and inferior)* terms [43] and expands them across the binary relations of see also (such as
good—favorable, obedient) and direct antonyms[2] (good—bad) in WordNet.
All synsets gathered in this step are used as a template for discovery of other synsets,
grading them with a value of emotion between negative and positive. Such discovery
runs on WordNet repeatedly to cover all synsets. As a result, each synset is graded
with a value of positivity, negativity and objectivity, where the latest class contains
words with no emotional value.

Another approach that can be used in text processing, and therefore in sentiment
analysis, is based on word embedding. For example, Word2Vec implementation [44]
is able to represent each word as a vector of real numbers, based on a given dataset.
The interesting consequence is that simple mathematical operations on such vectors
are able to provide information about relations between these words. This approach is
even able to predict a word from the context. The vector representation of individual
words can be then used to learn and deduce their sentiment directly although a more

---

[2]Indirect antonyms cannot be used, as they do not express opposite sentiment, as for example in
relation child—parent.

complicated sentiment analysis is usually performed. One example of a sentiment analysis system that used word embedding (and Word2Vec as a baseline) is described in [45].

1.3.2.2 Machine Learning Approach

Even though many of the previously mentioned lexicons and methods used machine learning for their construction, classifiers can be trained on larger segments of text as well. One possible approach may use an extension of Word2Vec to describe whole sentences and documents by relatively short vectors of numbers [46] and train sentiment on such vectors.

A generally used approach then employs a simpler bag-of-words representation of the document. In such a representation, each document is transformed into a vector with the same length as the number of words in the considered dictionary.

Traditional document processing uses the TF-IDF scheme, where the value of each word is proportional to the number of occurrences in the considered document and decreasing with respect to the fraction of documents, where the word appears. The probably most common definition of the *term frequency—inverse document frequency (TF-IDF)* weighting scheme is:

$$\text{tf}(t, d) \cdot \text{idf}(t) = f_{t,d} \cdot \log \frac{N}{1 + n_t}, \tag{1.1}$$

where $f_{t,d}$ denotes the frequency of the term t in a given document d, N is the number of documents and n_t is the number of documents containing term t.

Another approach that seems to be superior in sentiment analysis [47] uses only a binary representation to store whether the word is present in the document or not.

In many natural language processing approaches, stop-words (most common words in the language with little semantic information) are removed from such dictionary, yet they might be important for correct detection of some features in semantic analysis, such as negations.

A basic sentiment analysis system can be constructed for example with a NLTK (Natural Language Toolkit), in particular with a training utility (such as nltk-trainer) and Sentiment Polarity Dataset or the VADER Sentiment Lexicon [48], which are both available in the NLTK. Possible way how to create sentiment analysis tool from these components is described partially in the book [49].

An example of such a system is the Mashape Text-processing and its Sentiment Analysis tool.[3] The system is designed as a two-level classifier, and the output of the system consists of posterior probabilities of the classes according to each of both classifiers. The first level returns the subjectivity score of the text and the second level returns the polarity of the subjective text. In case of the above mentioned review of the movie *The Revenant* (Fig. 1.4) the returned subjectivity is 50% and the polarity

---

[3] Available at http://text-processing.com/demo/sentiment/.

results in 80% positive. The original star rating from the author is 70%, so the result of this system is slightly over-positive. This system is however unaware of negations and other valences. Therefore, a phrase "not bad" is classified as 80% polar and 70% negative, based solely on a presence of negative words "not" and "bad".

Just a simple extension to the method of data vectorization can however improve a lot the overall results and robustness of the classifier. In particular, refer to the Sect. 3.3.3 for more information on sentiment analysis with Bayes classifier.

Although the simplification to a bag-of-words enables much better processing of text documents, sentiment analysis systems do not have to process all words in the documents either. As most of the affect information is stored in verbs, adverbs and adjectives, the AVA framework [50] proposes to split verbs into four groups: positive no doubt (e.g. accepts, appreciates), positive doubt (admits, feels), negative no doubt (opposes, rejects) and negative doubt (claims). Adverbs connected to such verbs then have an effect to strength: affirmative (e.g. absolutely, certainly), doubting (possibly, roughly), intensifying (extremely), diminishing (barely) or negative (hardly). Adjectives are connected in this model to one or multiple adverbs.

The AVA framework provides a numerical value of sentiment for each adverb-verb-adjective combination on itself, however a classification layer is proposed to return only the information on positivity of the whole input text. Other authors [51] use the combination of verbs, adverbs and adjectives only as a multi-word extension to standard classifiers.

Another very different approach uses convolution for text processing. Documents are then processed as stream of data, where not only the current word is considered, but also a small "window" of words around, thus capturing a context as well. To enable convolution over the whole document, both ends need to be padded by neutral elements.

Sentiment classification can be then based on features extracted from such convolution, for example from internal values of a convolution neural network [52].

1.3.3   Open Challenges

Until now, we assumed that the whole text is in English, grammatically correct, contains no jargon and irregular concatenations, contains an opinion of one person, and that it discusses only one entity. But in many cases, the discussed named entity is connected only with a particular segment of the input document or even part a of a phrase. For example: "Product A is much better than product B" should be understood as a positive sentiment assignment to A and negative or neutral to B.

The opinion can also combine text from multiple opinion bearers. Take for example a quotation—the cited text might be overall positive, yet the author may criticise its content. The system also needs to recognise which parts contain objective information and which describe the reviewers' opinion. Subjective word and phrase usage also depends on language, as described on an example of French and Dutch in [53].

Another issue that we need to be aware of is honesty and conditional clauses. In case of sarcasm, for example, the statement may be over-positive, yet holding a very negative opinions [54]. Such as: "Lost my keys. What a lovely day!" And while we can take use of facial expressions in real life and prosodic cues and changes in resonance in spoken text [55], there are usually very little markers of sarcasm in a written document. Relying on detection of overstatements is also not viable, as sometimes the overreaction may be truthful.

Conditional clauses may be also used to express negative emotions, even if there are no negative words contained. For example: "If you want a good product, don't buy this." Even though there is only one positive word "good", the sentence is negative about the product. Such conditional clauses are possibly recognizable in case of some fixed patterns [56], yet are still an open field for further research.

1.3.4 Affect Analysis in Multimedia

With advances in communication technology, mobile computing devices and digital audio and video capture, personal opinions can be shared not only through text, but also by an audio or video. The detection of affect in audio may be helpful for example in the evaluation of call centre recordings as discussed in [57]. In sentiment analysis from video, a whole video-blogging genre dedicated to expressing spontaneous emotions on products are the "unboxing" videos, where a person unpacks and briefly reviews the product—mainly the aesthetics, user experience and primary functions—without much of a prior knowledge.

As discussed in [58], the multimodal sentiment analysis can benefit from the detection of many more features—facial expression, body movement, intonation, selection of background music, etc. A study conducted two years later [59] then extracted features for each modality (facial expression from video, semantics from audio and text) and classified them into one of six basic emotions (surprise, joy, sadness, anger, fear and disgust). Based on the individual features, outputs of the individual classifiers were fused and new multimodal model has been proposed.

To introduce very briefly the features used for the description of multimedia content: Affect in human speech can be recognised by changes in prosody (tonality of the speech), amount and length of gaps between individual phonemes, amount of incomprehensible words of fillers (like, so, [ε :], ...). Video data is then processed frame-by-frame with the detection of face features (eyes, eyebrows, nose, mouth, smile lines, etc.) and distances among them.

1.3.5 Sentiment Analysis as a Classification Task

We hope it is now obvious that the automated sentiment analysis is in fact a classification task. Text document is mapped through a set of rules or a trained system to individual classes, typically representing positive and negative sentiment or coarse list of emotions. For such evaluation, either a fixed classifier based on preselected dictionary entries may be used, or for better precision, a particular sentiment analysis classifier can be created for required task and data type.

As in other uses of classification methods, the result of the classifier is determined not only by its type and setup, but more importantly by the feature space we are building the classifier upon. In case of sentiment analysis, the naive construction of such feature space (for example the simple bag-of-words approach) may not yield satisfactory results, as we would like to base the classifier not only the words themselves, but also on their context—position in sentence, negations, occurrence of conditionals and sarcasm.

1.4 Example-Based Web Search

When the users know what they are looking for, they have, in general, two distinct ways how to perform a search. The user either describes the object of interest in terms that would be hopefully understood, or presents an example. When used on the Internet, the first approach leads to text processing of the user query and search or recommendation among other indexed text documents, possibly connected to a digital representation of an object, such as sound recording, pictures, videos, 3D models or others. The results may also be referring to real objects, available for rent or purchase.

In many cases, this approach is rather inconvenient. It relies on a precise description of the object by a set of words, and thus requires a prior knowledge of the object domain and its important features. Also, the object needs to be described in an as similar way as possible to the description of the authors of the index. The authors may, however, have a deeper knowledge of the object or its context, and describe it only by a precise name, using jargon or technical properties unknown to general public. Other objects might have a subjective description and thus be prone to a possibility of different characterisations by other users or index creators.

Also, the user has to know at least something about the searched object. For example, to buy a spare part, the user needs to know either some identifier, an original location of the part or at least a purpose of it. Even then, the user may need to consult the technical documentation and locate the part identifier. However, such detailed documentation might be unavailable. In such cases, giving the broken part to a specialised technician as a reference of what the user is looking for, is much more convenient.

When we transfer such a search method to the online environment, a digital representation of the real object needs to be constructed. However, instead of using words susceptible to a lot of imprecision, methods of example based search tend to use robust features that are invariant to common transformations.

Before we start describing the individual methods involved in example based search and their relation to classification, we need to define more precisely our scope. Some literature [60] describes user critiquing and turn-based search as a type of example-based search. Although the features used in such systems are similar and the user may feel that the final choice is influenced by his selection of presented examples, the internal engine is closer to a recommender system, which is described in Sect. 1.2.4.

1.4.1 Example-Based Search and Object Annotations

To fully understand the common uses of example based search, we have to keep in mind that the digital objects (such as images, sounds or even 3D models) are usually accompanied by a set of technical and human-created metadata, usually stored in a semi-structured manner.

In many cases, the unstructured name of the subject may be sufficient for indexation. When the user has enough information, the correct name can be deduced by an approximate search. For instance, search for images containing "screw head cross" reveals the marketing name of "Phillips screw." Once the user gathers the appropriate name of the object, he or she might continue in the text-based interaction and search for other instances by their exact name. Such a simple information can also be stored directly in the filename of the digital representation of the object.

A more sophisticated technical metadata storage, especially used in digital photography, is EXIF [61]. The metadata is stored directly in the header of the image, containing especially information on the image acquisition circumstances. Usually including, but not limited to: date and time, capture device type and orientation, shutter speed, aperture number, flash information and focal length.

With the expansion of digital multimedia objects, an *eXtensible Metadata Platform (XMP)* have been proposed by Adobe and standardised [62]. It allows to add structured metadata, both understandable by human and computer, and store them in a data packet either attached to the original file or stored separately. Information stored in the XMP range from the technical metadata to information regarding the processing flow of the multimedia content, such as information about applied colour filters for still images or a scene and a take number for cinematographic use.

This platform can be extended for any particular use, which may limit the intelligibility of the stored metadata by different software tools. To mitigate this, XMP uses a subset of Resource Description Framework (RDF) and usually limits the object properties by well-known ontologies.

A widely used ontology, *Dublin Core* [63], defines "fifteen generic elements for describing resources". These properties are then used for a coherent approach to metadata annotations for text documents as well as multimedia clips. In the full RDF scheme, the subject is then connected to other objects and subjects. Such entities can

@prefix dc: <http://purl.org/dc/terms/>
@prefix dbr: <http://dbpedia.org/resource/>
@prefix dbo: <http://dbpedia.org/ontology/>
@prefix dbp: <http://dbpedia.org/property/>
@prefix loc: <http://id.loc.gov/vocabulary/relations/>

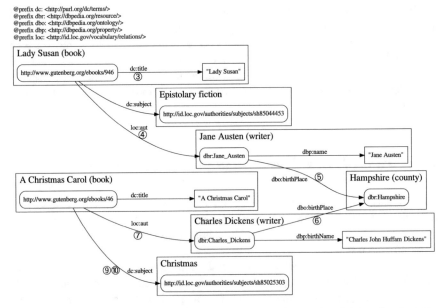

Fig. 1.5 Fragment of a semantic web with search for a book about topic "Christmas" authored by a writer born in the same county as the author of "Lady Susan". Circled numbers correspond to lines in Fig. 1.6

be from other databases and linked through different ontologies, thus connecting to a semantic web [64]. In the example shown in Fig. 1.5, a book digitised by Project Gutenberg can be connected to its name through Dublin Core property Title and to DBpedia entry of the author through Library of Congress ontology.

When objects become connected to the semantic web, graph searching and traversing methods can be utilised for a discovery of objects similar to the provided example. Moreover, because several ontologies include a concept of abstraction or relations of similarity, it may closely resemble a human perception of conceptual distance between the objects. In our example, Jane Austen was born in Steventon and Charles Dickens in Landport. In such case, we might get the object representing the county of Hammington on DBpedia through `dbo:shireCounty` relation for the city of Steventon, respectively through a transitive query on `skos:broader`[4] relations from the concept of Areas of Portsmouth, one of which is Landport.

However, the generalization to the county of Hampshire is, in our case, also present directly as another `dbo:birthPlace` relation for both author objects. Thus, the SPARQL query on DBpedia database can be as simple as presented in Fig. 1.6.

---

[4] `skos` prefix denotes the Simple Knowledge Organization System Schema by W3C.

```
① SELECT DISTINCT ?book2
② WHERE {
③   ?book1 dbp:name "Lady Susan"@en.
④   ?book1 dbo:author ?author1.
⑤   ?author1 dbo:birthPlace ?place.
⑥   ?author2 dbo:birthPlace ?place.
⑦   ?book2 dbo:author ?author2.
⑧   ?book2 a dbo:Book.
⑨   ?book2 dct:subject ?subject.
⑩   ?subject skos:broader* dbc:Christmas
⑪ }
```

Fig. 1.6 Example SPARQL query for books (line 8) containing subject of Christmas (9–10) by authors with shared birth place (5–7) to author of Lady Susan (3–4)

Methods of search that use ontology-driven semantics are proposed for text and media documents [65], image collections [66], biomedical data [67] and other domains. Especially in images and other media documents, description of an event can be represented by a set of entries from a *WordNet ontology* [41] and be thus insensitive to use of synonyms in the otherwise unstructured textual description.

The combination with example based search is, however, still very much dependent on a possibility of automatic semantic description. Only if we can extract semantic information from all objects, including the example, ontologies can be used for normalization of the descriptors and enable search based on the semantic information gathered from the example. Therefore, we will focus mainly on feature extraction and automatic object description in the following subsections.

1.4.2 Example-Based Search in Text Documents

Although this might seem controversial, search in text documents by a text query might also be considered an example based search—we use the same modality for the query as is the indexed modality of the objects. Also, understanding of the fundamental concepts involved in the text-based search might provide a better insight into the indexing and search of more complex objects.

To enable a simple full-text search, the documents are commonly processed as a bag of individual word lemmas (basic forms), excluding the stop words (most common words carrying no semantic importance). From such information, a reversed index is created, where each lemma points to a set of documents, where the word occurs. Such simplification into a bag of word lemmas then leads to an invariance of reordering of the words in the document, changing aspect or conjugation.

The input query is then analogically processed into a set of lemmas, corresponding sets of documents are gathered from the index and ordered by the amount of matching words or another measure (such as the relevance of the document). More complex search agents are also able to recognise and resolve possibly misspelled words or negatives in the query.

Other indexing strategies may preserve the information about relative or absolute positions of the words in the document. This way, the order of words in the query may also be considered during sorting of the search results.

The historical prevalence of text-based interaction is understandable. Text, in particular as a bag of lemmas, needs a much smaller number of features for further processing and yields a reasonable-sized index. Such reduction is favourable due to the so-called "curse of dimensionality", a term used for the need for multiple data points in every possible feature value combination to guarantee a possibility of prediction. Vaguely said, the more dimensions we would like to use for distinguishing objects apart, the more instances we need to train the algorithm.

1.4.3   Example-Based Search in General Objects

As could be seen already in the previous subsection, an example based search method is centred around the method of data indexing. Once we propose a robust index construction method, the search itself is usually trivial.

Until now, we have been discussing mostly the use of classifiers on text data or data directly convertible into a numerical representation in a relatively low-dimensional feature space. With the increasing use of mobile devices and consumer electronics equipped with microphones, image sensors and touch screens, different means of interaction with the search engines and other services do appear. Therefore, so does arise a necessity of appropriate indexing methods.

The index construction is commonly based on the extraction of object features in such a way that the common object transformations do not alter the extracted features significantly. For example, photographs of a landmark may be captured from different angles and with different rotations of the camera, but we would like to index all photos of the same landmark very close to each other. Audio recordings may be indexed with a use of descriptors invariant to some pitch shifting, cropping of either end or distortions. Proteins may be indexed with regards to the similarity in amino acid sequences or structures.

The index also usually uses as many information from the original object as possible to provide a sufficient resolving power. For example, if we would like to recognise the Eiffel Tower in Paris from the copy in Las Vegas or any other copy or derivative, we have to consider also the surroundings of the object.

On the other hand, if we are looking for objects similar only in some aspects, such as painting style, music genre or document type, features encoding only such properties should be used for the construction of the index.

In the following subsections, these concepts will be expanded to more complex objects, namely drawn symbols, still pictures, sound and video. However, only the efficient and correct creation of the index (and actual possibility of the user to enter the query data) are the limit of example based search.

1.4.4 Scribble and Sketch Input

The simplest visual input is possibly a small set of scribbles—continuous two-dimensional lines, which can be represented only by the changing angle in given distances.

Although this input method has been pioneered by Palm in a generic text input system Graffiti, it has been proven inferior to the virtual keyboard [68], even if the system was enhanced later [69].

The scribble-based user input is however of great use if more complicated symbols have to be inserted and the user lacks appropriate input method or knowledge to use it. Examples of such input systems range from recognition of space separated symbols, such as in Chinese and Japanese scripts or discrete characters and numbers in other scripts, to the recognition of whole cursive script written words [70].

Among more recent uses of scribble based input, *Detexify* [71] is an online system for recognition of handwritten symbols that directly offers a command to reproduce the symbol in the LaTeX environment.

Technically, most of these systems are based on a Dynamic time warping method [72], where the user query is compared to a pre-trained set of scribbles in a somewhat flexible manner, and the entry with the smallest deviation is returned.

A bit more complex user input is a sketch of an object. As the number of scribbles increase, it is less advantageous to compare them in all permutations with the scribbles stored in a database. However, it might be still a good idea to keep the information in individual image vectors and their order as they represent the edges of the symbol, object and texture directly.

Various types of neural networks are also used in scribble recognition, especially the ones that incorporate the temporal information, such as time-delay neural networks and recurrent neural networks, as discussed in [73]. A similar approach can be then used for not only the online symbol and handwriting recognition but also for online doodle recognition, as demonstrated on a popular game "Quick, Draw".[5]

In many cases, the vector information is not available. Recognition of hand-drawn symbols in raster images is then, for example, a matter of stroke direction detection (using edge detection mask) and classification against a stored set of images [74].

In the case of complicated objects (for example, human faces) the sketches are influenced significantly by their author. Moreover, without the vector information, search algorithms have to utilise some of the sketch synthesis and recognition methods to match the style of input and database entries [75]. If considering such operations, usually a bitmap query is used, thus leading to processing similar to that of photographs.

---

[5]By the time of writing this book available at: https://quickdraw.withgoogle.com.

1.4.5   Multimedia and Descriptors

Sound, images and video contain a lot more raw data to process than text documents and scribbles. For example, a standard page of text document takes 1 800 Bytes. Sound captured with a sample rate of 48 000 Hz, 16-bit depth (2^{16} volume levels) and with two channels takes up 675 MB per hour. One FullHD frame (1920 × 1080 pixels, three channels of 8-bit depth without subsampling) takes almost 6 MB. An hour long video with 25 frames per second then takes a whopping 534 GB. Multimedia files commonly use a lossy compression to deal with such high amounts of data, usually based on some model of human perception (a detailed description of such methods can be found in [76]). Such compression reduces the amount of information that needs to be stored or transferred but may distort the original signal in a way that is possibly unfavourable for further processing, such as classification [77]. For example, filtering out high frequencies results in reduced presence and legibility in the audio signal and blurring or distorting edges and colour in image data.

In addition to the size, simple reductions of the multimedia feature space (sound levels and pixel colour) do not maintain a semantic information of the content. However, this is similar to text processing, where individual letters do not carry much information, but words and phrases do. Therefore, descriptions of a higher (semantic) level need to be extracted from multimedia content to provide the requested indexing possibility.

The description level should not be treated as a strictly set property of a feature. Low-level features can also be utilised for construction of high-level object descriptions.

For instance, colour in digital images is considered a low-level feature. If the descriptor uses only a simple transformation of input data, we also tend to describe the resulting description as of a low level. Such features are usually specified by a significant dependency of the incoming data and either a large dimensionality or low discrimination power. As an example, a descriptor returning n most common colours in a picture produces a low-level description of the image.

The descriptors of a higher level, on the other hand, are designed to be invariant to common variations and changes while preserving the discrimination power. For example, instead of indexing the individual colours, low-dimensional representations of local colour histograms can be stored and later compared with a measure that is invariant to global colour shifting. The details of a histogram distance computation can be found, for example, in [78]. The whole method is described in [79].

Other features, gathered from the input data, may be especially suitable for creation of high-level descriptions. Few examples of such features follow.

1.4.5.1   Object Edges

One of the features that can be extracted from picture data is object edges. The Canny edge detector, for example, uses several processing stages to gather only stable and

(a) Original image

(b) Canny edge detector

(c) SIFT

(d) SURF

Fig. 1.7 Still picture processing

connected edges from the picture [80]. The result of this detector can be seen in Fig. 1.7b.

Such simplification then leads to a similar approach as the hand-drawn sketches. Object edges are compared adaptively to detect similar objects in other pictures. However, in the case of photographs, we need to pay special attention to object occlusions and framing. As some parts of the object may not be visible, the edge encoding and comparison procedures need to be more flexible.

The shape of the edge may also be distorted due to a projective transformation.

1.4.5.2   Interest Points and Local Textures

A different approach for extraction of high-level data from still images uses a detection of "interest points" (corners or blobs) and local description of the picture around them. In practice, the detected points are usually located near significant gradient changes (edges, textures), and the descriptor part records the influence of preselected functions on the local texture of the image.

Two visualisations of such image descriptors (SIFT [81] in Fig. 1.7c and SURF [82] in Fig. 1.7d) show the detected point (centre of the circle), the main direction of the gradient (straight line) and the size of the considered neighbourhood for the creation of the descriptor (circle).

Due to the fact, that these image descriptors are normalized in scale and rotation, many of the indexing and searching algorithms based on these image descriptors are inherently rotation and scale invariant; which may be desirable.

Trivial image search algorithm may be then constructed with an index of such interest points, matching against the interest point descriptors from the query image. Such approach is, however, usually not good enough.

To this end, image features are clustered into "visual words" [83] and whole images described by the histogram of such visual words.

Indexing of such information then creates a simple, yet powerful tool for search of visually identical or similar images. Just to recall, as the online images are usually connected to some document or at least an alternative description, the same search engine may be able to not only find the similar pictures but also to describe them.

1.4.5.3   Histogram Descriptors

Other commonly used descriptors are the Histograms of oriented gradients (HoG) and Histograms of oriented optical flow (HoF).

The HoG method accumulates the orientation of gradients in the image divided to smaller sub-images, so called "cells". The combined histogram entries are then used as feature vectors describing the object in the scene. In combination with appropriate contrast normalization and histogram collection over a variable window, even a linear Support Vector Machine can be used for pedestrian detection in static images [84].

For video content, Histograms of oriented optical flow approach combines the description gathered by HoG from the primary image with optical flow obtained from consecutive image frames. The first approach of Motion Boundary Histograms [85] was again used for pedestrian detection. In later research this descriptor was used in event detection [86, 87] and other action classification tasks [88].

1.4.5.4 Sound

If we omit the search by voice input, which is processed by an automated speech recognition system, and then passed to the search engine as a regular text query, search by sound example is used only in two varieties.

Either the user query is the sound itself, such as a song, and the question is "What am I listening to right now?". Alternatively, the query is very imprecise, possibly just a hum of a melody with a question from the user "I have this stuck in my head, what is it?" In both cases, the search engine has to be insensitive to a possibly bad quality (of reproduction and capture) and account for access to only a short section of the song as a query, usually tens of seconds.

In both cases, the audio signal is not used as a raw data but rather described by a broad set of attributes, both spectral and temporal, such as Log-frequency power spectra, attack and temporal centroid; as outlined in the MPEG-7 standard [89].

The query by precise song can be handled, for example, by the Shazam application [90]. As presented in the article, short segments of the input sound are extracted and matched against the database of original recordings. Even though the signal-to-noise ratio is usually poor in bars and clubs, even few matched tokens are allegedly enough to detect the right song.

This approach can be then extended to a search for similar music [91] or the cover versions and remakes of the given examples [92].

The situation changes when a segment of original music is not available for query, only a partial memory of the melody. One possible approach is to extract the main melody of both the original song and the hummed query [93] and match them using a method to certain extent invariant to variable time stretching and pitch shifting.

Instead of a relatively complicated melody extraction from the original polyphonic music track, the index can be created from data similar to the future queries, i.e. other users humming the main melody of a notoriously known section of the song. Data processing can be then similar to the before mentioned original music recording, involving the MPEG-7 descriptors and token matching.

Some of these systems are described in [94], in particular, midomi.com.

1.4.5.5 Fusion of Multimedia Descriptors

To obtain more precise results in complex tasks, such as event detection from multimedia content, a set of various features with corresponding classifiers can be combined. Such approach enables us to utilise visual, aural and other information fully.

The more traditional approach, in this context known as late fusion, uses a weighted average of posterior probabilities gathered from individual classifiers. For example in [95], results from two distinct visual feature, three motion feature and one audio feature classifications are combined with two visual concept models and results from automated speech recognition, followed by a text classification. This particular approach, therefore, combines results from low-level visual features (extensions of SIFT) with results obtained by matching of semantic visual concepts.

One of the visual concept descriptors is, for example, Action Bank [96], which represents the considered video as a feature bank vector of length equal to the number of action detectors in the considered bank multiplied by the number of scales on which the detectors are executed times three levels of the octree. Average accuracy increases with the higher number of action detectors, however, increases the computational cost. The resulting feature vector is then commonly classified by a support vector machine.

Another approach, presented in [97], utilises both fusion of classification results (late fusion) and a combination of individual extracted features (early fusion). The motivation of the late fusion is the same—to combine results of multiple classifiers running on different features from all available modalities of multimedia content. The early fusion stage then provides information from more than one feature extractor (or modality) to the classifier, enabling it to utilise more diverse information about the multimedia of concern. Combination of multiple feature vectors (or kernel spaces), however, expands the dimensionality and may lead to poor learning performance if not enough training examples are available. Also, variously encoded features may be hard to combine into a single feature vector at all. The significant advantage of an early fusion scheme is, however, in the elimination of poor performing classifiers based on features not significant for the particular task and input data (e.g. automated speech recognition in a silent clip).

1.4.6   Example Based Search as a Classification Task

Example based search is closely related to classification. It tries to find the closest known objects to the incoming data. Just to remind, classification attempts to deduce the most likely class for the incoming data. Therefore, both of these processes can use a very similar data representation—data points in a feature space.

The main part of the example based search methods is therefore centred around the extraction of the features. The same features are also extracted from the provided example and matched against the data points already present in the database of objects.

In the simplest case, the similarity of data points can be transformed into a single distance measure. Changes in individual features may have a different effect on a similarity assessment, yet this effect is usually known by the design of the descriptor. The data points are then sorted according to the distance measure.

However, in many cases, the individual features are not directly comparable. Also, no simple feature space transformation leads to a direct comparability. For example, the direct comparison of image features would be possibly suitable for detection of image duplicates, but would not have enough power to detect a presence of the same object.

Dictionary methods can be used to achieve a higher insensitivity. Each object can be then classified by a "bag of features", possibly partially sensitive or insensitive to the location of the original descriptors. Such representation of objects is then directly comparable, as the resulting vector contains information on the presence of a given feature from the dictionary and similar objects should include similar features.

For custom comparison of complex objects, classification methods can be utilised directly for deduction of similarity. For example, the classifier may return information whether two given objects belong to the same class and therefore should be treated as similar, or not.

1.5 Malware Detection

The detection of malicious software (malware) is a young but very important area of research. It is focused on discovering any piece of malicious code, attachment or document stored or executed on the considered device. Its importance is still on the rise because a majority of the inhabited land is currently covered by internet connections. The omnipresence of internet helps to spread information quickly but on the other hand, it also helps to spread all possible kinds of malware. Furthermore, the availability of internet connection combined with the growing number and complexity of connected electronic devices significantly bolster the thread landscape every year. This is also reflected in the rise of malware targeted to mobile operating systems such as Android.

Let us quote two security reports from 2017 to illustrate how great the problem actually is. The Cybersecurity Ventures [98] predicted that cybercrime will cost the world $6 trillion annually by 2021, up from $3 trillion in 2015. According to them: "This represents the greatest transfer of economic wealth in history and will be more profitable than the global trade of all major illegal drugs combined". This prediction is also supported by the Accenture security report for 2017 [99]. According to this report almost 40% of the loss caused by cybercrime is caused by malicious software alone.

Those numbers illustrate why malware detection and analysis is number one priority of many security and anti-virus companies. Employed methods ranges from completely human driven analysis and manually produced characteristic patterns, aka signatures, to fully automated approaches driven by machine learning. Methods used for the automatic malware analysis are frequently divided into static malware analysis and dynamic malware analysis. However, the boundary between them overlaps more and more as techniques such as the execution graph analysis are used in both. For the sake of simplicity, everything done without executing the binary will be referred to as static malware analysis and methods which require file execution will be referred to as dynamic analysis. The hybrid malware analysis which uses features obtained by static analysis together with features observed while running the binary in controlled environment is discussed in its own subsection followed by a short description of the most common malware families.

1.5.1   Static Malware Analysis

Static malware analysis treats binary files as an arbitrary data file from which a multitude of features can be extracted, whilst its execution is completely omitted. First approaches of this kind were based on manual identification of a set of instructions responsible for malicious actions and not used by the legitimate programs [100]. These sequences of instruction are often called tell-tale. This simple approach of course has a very low false positive rate but on the other end the investment of human resources in order to reliably detect at least the already known malware types is huge and doesn't scale at all.

Furthermore, malware authors soon adopted techniques which yielded a tell-tale approaches ineffective. They are called obfuscation and polymorphism. Obfuscation is a method when even the simplest code is written in a way where it is very hard to understand what it actually does. Polymorphism is an ability of piece of code to change itself. This two methods combined together created an effective way of avoiding a signature based detection. In theory, reverse analysis of obfuscated and polymorphic code is NP-hard [101]. Therefore, recently the analysis of binaries moved to modelling high-level building blocks, such as sequences of system calls, and estimating their actions [102, 103], rather than modelling the flow of individual instructions. The rationale for modelling high-level building blocks is that it is much harder to hide their true purpose. The most commonly used representations of binary files that respect the high-level functions paradigm are function call graphs and n-grams. A completely different approach to binary files representations which appeared quite recently is to treat a binary as an image and analyse it using convolutional neural networks [104].

1.5.1.1   Function Call Graph

One of the most common representations of software are control flow graphs (CFG) or simply call graphs. Nodes in CFG represent basic building blocks of a binary, typically sequences of instructions without jumps, and directed edges capture dependencies (jumps) between those sequences. It can be constructed dynamically or statically. In a dynamic setting, the binary is executed and the graph captures the flow inside the binary during its execution. In a static setting, all instructions are analysed without actually executing the program and all possible paths are then presented in the graph.

The second popular representation is a function call graph where each node represents an internal or external function and edges again represent the dependencies between them. An example of such graph is depicted in Fig. 1.8.

Both call graph variants can still be affected by a combination of polymorphism and obfuscation. Therefore, Bruschi et al. [106] proposed a code normalization technique to reduce their effect. Also in [107], the authors presented a method to overcome standard obfuscation techniques, this time by creating a finite-state automaton from an annotated CFG. In [108], the authors assume that malware binaries from the same

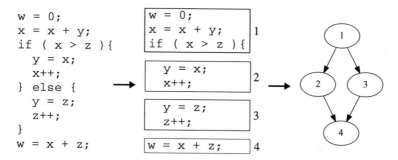

Fig. 1.8 Example of a simple function call graph [105]

family still should have similar structure despite polymorphism and obfuscation. Therefore, they proposed to cluster the extracted call graphs using the graph edit distance. Then, the clusters should represent individual malware families. In [109], a generic framework that extracts structural information from malware programs as attributed function call graphs is presented. Additional features are encoded as attributes at the function level. Then a similarity metric between two function call graphs is learned to ensure that malware binaries from the same family have high similarity and binaries from different families are as dissimilar as possible.

Unfortunately, malicious actors quickly adapted and started to use other tools originally developed to protect intellectual property, encryption and packing. The main difference between these two is that in order to execute an encrypted binary, it has to be fully decrypted in the memory. The packed binaries, on the other hand, can be executed directly. Headers of the packed binary and also first instructions containing so-called unpacker are stored as plain text. Therefore, when executed, the unpacker decrypts only the requested small part of the binary, runs the code, remembers its result and clears the memory. The overall effect is that only a small piece of a binary is decrypted at a time and it is almost impossible to discover the whole purpose. A second effect packing has on analysis is that, if done properly, it can hide all imports. An analyst is then left with only seeing imports of the unpacker itself.

Malicious binaries are very often packed or encrypted and all previously mentioned malware detection approaches are vulnerable to encryption and packing of binaries. To address this problem, Cesare and Xiang in [103] extended the idea of [107] and proposed an approach for automatic analysis of packed binaries and extraction of an approximative function call graph and matching individual sequences of extracted code to static signatures stored in database.

1.5.1.2 Using *n*-grams

The other representation commonly used for static malware analysis was inspired by natural language processing. In document classification, the occurrence or frequency of each word is used as a feature in so called bag-of-words (BoW). A direct application of BoW principle in static malware analysis wouldn't work because the same function with different parameters may behave differently. Therefore, instead of using the function names as words, the authors of [110] proposed making n-grams from the whole source code without any preprocessing. Binaries are then classified using the k-nn algorithm. A similar approach was actually proposed 15 years earlier in [111], but without any experimental evaluation.

The most serious drawback of this approach is the exponential growth of the number of features with increasing n. To address this issue, [112] proposed using only the most frequent n-grams from malicious and benign binaries. By selecting the different amounts of n-grams, one can balance the computational costs and detection capabilities. Stiborek in his Ph.D. thesis [113] showed that by clustering of the n-grams one can use aggregated information of all features while significantly reducing the computational costs.

1.5.1.3   Images

In 2011, Nataraj et al. [114] proposed a novel way of representing binary files called binary texture, in which binary files are represented as grey-scale images. They also observed that for many malware families, the images belonging to the same family appear very similar in layout and texture. Motivated by this observation, they used standard features for image processing, combined with the k-nn classifier, and achieved very promising results on a big and diverse set of malicious binaries. In their subsequent paper [115], the same authors showed that this approach is also resilient to anti-detection features like packing and section encryption.

Another similar representation was presented in [104]. The proposed method generates RGB-colored pixels in image matrices using the opcode sequences extracted from malware samples and calculates similarities for image matrices. Originally, it was designed for dynamic analysis of packed malware samples through applying them to the execution traces. But in the end, it can be applied to dynamic and static analysis as well.

1.5.2   Dynamic Malware Analysis

As we have seen in the previous section, static analysis evolved into a strong tool for malware discovery. But this attacker/defender race is in fact a fast co-evolution. As the answer, malicious authors started to employ legitimate tools such as packing and encryption in order to overcome the static analysis. In the previous section, we mentioned few recent works that attempted to tackle with packed binaries but in many cases, executing the binary is still the only way to discover its true purpose. And this is exactly the place for dynamic malware analysis. Dynamic analysis doesn't try to understand or find signatures within packed or encrypted binaries, it rather lets them run in a controlled environment (sandbox) and analyses their interactions with the operating system.

The behaviour of a binary file under investigation can be observed trough system calls or higher level actions. System calls are, in modern operation systems, the only way for applications to interact with hardware. The higher-level actions include writing to a file, creation or modification of a registry key, starting a new process, and the like.

1.5.2.1 System Calls

There are multiple ways how to process logged system calls. Lanzi et al. [116] used n-grams extended by the information about operations involving registry keys and files initiated by the binary under investigation. A problem with n-grams, also mentioned in the context of static analysis, is the exponential growth of the feature space size with increasing n. Many ways how to reduce the dimensionality were proposed, e.g., singular value decomposition or linear discriminant analysis [117, 118]. The Pfoh et al. [119] suggested using a string based kernel together with the SVM classifier instead of a bag-of-words based similarity. They classified system call traces in small sections while keeping a moving average over the probability estimates produced by the SVM. With this approach, they achieved an appealing efficacy almost in real-time.

Very interesting is also a paper of Rieck et al. [120]. The authors used normalized histograms of n-grams as their feature vectors. This vectors are then clustered using hierarchical clustering algorithm. Prototypes are extracted from each cluster, employing a linear time algorithm by Gonzalez [121] that guarantees solutions with an objective function value within two times the optimal solution. Each of these prototypes then represents a malware family and every newly observed sample is compared to them. Another point we have to take into consideration is that malicious binaries often execute additional system calls to hide their true purpose. Furthermore, those additional system calls can be easily reordered or randomly left out as they are irrelevant for the outcome of malicious activity. Kolbitsch et al. [122] developed a graph representation of system calls which captures system call dependence and ignores irrelevant system calls. In that so-called binary behavioural graph there is an edge between two nodes only if node b uses as its input some data from the output of node a. This representation alongside with its detection performance is also well designed to identify parts of code, called slices, responsible for the key functionality. The slices approach was exploited in a later paper of the same authors [123] concerning multi class classification. The slices extracted from known malware families are used as a dictionary and slices extracted from new samples are compared to them.

A very similar approach was used in [124] to classify malicious binaries into known malware families, but in that case, a maximal common subgraph was used as a similarity metric. In a later paper, Park et al. [125] proposed a method that creates a binary behaviour graph based on system call traces of each binary file. From those graphs, a super graph representation of the whole malware family is then created.

The super graph contains a subgraph called hot path, which was observed in the majority of analysed malware samples of a particular family.

1.5.2.2 Higher-Level Actions

The adoption of evasive techniques, such as code injection, shadow attack or section encryption, has started a shift from monitoring sequences of system calls to monitoring higher-level actions. First attempts with higher-level actions were made by building hierarchical behaviour graphs to infer higher-level behaviours from combinations of system calls [126]. They were designed to detect alternative sequences of events that achieve the same high-level goal.

A different approach was presented by Rieck et al. [127]. They employed a CWSandbox [128] to log high-level actions using a technique called API hooking [129]. Their features are frequencies of strings contained in those logs. To enable generalization they use not only a higher level action with all its parameters, e.g. `copy_file(source=file1, target=directory)`, as one word but they also create other p words, where p denotes the number of used parameters, by always removing the last parameter. Such representation has a huge number of sparse features. In their experiments, 20 million features were extracted from 6000 samples.

Stiborek et al. [130] defined similarities specifically designed to capture different aspects of individual sources of information (files, registry keys, network connections). The problem of huge dimensionality was solved by clustering the features. The similarities to the resulting clusters then serve as new features. The authors also presented a multiple-instance learning paradigm into the malware analysis.

1.5.3 Hybrid Malware Analysis

The fact that advanced malware has developed mechanisms such as obfuscation or polymorphism to avoid detection by static analysis and also execution-stalling techniques to avoid detection while in sandbox has motivated Anderson et al. [131] to create a unified framework combining both types of analysis. Quoting the authors:

> A malicious executable can disguise itself in some views, disguising itself in every view while maintaining malicious intent will prove to be substantially more difficult.

They used bi-grams, opt-code sequences, control flow graph, dynamic instruction traces, dynamic system call traces and features extracted from the file itself, such as entropy, number of instruction and the like as a feature set. Then, a multiple kernel learning [132] was utilised to find a weighted combination of those data sources. The presented approach improved the detection capability of trained SVM classifier and helped to reduce the false positive rate.

Another hybrid analysis framework was introduced by Santos et al. [133] later that year. In this paper the opt-code sequences are extracted and their frequencies are used as static features. The manipulations with files and registry keys, actions regarding network connections, retrieval of system information and sandbox errors serve as their dynamic counterpart. More than 140 000 binary features were extracted, but only 1000 most important were used. The results confirmed that combining both types of features improves detection capability.

1.5.4 Taxonomy of Malware

The detection on the different OS systems is almost the same, but the threat landscape is very different. According to the McAfee quarterly report from spring 2018 [134], Windows executables are still predominant source of infection, followed by pdf and Android apk files. In the 2017, more than 180 millions of malicious samples for Windows, more than 7 millions of malicious samples for Android and only about 260 thousand malware samples for Mac OS were discovered. The diversity and low market coverage of various unix/linux distributions lead the malicious actors to almost completely ignore them as a targets.

Malware categorisation and naming is actually a much harder problem than it may seem. Main reasons being that malware is typically named and categorised according to the behaviour which lead to its detection. But different anti-virus systems use different detection algorithms and may detect malicious binary in different ways. Attempts to unify naming conventions dates back to early 1990s and to the alliance of anti-virus specialists called CARO (Computer AntiVirus Researcher's Organization). They have created a malware naming scheme. An uprise of new device platforms together with growing number of anti-virus vendors on the market and increasing sophistication of malicious programs lead to the situation when the scheme has ceased to be used [135].

From time to time, a new initiative to create a unified detection scheme appears. The latest project for providing unique identifier to a newly discovered malware was a project called Common Malware Enumeration [136]. A unique identifier of threats for anti-virus vendors is still far from being a reality. But currently, most of them at least follow the common naming scheme similar to this:

```
[Prefix:]Behaviour.Platform[.Name][.Variant],
```

where gen (generic) or heur (heuristic) are the most common prefixes, meaning that the malware binary was discovered by some kind of generic detector or heuristic process. Behaviours such as: Trojan, Adware, Worm, etc. indicates the type of a threat. The platform may represent operating system Win32, Android or a particular application like Apache server. The name is typically used only for well established malware families like Conficker, CryptoWall or Sality while being left out for newly discovered threats. The variant is most often a number of version or a single character denoting the same. In rare cases it can be indicator of specific domain or string found in source code.

The following paragraphs provide a brief overview of the most common malware types. We could provide deeper overview of the threat landscape with much more details but it would easily fill a separate book which would be outdated even before publishing. Therefore, we focus on the general and well established malware groups and the main differences between them:

- high-severity

  - viruses,
  - worms,
  - trojans,

- low-severity

  - adware,
  - pornware,
  - PUA/PUP.

Viruses are malicious programs that self-replicate on the local machine. They copy their parts into different folders and registry keys and spread on any available attached media, like USB sticks or mounted remote hard drives. Malicious actions done by viruses are various ranging form sending spam, exfiltrate sensitive data to taking control over the infected computer. Furthermore, many current viruses are highly modular and can provide additional functionality on demand. Because of the broad spectrum of the malicious intents of viruses, they are more often divided according to the method used to infect a computers:

- file viruses,
- boot viruses,
- macro viruses,
- script viruses.

Worms spread trough computer networks. Their dominant strategy is sending emails where the worms are in the form of an attachment or as a link to a malicious web resource. A smaller group of worms uses instant messengers or social networks to send malicious links. Those emails and social media messages are often accompanied with some social engineering campaign, to convince a user to really click on the link or open the attachment. There are also worms that are spreading inside network packets to infect a computer directly by exploiting network configuration errors, and the like. Worms, similarly to viruses, can be devided according to the infection method:

- mail worms,
- instant messenger worms,
- P2P worms,
- net worms.

Trojans, unlike viruses and worms don't spread nor self-replicate. They are malicious programs which perform unauthorised actions without user's knowledge. Their name comes from the Trojan horse and similarly to the famous wooden horse, modern trojans hide a highly intrusive content. Trojans are usually divided according to the type of malicious action they perform. The major types are:

- backdoor,
- banking trojan,
- click fraud,
- cryptocurrency miner,
- dropper,
- exploit,
- information stealer,
- ransomware.

Backdoors provide remote control of a user's device. They can install additional modules, send or receive files, execute another programs, log user's activity and credentials. *Banking trojans* are specifically designed to steal banking accounts, e-payment systems credentials or credit card information. *Click fraud* trojans intent is to access internet resources. It can be achieved by hijacking browser or replacing system files. It may be used to conduct distributed denial of services (DDoS) attacks, increase the number of visits of certain pages or generate clicks on particular online ads. DDoS attacks are launched from a huge networks of infected computers, often called botnet. In this type of attack, a huge number of computers simultaneously request resources for some type of service, until they are completely depleted. This often ends by the service being crashed or unavailable for its purpose *Cryptocurrency miners*, as their name suggest, are used to mine various cryptocurrencies. They does not present a threat to users sensitive data but they steal computational power of his/her device. They are typically file-less and run directly within a web browser. *Droppers* are used to secretly install malicious programs to victim's computer and launch them without any notification. Malware authors and net-worms often breach a users device by taking advantage of a vulnerability in software running on that device. A program designed to do so is called *exploit*. Authors often implement a tool which takes advantage of multiple vulnerabilities, *exploit kit*. Malicious programs designed to steal users logins, credentials and information about his/her computer, such as operating system, installed software, browser and its extensions, are called *information stealers*. The last type of trojan we will discuss in this section is called *ransomware*. Ransomware is recent addition to the trojan family which, instantly after infection, encrypts user's disk and demands a ransom. The ransom is typically several hundreds dollars in some obscure cryptocurrency, most recently monero and z-cash seems to be favourite currencies due to their private blockchain.

Adware serves as a collective name for the malware family designed to display advertisements. Various approaches ranging from silently injecting additional banner or two into a web browser (add injectors) to violently jumping windows with endless adds (malwertising) are employed. In addition, adware can also gather sensitive information about users to customise presented adds. They can also tamper results of search engines to deliberate redirect users to particular web pages (fake search engines). Those pages can be both legitimate or malicious.

Pornware's functionality is very similar to the adware but instead of showing advertisements it forces users to visit malicious or paid porn servers.

PUA/PUP is an acronym for a possible unwanted application/program. Typically, it is installed together with some freeware or shareware and in most cases with user's agreement. They seemingly serve a legitimate purpose while being some sort of an adware/trojan in disguise. A browser toolbar with fake search engine can serve as a good example.

1.5.5 Malware Detection as a Classification Task

Similarly to other applications discussed in this chapter, malware detection accomplishes a mapping of multidimensional vectors of features, in this case features charcterizing the considered software, to one of particular classes. As was explained above, the features can be of two principally different kinds:

- Features resulting from a static analysis of the software file. Examples of such features are the structure of the file, structure and size of its header, the availability of a digital certificate, the presence of packing, encryption and overlying, information about the entry point, about dynamic libraries, imported and exported functions and about calls to API functions of the operating system present in the code, to recall at least most important ones.
- Features, resulting from a dynamic analysis of the software, i.e., from running it in a controlled environment (sandbox). As was recalled in Sect. 1.5.2, these are system calls, or higher level actions, such as writing to a file, creation or modification of a registry key, or starting a new process.

The set of classes includes in the most simple case only malware and benign software. Sometimes, PUA/PUP is added as an an additional class between both of them, and finally, it is possible to discriminate between various subsets of malware as described in the previous subsection.

Binary classification is dominant in malware detection, but the multi-class classification is slowly being included. The users are not satisfied to know only whether a code is suspicious or malicious, they want to know how serious the infection is. There is clearly a difference between seeing an additional advertisement banner here and there and getting all your precious data encrypted or bank account stolen.

The classification of malware samples is very challenging task, the main reason being a quickly changing behaviour of malicious and also benign software. Therefore, pre-trained models cannot be used or have to be regularly updated. A second very serious reason is that obtaining a large enough, up-to-date and well labelled dataset is almost impossible. The publicly available datasets are typically very small, unreliably labelled and soon outdated. It may be surprising that while in the wild, benign files are still the majority class, malware is dominant in the training datasets.

1.6 Network Intrusion Detection

An *Intrusion Detection System (IDS)* is a hardware or software implementation of a network security monitoring technology, which is designed to be a passive or listen-only device. The IDS monitors traffic and reports its results to an administrator, but cannot automatically take action or block incoming/outcoming traffic. The active variant of IDS which is able to take an action and therefore, prevent malicious actions to happen, is called *Intrusion Prevention System (IPS)*.

Another difference between the IDS and IPS solution is its location in a network. IPS solutions are placed directly inline to be able to filter or block network traffic in real-time, whereas IDS typically sits out-of-band where it analyses a copy of the inline traffic stream. IDS was originally developed this way because at the time the depth of analysis required for intrusion detection could not be performed fast enough to keep pace with components on the direct communications path of the network infrastructure.

Nowadays, both approaches IDS and IPS are often combined. Furthermore, in the context of this book, which is in this section the classification of network traffic, both solutions are equivalent. Therefore, we will use only the acronym IDS.

There are many types of IDS solutions differing by what kind of data they analyse and how. From the data perspective, we can differentiate network-based (NIDS) and host-based (HIDS). An IDS that analyses network traffic is an example of NIDS, while an IDS that monitors important operating system files is an example of a HIDS. According to the detection method the IDS are *signature-based* or *anomaly-based*, both will be in more detail described below. The signature-based IDS typically use predefined set of rules matching signatures of known threats. Signature-based IDS are excellent in detecting well known attacks, but are vulnerable to modified or novel attacks, also known as zero-day attacks. On the other hand, the anomaly-based IDS typically have multiple statistical or machine learning detectors, which can discover even previously unseen threats at the cost of lower precision on the already seen ones. Most current solutions successfully combine HIDS and NIDS but surprisingly a combination of anomaly-based and signature-based IDS is still rare, despite a scientific effort in this field [137, 138].

1.6.1 A Brief History of IDS

The first IDS started to emerge in early 80's after the [139] was published by the pioneer in information security and member of the Defense Science Board Task Force on Computer Security at the U.S. Air Force, James P. Anderson. A first prototype based on his ideas, called the *Intrusion Detection Expert System (IDES)*, was developed by Dr. Dorothy Denning in 1984. What started as a simple signature-based IDS, evolved over the years into a mature IDS with combination of signature list and strong anomaly detection core [140]. It was used for tracking and analysing audit data containing authentication information of users.

The work of Dorothy Denning and Peter Neumann also led to the design and development of an audit-trail analyser for remote logins into MILNET/ARPANET, providing both live detection and after-the-intrusion analysis [141].

A few years later, another working prototype [142] was developed at the University of California as a security solution for the U.S. Air Force. As the inventors said in an interview, "searching through this large amount of data for one specific misuse was equivalent to looking for a needle in a haystack", so they called their IDS *Haystack*. Haystack worked as a signature based IDS with a predefined set of rules designed for monitoring server machines. Its successor *Distributed Intrusion Detection System (DIDS)* [143] was extended to be able to monitor servers as well as client workstations. The authors of Haystack, and later DIDS, formed the commercial company Haystack Labs and were the first who started to sell IDS solutions to a wide public. Their first commercial product line was a host-based IDS called the Stalker.

Both IDES and Haystack were originally designed as purely host-based, meaning that they monitored only client workstations or servers. The first who came with the idea of a network-based IDS was Todd Heberlein in 1990 [144]. Heberlein with his team implemented this new idea in the NIDS called *Network Security Monitor (NSM)*. NSM was successfully deployed at major government installations. That generated more interest and consequently more investments in to the field of intrusion detection. Heberlein's idea also inspired the Haystack Labs team and together they introduced the first hybrid IDS (host-based together with network-based). The work of the Haystack project and the introduction of the Network Security Monitor revolutionised the IDS field and brought it into the commercial world.

In 1990, Lunt proposed adding an artificial neural network as a third component to the IDES [145]. In the following years, IDES was transformed into its successor called Next-generation Intrusion Detection Expert System (NIDES) [146]. NIDES was able to process both host-based and network-based data. It was distributed in that sense that it had multiple data collectors and one or multiple dedicated analytic nodes. Furthermore, it was highly modular, enabling the anomaly-based engine to run alongside signature-based and first artificial intelligence based modules.

Real commercial success for the IDS/IPS product families started around 1997, when Internet Security Systems, the security leader of that time, introduced a network intrusion detection system called RealSecure. After that, all big players in networking or security like Cisco, Symantec and others introduced their own IDS solutions.

The year 1999 brought two intrusion detection systems used up until now. First, the Snort [147] is an example of a pure signature-based network intrusion detection system. The Snort engine is currently free open-source, but unfortunately, rule sets are paid. Second, the Bro Network Security Monitor [148] is also an open source, yet delivered under BSD license. It can be used as a network IDS and/or for collecting network measurements and conducting forensic investigations.

Despite the two previous examples, most later solutions relied heavily on machine learning, such as supervised and unsupervised anomaly detection and classification. A vast amount of machine learning based methods were proposed to serve as a part or whole IDS, e.g. artificial immune systems [149], neural networks [150, 151], self organising maps [152] support vector machines [153, 154], random forests [155], genetic algorithms [156] and many others. For more references see e.g [157, 158].

For more details about history and evolution of IDS/IPS, see [159, 160].

1.6.2 Common Kinds of Attacks

A reliable IDS has to collect enough evidence to convince network administrator/ security analyst and provide a description of what actually happened. A first step into this is a multi-class classification. Knowing what type of attack it is can really help in mitigating the threat and/or with remediation [137, 161, 162]. According to the McAfee Labs quarterly threat report [163], the most common attacks in 2017 were:

- attacks against browser,
- password brute forcing,
- denial of service,
- worms,
- malware (cf. Sect. 1.5),
- web attacks,
- scanning.

Attacks against browser are on upraising trend since 2014. The reason for such popularity is that users may use a multiple different operating system such as different versions of Windows, macOS, Linux, Unix or Android, all with different set of security patches and therefore different sets of vulnerabilities. But all users do use an internet browser. Current internet browsers (Firefox, Google Chrome, Opera, Internet Explorer,...) have by default enabled the automatic update functionality, therefore the user landscape is much more homogenous. The two typical scenarios of attack agains browser are AdInjector plugin and watering hole. The AdInjectors typically exploit a vulnerability in the current version of a particular internet browser or its extension and install another plugin which shows additional advertisement to the user and/or tampers the search engine results. Watering hole attack is defined as infecting a popular web site visited by many users with a malicious code. Every visitor of that page can be potentially infected.

Brute force attacks simply attack any part of a network visible from outside through password guessing. There are two types: random guessing and dictionary brute force. Random guessing just tries all possible combinations of letters and numbers, optionally some special characters such as dots are included, until a match occurs. The dictionary attacks use the publicly available dictionaries of the most commonly used passwords (such as "admin" for the user "admin") and optionally try some simple modifications of them. Brute force attacks are the motivation for password policies in most companies.

Denial of service (DoS), or currently distributed denial of service (DDoS) is focused on resource depletion. There are many different kinds of DoS, but what all of them have in common, is flooding the targeted service with a huge amount of traffic. That service is than slowed down or stops working entirely. The distributed version is much harder to stop as the traffic is made by up to hundreds of thousands of unique sources. In some cases the DDoS attack is used as a smokescreen for some sophisticated attack.

The worms were a prominent sources of infection in the early days of internet and are on uprise once again. The main characteristics of worms is they can multiply and spread themselves once executed. The typical source of initial infection is via e-mail attachments and malicious URL. The most common type of internet worms nowadays is the already mentioned ransomware. Once such worms get into a network, they infect nearly all devices in it.

Web services and databases are also a target of network attack. The most prominent ways to compromise web applications are trough Cross-Site Scripting (XSS), SQL injection and path traversal.

Scans are rather a pre-attack reconnaissance than an attack itself. Two basic types of scans are differentiated: horizontal scans and vertical scans. A horizontal scan denotes scanning a group of devices on one particular port, e.g. 80, 443, 23 etc., whereas a vertical scan is the mapping of one device across a wide range of ports.

1.6.3   Host-Based Intrusion Detection

Host-based intrusion detection systems (HIDS) analyse audit data of the operating system (OS) or/and applications logs with the goal of identifying malicious activity. The operating system level intrusion detection usually relates to low-level system operations such as system calls, the modification of crucial system files and user actions like logins attempts. As those operations are low-level (close to a kernel), they are usually reliable and difficult to tamper, unless the system is compromised at the kernel level. However, the audit data provided by the OS are not designed to be used by IDS, therefore, they contain irrelevant information and often lack some valuable information as well. The question what kind of information should be provided by OS and logged is a hot topic since 1993 [164]. The original idea has been refined several times since then e.g. by Daniels [165].

The signature-based techniques, called misuse detection in older literature, perform a real time comparison of the ongoing audit data stream against the list of threat descriptions (signatures) and trigger an alarm whenever a signature is matched. Those threat descriptions are written in so called attack language. An attack language provides mechanisms and abstractions for identifying the manifestation of an attack. Unfortunately, almost every IDS vendor has its own attack language and only few of them support more than one. As an example of more wide spread attack language may serve the P-Best [166], which was used in NIDES, or STATL [167], which supports state transitions and therefore, enables the state-full description of an attack.

The anomaly based approaches create models of normal behaviour, sometimes called profiles, and detect any deviation. Contrary to the misuse detection, everything that doesn't fit into those models is reported as malicious. One can easily imagine that signature based HIDS are doing a behavioural black-listing whereas the anomaly detection HIDS are doing behavioural whitelisting. The main advantage of the anomaly detection is that it can discover a zero-day or strongly modified threats but at the cost of occasional false alarms.

Anomaly models can be specified by a user or learned from data. In the former, models are manually written by the administrator based on his/her expert knowledge or by an analysis of the application code. The early systems were mostly based on specification. One of the first was [168], which was soon refined with more focus given to system calls [169]. An alternative approach is called sandbox and was presented in [170]. A sandbox is an artificial environment where computer programs can be run with constrained access to resources and all system calls, the used libraries as well as other descriptors are logged for further analysis. There are known cases of malware escaping or fooling the sandbox environment [171] but still the information obtained by sandboxing is trusted and nowadays it is a crucial part of all anti-virus products. Actually, sandboxes are currently used as an additional layer of security in web browsers, pdf readers, touch device OS and many other programs too complex to be bug proof.

An example of a learned HIDS is [172]. During the training phase, it collects sequences of system calls of benign processes, creating signatures describing normal behaviours. In the detection phase it compares system calls of all processes and triggers an alarm whenever no match is found within the list of benign signatures. This and even following approaches missed the history of program calls. An extension that included the context provided by the program call stack was described in [173]. It allowed to trigger an alarm not only on a wrong sequence but even on a bad order of otherwise benign sequences.

An application level intrusion detection uses information provided directly by applications in the form of logs, API or via integration. Application data are reliable, rich and very specific making it easy to identify application responsible for a particular event. Unfortunately, each application has its own logging format and provides a different information with a varying level of detail. Although the operating systems tend to store the application logs into the centralised logging directory many applications store their own logs elsewhere making it difficult to unify the data gathering process. The last disadvantage is that an event is logged only after it was already finished, making it impossible for IDS to play the role of an IPS.

The situation is getting better, at least with the different logging format, as more and more software vendors implement a logging format similar to the common log file format [174], the extended log file format [175] or at least allow users to specify its own log format. Both the common and extended log file formats were designed for web servers but they can be used for a much wider set of applications with only a minor changes. The common log file format has the following pattern:

```
host userID username date request status bytes
```

Let us consider a following example for illustration.

```
10.0.0.42 - bajeluk [10/Jan/2017:11:42:00 -0100]
"GET /PhD_thesis.pdf HTTP/1.0" 404 80
```

This entry logs an attempt of user bajeluk from his local IP 10.0.0.42 to get file PhD_thesis.pdf, unfortunately the file did not exist at that time (status 404 stands for file not found) and the response of the server contained 80 bytes. The extended log format is more flexible allowing additional fields or on the contrary less fields to be present.

One of the simplest signature-based systems that monitor application audit data is a UNIX simple watch daemon (Swatch). This tool monitors log files, filters out unwanted data and takes user specified actions based upon found patterns. Patterns are written in a form of regular expressions. The logSTAT [176] tool also works with application logs, but it uses a STATL attack language. Let us recall that the STATL attack langue is a state-full approach to describing more complex threats using state transactions.

An example of an anomaly-based application monitor that is learned from data is DIDAFIT (Detecting Intrusions in Databases trough Fingerprinting Transactions) [177]. DIDAFIT as an application specific monitoring system was designed to protect valuable databases by learning patterns of legitimate database transactions and alerting the administrator if an unusual or high-risk SQL statement occurred. In [178] yet another example of the application specific anomaly detection is described, this time targeted on web servers and clients for web based applications. The system analyses client queries and their parameters seeking a sudden increase of the usage frequency at unusual time and many other anomalies which may signalise an attack or misuse of a service.

1.6.4   Network-Based Intrusion Detection

A network-based IDS monitors one or more network segments and searches for a suspicious activity. The location of the IDS within the network depends on the desired speed of analysis. If our goal is a real-time analysis with the ability to block incoming/outgoing traffic, we employ the term inline mode and use an inline placement, otherwise the IDS is typically located in a dedicated part of the network and does an offline analysis.

An IDS in the inline mode directly monitors all incoming and outgoing traffic in real time and is able to act preemptively and successfully block recognised incoming attacks or misuse of a service. But processing all traffic in real time is of course resource demanding and thus an inline IDS can be a bottle-neck when guarding a large network segment. In fact, deploying an inline IDS on a highly saturated network can make the whole infrastructure susceptible to DoS attacks. Therefore, inline IDS are typically put behind the firewall, which does the initial pre-filtering, lowering the amount of analysed traffic.

The IDS in offline mode can be implemented in two ways, using port mirroring or via a test access point (TAP). A port mirroring functionality may be found under a variety of names depending on the vendor, e.g., Switched Port Analyzer (SPAN), used by Cisco, or Roving Analysis Port (RAP), used by 3com. It is a port in a network switch where the whole traffic is visible. Therefore, it is an ideal place to plug-in an offline IDS or any other kind of network monitoring device. Although an IDS in offline mode does not need to process data in real time, there is still a vulnerability to the DoS type of attacks. As port mirroring consumes additional resources, a switch may start to sample the traffic cloned to the SPAN port or disable the port until enough resources are available when it is too busy. On the other hand, this solution is cheap as the SPAN port is already there and ready to use on most of the network switches.

An offline mode using TAPs is more robust and flexible than using port mirroring as it can be placed anywhere in the network, not only on switches with corresponding functionality. TAP devices are designed to handle fully saturated network lines without sampling or loosing packets. Depending on the type of final analyses, it may be seen as another advantage that TAP devices mirror the data perfectly including physical errors. On the other hand, a TAP device brings additional cost to the network infrastructure (Figs. 1.9, 1.10).

From the above paragraphs, it should be evident that apart from location, a fast enough data collection is another crucial element. For small network segments, it is adequate to install SW data collector on crucial routers. SW solutions are cheap and some of them are even open-source or at least free to use. On the other hand it puts a computational burden on the routers. Therefore, if we move to high velocity, highly saturated or just larger network segments, it is necessary to use dedicated hardware such as Flowmon probe [179]. Currently, such hardware is able to handle fully saturated 100Gb optical network, that is at the time of writing this book enough even for the backbone traffic.

According to the used detection method, NIDS similarly as HIDS can be divided into the signature-based and anomaly-based. Both signature-based and anomaly-based detection approaches are most frequently build around analysis of the uniform resource locator (URL). The URL is in fact a structured object consisting of protocol, hostname, path and query, following the scheme:

```
[protocol]://[hostname]/[path]?[query]
```
In the example:
```
https://en.wikipedia.org/w/index.php?title=URL&action
=info
```

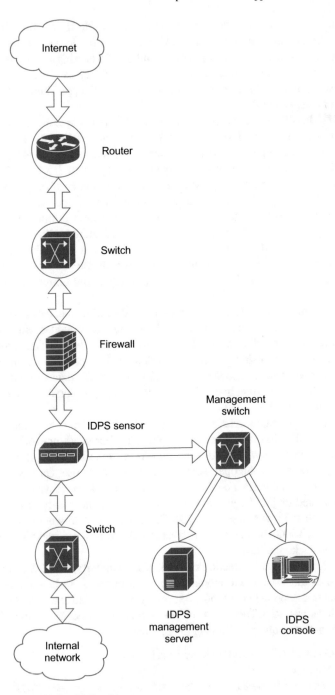

Fig. 1.9 Example of inline IDS architecture

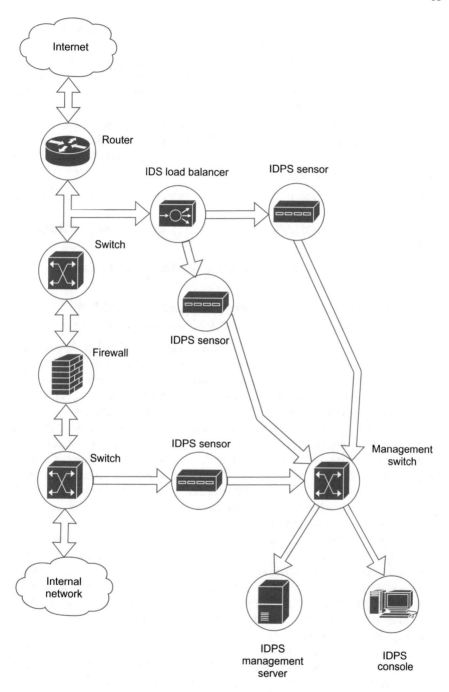

Fig. 1.10 Example of offline IDS architecture, showing both SPAN port and tap alternatives

The `https` protocol is a secure version of the hypertext transfer protocol, `en` is a subdomain, `wikipedia` a second level domain (SLD), `org` is a top level domain (TLD). Subdomain with SLD and TLD together are typically called host or hostname. The path part, `w/index.php` in the above example, marks the location of the requested file. And the last but in many cases most interesting part the query, `?title=URL&action=info`, stores addition information or parameters of the request. The query can be used by adversaries to exfiltrate personal data such as: browser version, installed extensions, size of a display and its resolution, operating system and much more. The exfiltrated message is often encoded as a base64 string as illustrated in the following example:

P2FwcElEPTEwMCZzeXN0ZW1JRDO3MzEmYnJvd3NlclR5cGVJRDOxNCZ
PU01EPTY
mc2NyPTE5MjB4MTA4MCZ3bmQ9MTkyMHg5NjAmcmFuZDOwLjk4NTQ
5OTE5MTE=

This visually random string in fact contains a lot of useful information when decoded:

?appID=104&systemID=731&browserTypeID=14&OSID=6
&scr=1920x1080&wnd=1920x960&rand=0.9854991911

The encoded query contains the information about application that sent the request, about operating system, browser, screen and window resolution along with some custom fields. Those fields are most probably used for some kind of targeted advertisement, but it also can be an initial information gathering before launching a malware campaign target for an OS version with a specific vulnerability.

There are many features that can be extracted from network traffic such as protocol, source and destination IP addresses and ports, counts of transferred bytes, elapsed times, to name a few, the opinions as to their usefulness strongly differ. The most research reported in the literature was done using time-based features (length of connection, inter-arrival times,...) and transferred bytes. The results in the scientific papers may seem promising, but in the real world scenarios, they typically fail. This is because a real world environment is much more heterogenous and noisy. For instance, the inter-arrival times depend on many influences (such as the number of devices on the route, the current work-load of those devices, the network architecture, etc.), which can completely invalidate any timing data. The count of transferred bytes is more useful when aggregated over a longer period to detect data exfiltrations, rather than used as a feature for classification of individual transactions. Of course, there were cases when attackers managed to produce packets of exactly the same size, but such exploits died quickly.

According to our experience, the most valuable features for intrusion detection are currently URL and referrer features for the HTTP and TLS based features for the HTTPS. This is also reflected by the fact that commercial signature-based detection algorithms usually rely on patterns in the URL. The simplest methods block a connection or trigger an alarm when a visited domain/hostname is marked as malicious, bogus or fraudulent in their blacklist. Unfortunately, current attackers register sub-domains/domains automatically and in huge amounts and change them as often as possible, making the blacklisting approach impractical and ineffective. In the most harsh malicious campaign we have seen, domains were rotated after several hours, but a typical rotation period spans from few weeks to months.

More sophisticated signature-based approaches also include patterns on the path and/or query part of the URL. Still there are many options how to avoid detection, e.g. often change the parameter names and order, name the parameters randomly and relay only on their order, add parameters with random values or simply encrypt the whole query string. But the last nail into the coffin for signature based detection is the upraising usage of the secured version of the HTTP protocol, HTTPS. The https protocol ensures users security and privacy by encrypting their traffic, but on the other hand, it hides the most important features, URL and referrer, from the analysts. In the best possible scenario, there is at least hostname which can still be used for blacklisting, but typically, there are only IP addresses. Blocking a whole IP address range instead of a particular domain is risky as multiple domains are typically hosted within the same IP range, especially now when huge cloud providers as Amazon and Google change IP addresses for particular services regularly due to load balancing. Another property worth mentioning about https traffic is that it enables multiple connections to be aggregated under one https tunnel, effectively making any information about transferred bytes, timings and other features unusable. If this is still not enough, there is domain fronting. Domain fronting is a technique that allows visit almost any site while still being hidden within a completely legitimate https tunnel to, e.g., google.com. For more information, check [180].

The anomaly based detection can help with solving some of the above stated issues of signature based detection. There are two basic approaches—the deep packet inspection (DPI), and the NetFlow/Proxy log analysis. DPI does an analysis of the payload of the transferred packets. Due to this unique information, it exactly knows what is transferred, thus its precision can be really high but at the cost of overwhelming computational costs. Therefore, it can be used on small networks only and of course, it can be easily avoided using the https protocol as in that case the payload is encrypted. A possible remedy is to intercept traffic at the network gate, to decrypt it.

The NetFlow/WebFlow analysis, on the other hand, doesn't know anything about payload, nor even the individual packets itself. The NetFlows were designed by Cisco for the analysis in large and fast networks. A simplistic definition of a NetFlow (network flow) can be that it is the unidirectional sequence of packets that share protocol, source and destination IPs and ports. The exact definition of the net flow differs version by version mostly by the exported fields. The exact definitions of the most common versions 5, 9 and IPFIX can be found in [181–183]. Proxy logs are very similar but are collected at the application layer instead of the network layer.

For both the NetFlows and Proxy logs, the statistics on network traffic are collected by the routers and switches or proxies and later exported. The export is typically done every 5-minutes, therefore, the analysis cannot be performed in real-time or used as a prevention. It is, however, well suited for discovery of the ongoing infections in the network environment. The obtained statistics are very useful for discovering command and control channels of the botnets, fraudulent transactions of banking trojans, malicious advertisement or click-fraud, to name some examples. On the other hand, they are typically insufficient to discover upraising threats called ransomware. Ransomware encrypts files on the attacked computer and demands a ransom for providing a decryption key. Such malware has typically very few network connections as it only needs to communicate with its master server once, therefore, the only suitable way is to discover it with a DPI or HIDS.

Anomaly-detection-based NIDS and flow analytics are currently the best performing methods, and they still improve in their precision and recall. Finally, they are becoming accepted by a wide public, and more and more companies are starting to use a machine learning solution for network security. Nevertheless, the progression of latest machine learning discoveries is still introduced very slowly into practice. The main reason is lack of a common, authentic and large enough dataset with detailed and accurate labels. Up to date, most research was done on datasets like KDD99, despite it is almost 20 years old and contains mostly extinct threats, whereas the network background completely changed since the time it was crated.

Contrary to the current state, a high number of false alarms in early stages of anomaly-based IDS nearly caused their rejection due to a high mistrust. The main reason why anomaly-based approaches are still being overlooked despite a substantial improvement in machine learning methods is that rules for already known attacks can be written very specifically, whereas anomaly detection needs to be general and report any suspicious activity. It is much easier to describe a particular behaviour than to create a detector that would fit into every network environment. It is very hard to design features general enough to be applicable to most networks yet specific enough to detect suspicious behaviour. It is even harder because, usually, the anomaly model cannot be trained from known malware samples as they originate from a different network environment. Hence, such models have to be trained directly within the target network and kept up to date. Therefore, it is very difficult to have a ready-to-deploy solution that is based solely on anomaly detection. Furthermore, the results of anomaly detection have a context-dependent nature: what can be a serious anomaly in one network may be considered as a legitimate trafic in other networks.

Even within one network there can be segments with a really different behaviour. For example, imagine a hospital network. The network may have, e.g., three segments with a very different privileges and expected normal behaviour. The first segment would be a public segment that provides a Wi-Fi to patients, visitors, or anyone within the reach of wireless access points. In this segment, the heterogenous traffic with a lot of suspicious behaviour should be expected. The second segment would serve to administrative personal with access to file-servers with files ranging from accounting to public relations. Most of the traffic in this network segment would be trough verified software channels and restricted to working hours only. The third and most critical segment would be available only for physicians granting them access to sensitive data on patients. In this segment, you would expect the highest risk and even a slight discrepancy from the expected behaviour should be reported to the admin.

From this simple example you can imagine that different policies are necessary for rising an alarm even in case of a small network located in one or few geographically close buildings.

To wrap up the above paragraphs, anomaly detection is really able to detect novel and modified attacks, malware and even misbehaving devices in the network. Typically, it detects a majority of suspicious behaviours present in the network, but at the same time, it tends to have a higher false positive rate or higher maintenance costs for more specialised anomaly models. In the current situation, when typical attacker changed from a lonely computer enthusiast looking for challenge to professional organisations full of well trained experts and when the number of devices and consequently the number of vulnerabilities is growing so rapidly that there is no way we can take care of all of it manually, it is more obvious than ever before that smart machine learning solutions are our only option to ever win the never ending race with attackers.

1.6.5 Other Types

One of the latest inventions are collaborative intrusion detection networks (CIDN). CIDN are overlay networks that connect Intrusion Detection Systems and provide information sharing such as alerts, blacklists, signatures, etc. There are multiple advantages of using CIDN over an isolated intrusion detection system, which strictly relies on updates from its vendor, whereas multiple IDS can share knowledge base and react more accurately and faster on novel threats. Thanks to the knowledge sharing, it can better correlate detected threats and discover ongoing large-scale attacks. Furthermore, it provides a robustness against adversaries trying to compromise an IDS as its final decision is based on multiple opinions. Therefore, if well configured, the CIDN should survive and work well even if its part was compromised. On the other hand, building a well tuned CIDN brings many challenges. The main problems are the communication overhead and computational resources spent on collaboration. The other question is the expertise of different CIDN nodes and how to decide whom to trust more and vice versa and of course, how to find a compromised or misbehaving node. The CIDN is a promising concept but there is still much work to be done before it becomes a common solution. For more information, we recommend [184].

The other interesting concept in the intrusion detection are honeypots. Honeypots are physical or virtual network nodes that look like a weak spot in the network and many times are pretending to have access to valuable data to lure the attacker. But actually they are thoroughly monitored, with highly restricted access to the other parts of the network if not completely isolated. Sometimes, honeypots have intentionally weakened protection and security settings. If a honeypot is scanned, attacked and breached an administrator has an immediate information about the malicious actions made on this node and can observe and recognize the attackers strategy. This strategy can be then included into the signature-based IDS as one of the rules and all security weaknesses used to breach the honeypot should be patched.

There has been an interesting project dealing with high-interaction honeypots called honeynet project [185].

1.6.6 IDS as a Classification Task

In the previous sections, we introduced many different types of IDS. They may differ in the monitored traffic, location in the network or triggering principles, but at the end, all of them have to decide if the monitored object is normal/benign or irregular/malicious. This represents a typical example of binary classification, and is the basis of all IDS solutions, no matter whether they are host-based or network-based, whether they are online or offline or whether they are signature-based or anomaly-based. The overwhelming majority of approaches still rely on features hand-designed by domain experts and only a few really novel approaches at least tried the automatic feature extraction, e.g., using autoencoders.

References

1. Hiskey, D.: How the word "Spam" came to mean "Junk Message" (2010). http://www.todayifoundout.com/index.php/2010/09/
2. European Parliament and the Council: Directive 2003/58/EC (2003). OJ L 221
3. Ministry of Justice of Canada: S.C. 2010, c.23. An Act to Promote the Efficiency and Adaptability of the Canadian Economy by Regulating Certain Activities That Discourage Reliance on Electronic Means of Carrying Out Commercial Activities, and to the Canadian Radio-Television and Telecommunications Commission Act, the Competition Act, the Personal Information Protection and Electronic Documents Act and the Telecommunications Act (2010)
4. Congress, U.S.: Controlling the assault of non-solicited pornography and marketing act. (CAN-SPAM Act) (2003)
5. O'Gorman, B., Wueest, C., O'Brien, D., Cleary, G., Lau, H., Power, J., M., C., Cox, O., Wood, P., Wallace, S.: Internet security threat report. Technical Report, Symantec (2019)
6. Gray, A., Haahr, M.: Personalised, collaborative spam filtering. In: Proceedings of CEAS (2004)
7. Boykin, P., Roychowdhury, V.: Leveraging social networks to fight spam. Computer 61–68 (2005)
8. Krasser, S., Tang, Y., Gould, J., Alperovitch, D., Judge, P.: Identifying image spam based on header and file properties using C4. 5 decision trees and support vector machine learning. In: 2007 Information Assurance and Security Workshop, IAW'07, pp. 255–261 (2007)
9. Vergelis, M., Demidova, N., Scherbakova, T.: Spam and phishing in Q3 2018. Technical report, AO Kaspersky Lab (2018)
10. Gourville, J., Soman, D.: Overchoice and assortment type: when and why variety backfires. Mark. Sci. 24, 382–395 (2005)
11. Alspector, J., Koicz, A., Karunanithi, N.: Feature-based and clique-based user models for movie selection: a comparative study. User Model. User-Adapt. Interact. 7, 279–304 (1997)
12. Robinson, G.: Automated collaborative filtering in world wide web advertising. US Patent 5,918,014 (1999)

13. Davidson, J., Liebald, B., Liu, J., Nandy, P., Van Vleet, T.: The YouTube video recommendation system. In: Proceedings of the Fourth ACM Conference on Recommender Systems, pp. 293–296 (2010)
14. McGinty, L., Smyth, B.: On the role of diversity in conversational recommender systems. In: International Conference on Case-Based Reasoning, pp. 276–290 (2003)
15. Resnick, P., Varian, H.: Recommender systems. Commun. ACM **40**, 56–58 (1997)
16. Schafer, J., Konstan, J., Riedl, J.: Recommender systems in e-commerce. In: Proceedings of the 1st ACM Conference on Electronic Commerce, pp. 158–166 (1997)
17. Terveen, L., Hill, W.: *Human-Computer Interaction in the New Millenium*. Addison-Wesley, Reading (2001)
18. Auer, S., Bizer, C., Kobilarov, G., Lehmann, J., Cyganiak, R., Ives, Z.: DBpedia: a nucleus for a web of open data. In: The Semantic Web: 6th International Semantic Web Conference + 2nd Asian Semantic Web Conference, pp. 722–735 (2007)
19. Resnick, P., Iacovou, N., Suchak, M., Bergstrom, P., Riedl, J.: GroupLens: an open architecture for collaborative filtering of netnews. In: Proceedings of the 1994 ACM Conference on Computer Supported Cooperative Work, pp. 175–186 (1994)
20. Satzger, B., Endres, M., Kießling, W.: A preference-based recommender system. In: E-Commerce and Web Technologies, pp. 31–40. Springer (2006)
21. Chirita, P., Nejdl, W., Zamfir, C.: Preventing shilling attacks in online recommender systems. In: Proceedings of the 7th Annual ACM International Workshop on Web Information and Data Management, pp. 67–74 (2005)
22. Lam, S., Riedl, J.: Shilling recommender systems for fun and profit. In: Proceedings of the 13th International Conference on World Wide Web, pp. 393–402 (2004)
23. Adomavicius, G., Tuzhilin, A.: Toward the next generation of recommender systems: a survey of the state-of-the-art and possible extensions. IEEE Trans. Knowl. Data Eng. **17**, 734–749 (2005)
24. Burke, R.: Hybrid recommender systems: survey and experiments. User Model. User-Adapt. Interact. **12**, 331–370 (2002)
25. Kabiljo, M., Ilic, A.: Recommending items to more than a billion people. https://code.facebook.com/posts/861999383875667 (2015)
26. Ricci, F.: Mobile recommender systems. Inf. Technol. Tour. **12**, 205–231 (2010)
27. Commission Nationale de l'Informatique et des Libertés: The French Data Protection Authority Publicly Issues Formal Notice to Facebook to Comply with the French Data Protection Act Within Three Months. https://www.cnil.fr/en/french-data-protection-authority-publicly-issues-formal-notice-facebook-comply-french-data (2016)
28. European Commission: Information providers guide, the EU internet handbook: Cookies. http://ec.europa.eu/ipg/basics/legal/cookies/index_en.htm (2016)
29. Horrigan, J.: Online shopping. Technical Report, Pew Research Center. Internet & American Life Project (2008)
30. ComScore: Online consumer-generated reviews have significant impact on offline purchase behavior. http://www.comscore.com/press/release.asp?press=1928 (2015)
31. Rainie, L., Horrigan, J.: Election 2006 online. Technical report, Pew Research Center. Internet & American Life Project (2007)
32. Su, F., Markert, K.: From words to senses: a case study of subjectivity recognition. In: Proceedings of the 22nd International Conference on Computational Linguistics, vol. 1, pp. 825–832 (2008)
33. Subasic, P., Huettner, A.: Affect analysis of text using fuzzy semantic typing. IEEE Trans. Fuzzy Syst. **9**, 483–496 (2001)
34. Godbole, N., Srinivasaiah, M., Skiena, S.: Large-scale sentiment analysis for news and blogs. In: International Conference on Weblogs and Social Media, pp. 219–222 (2007)
35. Presser, M., Barnaghi, P., Eurich, M., Villalonga, C.: The SENSEI project: integrating the physical world with the digital world of the network of the future. IEEE Commun. Mag. **47**, 1–4 (2009)

36. O' Brien, S.: Humans and machines team up to predict brexit campaign result by analysing UK social chatter. Technical Report, Sensei Project (2016)
37. Sheth, A., Jadhav, A., Kapanipathi, P., Lu, C., Purohit, H., Smith, G., Wang, W.: Twitris: a system for collective social intelligence. In: Encyclopedia of Social Network Analysis and Mining, pp. 2240–2253. Springer (2014)
38. Donovan, J.: The Twitris sentiment analysis tool by Cognovi Labs predicted the brexit hours earlier than polls. https://techcrunch.com/2016/06/29/the-twitris-sentiment-analysis-tool-by-cognovi-labs-predicted-the-brexit-hours-earlier-than-polls/ (2016)
39. Pang, B., Lee, L.: Opinion mining and sentiment analysis. Found. Trends Inf. Retr. **2**, 1–135 (2008)
40. Leacock, C., Miller, G., Chodorow, M.: Using corpus statistics and WordNet relations for sense identification. Comput. Ling. **24**, 147–165 (1998)
41. Miller, G.: WordNet: a lexical database for English. Commun. ACM **38**, 39–41 (1995)
42. Baccianella, S., Esuli, A., Sebastiani, F.: SentiWordNet 3.0: an enhanced lexical resource for sentiment analysis and opinion mining. In: Language Resources and Evaluation Conference, pp. 2200–2204 (2010)
43. Turney, P., Littman, M.: Measuring praise and criticism: inference of semantic orientation from association. ACM Trans. Inf. Syst. **21**, 315–346 (2003)
44. Mikolov, T., Chen, K., Corrado, G., Dean, J.: Efficient estimation of word representations in vector space (2013). ArXiv preprint arxiv:1301.3781
45. Tang, D., Wei, F., Yang, N., Zhou, M., Liu, T., Qin, B.: Learning sentiment-specific word embedding for twitter sentiment classification. In: 52th Annual Conference of the Association for Computational Linguistics, pp. 1555–1565 (2014)
46. Le, Q., Mikolov, T.: Distributed representations of sentences and documents. In: International Conference on Machine Learning, pp. 1188–1196 (2014)
47. Pang, B., Lee, L., Vaithyanathan, S.: Thumbs up? Sentiment classification using machine learning techniques. In: Proceedings of the ACL-02 Conference on Empirical Methods in Natural Language Processing, vol. 10, pp. 79–86 (2002)
48. Hutto, C., Gilbert, E.: Vader: a parsimonious rule-based model for sentiment analysis of social media text. In: Eighth International AAAI Conference on Weblogs and Social Media, pp. 216–225 (2014)
49. Perkins, J.: *Python Text Processing with NLTK 2.0 Cookbook*. Packt Publishing, Birmingham (2010)
50. Subrahmanian, V., Reforgiato, D.: AVA: adjective-verb-adverb combinations for sentiment analysis. IEEE Intell. Syst. **23**, 43–50 (2008)
51. Gamallo, P., Garcia, M.: Citius: A Naive-Bayes strategy for sentiment analysis on English tweets. In: International Workshop on Semantic Evaluation, pp. 171–175 (2014)
52. Poria, S., Cambria, E., Gelbukh, A.: Deep convolutional neural network textual features and multiple kernel learning for utterance-level multimodal sentiment analysis. In: Empirical Methods in Natural Language Processing, pp. 2539–2544 (2015)
53. Boiy, E., Moens, M.: A machine learning approach to sentiment analysis in multilingual web texts. Inf. Retr. **12**, 526–558 (2009)
54. Hallmann, K., Kunneman, F., Liebrecht, C., van den Bosch, A., van Mulken, M.: Sarcastic soulmates: intimacy and irony markers in social media messaging. Linguist. Issues Lang. Technol. **14**, Paper 7 (2016)
55. Cheang, H., Pell, M.: The sound of sarcasm. Speech Commun. **50**, 315–346 (2008)
56. Narayanan, R., Liu, B., Choudhary, A.: Sentiment analysis of conditional sentences. In: Proceedings of the 2009 Conference on Empirical Methods in Natural Language Processing, pp. 180–189 (2009)
57. Devillers, L., Vidrascu, L.: Real-life emotions detection with lexical and paralinguistic cues on human-human call center dialogs. In: International Conference on Spoken Language Processing, Interspeech, pp. 801–804 (2006)
58. Cambria, E., Schuller, B., Xia, Y., Havasi, C.: New avenues in opinion mining and sentiment analysis. IEEE Intell. Syst. **28**, 15–21 (2013)

59. Poria, S., Cambria, E., Hussain, A., Huang, G.: Towards an intelligent framework for multi-modal affective data analysis. Neural Netw. **63**, 104–116 (2015)
60. Pu, P., Kumar, P.: Evaluating example-based search tools. In: ACM 5th Conference on Electronic Commerce, pp. 208–217 (2004)
61. Tesic, J.: Metadata practices for consumer photos. IEEE Multimed. **12**, 86–92 (2005)
62. International Organization for Standardization: Graphic Technology—Extensible Metadata Platform (XMP) Specification (2012)
63. American National Standards Institute, National Information Standards Organization: The Dublin Core Metadata Element Set (2013)
64. Berners-Lee, T., Hendler, J., Lassila, O.: The semantic web. Sci. Am. **284**, 28–37 (2001)
65. Castells, P., Fernandez, M., Vallet, D.: An adaptation of the vector-space model for ontology-based information retrieval. IEEE Trans. Knowl. Data Eng. **19**, 261–272 (2007)
66. Hollink, L., Schreiber, G., Wielemaker, J., Wielinga, B.: Semantic annotation of image collections. In: Knowledge Capture 2003—Knowledge Markup and Semantic Annotation Workshop, pp. 41–48 (2003)
67. Jonquet, C., LePendu, P., Falconer, S., Coulet, A., Noy, N., Musen, M., Shah, N.: NCBO resource index: ontology-based search and mining of biomedical resources. Web Seman. Sci. Serv. Agents World Wide Web **9**, 316–324 (2011)
68. Mac Kenzie, I., Zhang, S.: The immediate usability of graffiti. In: Graphics Interface, pp. 129–137 (1997)
69. Költringer, T., Grechenig, T.: Comparing the immediate usability of graffiti 2 and virtual keyboard. In: Human Factors in Computing Systems, pp. 1175–1178 (2004)
70. Tappert, C., Suen, C., Wakahara, T.: The state of the art in online handwriting recognition. IEEE Trans. Pattern Anal. Mach. Intell. **12**, 787–808 (1990)
71. Kirsch, D.: Detexify: Erkennung handgemalter latex-symbole. Ph.D. thesis, Westfälische Wilhelms-Universität Münster (2010)
72. Müller, M.: *Information Retrieval for Music and Motion*. Springer, Berlin (2007)
73. Keysers, D., Deselaers, T., Rowley, H., Wang, L., Carbune, V.: Multi-language online handwriting recognition. IEEE Trans. Pattern Anal. Mach. Intell. **39**, 1180–1194 (2017)
74. Ouyang, T., Davis, R.: A visual approach to sketched symbol recognition. In: IJCAI'09: 21st International Joint Conference on Artifical Intelligence, pp. 1463–1468 (2009)
75. Tang, X., Wang., X.: Face sketch synthesis and recognition. In: 9th IEEE International Conference on Computer Vision, pp. 687–694 (2003)
76. Jayant, N., Johnston, J., Safranek, R.: Signal compression based on models of human perception. Proc. IEEE **81**, 1385–1422 (1993)
77. Paola, J., Schowengerdt, R.: The effect of lossy image compression on image classification. In: International Geoscience and Remote Sensing Symposium—Quantitative Remote Sensing for Science and Applications, pp. 118–120 (1995)
78. Hafner, J., Sawhney, H., Equitz, W., Flickner, M., Niblack, W.: Efficient color histogram indexing for quadratic form distance functions. IEEE Trans. Pattern Anal. Mach. Intell. **7**, 729–736 (1995)
79. Deng, Y., Manjunath, B., Kenney, C., Moore, M., Shin, H.: An efficient color representation for image retrieval. IEEE Trans. Image Process. **10**, 140–147 (2001)
80. Canny, J.: A computational approach to edge detection. IEEE Trans. Pattern Anal. Mach. Intell. **8**, 679–698 (1986)
81. Lowe, D.: Object recognition from local scale-invariant features. In: 7th IEEE International Conference on Computer Vision, pp. 1150–1157 (1999)
82. Bay, H., Tuytelaars, T., Van Gool, L.: Surf: speeded up robust features. In: European Conference on Computer Vision, pp. 404–417 (2006)
83. Yang, J., Jiang, Y., Hauptman, A., Ngo, C.: Evaluating bag-of-visual-words representations in scene classification. In: ACM International Workshop on Multimedia Information Retrieval, pp. 197–206 (2007)
84. Dalal, N., Triggs, B.: Histograms of oriented gradients for human detection. In: IEEE Conference on Computer Vision and Pattern Recognition, pp. 886–893 (2005)

85. Dalal, N., Triggs, B., Schmid, C.: Human detection using oriented histograms of flow and appearance. In: European Conference on Computer Vision, pp. 428–441 (2006)
86. Chen, M., Hauptmann, A.: MoSIFT: recognizing human actions in surveillance videos. Technical Report, CMU, Computer Science Department (2009)
87. Chaudry, R., Ravichandran, A., Hager, G., Vidal, R.: Histograms of oriented optical flow and binet-cauchy kernels on nonlinear dynamical systems for the recognition of human actions. In: IEEE Conference on Computer Vision and Pattern Recognition, pp. 1932–1939 (2009)
88. Laptev, I., Pérez, P.: Retrieving actions in movies. In: 11th International Conference on Computer Vision (2007)
89. International Organization for Standardization: Information technology—Multimedia content description interface—Part 4: Audio (2002)
90. Wang, A.: The Shazam music recognition service. Commun. ACM **49**, 44–48 (2006)
91. Harb, H., Chen, L.: A query by example music retrieval algorithm. In: European Workshop on Image Analysis for Multimedia Interactive Services, pp. 122–128 (2003)
92. Tsai, W., Yu, H.: Query-by-example technique for retrieving cover versions of popular songs with similar melodies. In: Ismir, vol. 5, pp. 183–190 (2005)
93. Salamon, J., Gomez, E.: Melody extraction from polyphonic music signals using pitch contour characteristics. IEEE Trans. Audio Speech Lang. Process. **20**, 1759–1770 (2012)
94. Tao, L., Xinaglin, H., Lifang, Y., Pengju, Z.: Query by humming: comparing voices to voices. In: International Conference on Management and Service Science, pp. 5–8 (2009)
95. Myers, G., Nallapati, R., van Hout, J., Pancoast, S., Nevatia, R., Sun, C., Habibian, A., Koelma, D., van de Sande, K.E., Smeulders, A., Snoek, C.: Evaluating multimedia features and fusion for example-based event detection. Mach. Vis. Appl. **25**, 17–32 (2014)
96. Sadanand, S., Corso, J.: Action bank: a high-level representation of activity in video. In: IEEE Conference on Computer Vision and Pattern Recognition, pp. 1234–1241 (2012)
97. Lan, Z., Bao, L., Yu, S., Liu, W., Hauptmann, A.: Double fusion for multimedia event detection. In: Advances in Multimedia Modeling, pp. 173–185 (2012)
98. Morgan, S.: Cybercrime report. Technical Report, Cybersecurity Ventures (2017)
99. Richards, K., LaSalle, R., Devost, M., van den Dool, F., Kennedy-White, J.: Cost of cyber crime study. Technical Report, Accenture, Penomon Institute LLC (2017)
100. Lo, R., Levitt, K., Olsson, R.: MCF: a malicious code filter. Comput. Secur. **14**, 541–566 (1995)
101. Moser, A., Kruegel, C., Kirda, E.: Limits of static analysis for malware detection. In: Computer Security Applications Conference, pp. 421–430 (2007)
102. Ahmadi, M., Ulyanov, D., Trofimov, M., Giacinto, G.: Novel feature extraction, selection and fusion for effective malware family classification. In: Sixth ACM Conference on Data and Application Security and Privacy, pp. 183–194 (2016)
103. Cesare, S., Xiang, Y.: Classification of malware using structured control flow. In: Eighth Australasian Symposium on Parallel and Distributed Computing, pp. 61–70 (2010)
104. Han, K., Kang, B., Im, E.: Malware analysis using visualized image matrices. Sci. World J. **14**, 1–14 (2015)
105. Kiechle, D.: Fehlerraumapproximation durch verwendung von basisblock-fehlerinjektion. Master's thesis, Leibniz University Hannover (2018)
106. Bruschi, D., Martignoni, L., Monga, M.: Using code normalization for fighting self-mutating malware. In: International Symposium on Secure Software Engineering, pp. 37–44 (2006)
107. Christodorescu, M., Jha, S.: Static analysis of executables to detect malicious patterns. Technical Report, Department of Computer Sciences, University of Wisconsin, Madison (2006)
108. Kinable, J., Kostakis, O.: Malware classification based on call graph clustering. J. Comput. Virol. Hacking Tech. **7**, 351–366 (2011)
109. Kong, D., Yan, G.: Discriminant malware distance learning on structural information for automated malware classification. In: 19th ACM SIGKDD International Conference on Knowledge Discovery and Data Mining, pp. 1357–1365 (2013)
110. Santos, I., Penya, Y., Deves, J., Bringas, P.: N-grams-based file signatures for malware detection. In: 11th International Conference on Enterprise Information Systems, pp. 317–320 (2009)

111. Kephart, J.: A biologically inspired immune system for computers. In: Artificial Life IV: Proceedings of the Fourth International Workshop on the Synthesis and Simulation of Living Systems, pp. 130–139 (1994)

112. Reddy, D., Puajri, A.: N-gram analysis for computer virus detection. J. Comput. Virol. Hacking Tech. **2**, 231–239 (2006)

113. Stiborek, J.: Dynamic reconfiguration of intrusion detection systems. Ph.D. thesis, Czech Technical University, Prague (2017)

114. Nataraj, L., Yegneswaran, V., Porras, P., Zhang, J.: Malware images: visualization and automatic classification. In: 8th International Symposium on Visualization for Cyber Security, pp. 29–35 (2011)

115. Nataraj, L., Yegneswaran, V., Porras, P., Zhang, J.: A comparative assessment of malware classification using binary texture analysis and dynamic analysis. In: 4th ACM Workshop on Security and Artificial Intelligence, pp. 21–30 (2011)

116. Lanzi, A., Balzarotti, D., Kruegel, C., Christodorescu, M., Kirda, E.: Accessminer: using system-centric models for malware protection. In: 17th ACM Conference on Computer and Communications Security, pp. 399–412 (2010)

117. Canzanese, R., Kam, M., Mancoridis, S.: Toward an automatic, online behavioral malware classification system. In: 7th IEEE International Conference on Self-Adaptive and Self-Organizing Systems, pp. 111–120 (2013)

118. Canzanese, R., Mancoridis, S., Kam, M.: Run-time classification of malicious processes using system call analysis. In: 10th International Conference on Malicious and Unwanted Software, pp. 21–28 (2015)

119. Pfoh, J., Schneider, C., Eckert, C.: Leveraging string kernels for malware detection. In: International Conference on Network and System Security, pp. 206–219 (2013)

120. Rieck, K., Trinius, P., Willems, C., Holz, T.: Automatic analysis of malware behavior using machine learning. J. Comput. Secur. **19**, 639–668 (2011)

121. Gonzalez, T.: Clustering to minimize the maximum intercluster distance. Theor. Comput. Sci. **38**, 293–306 (1985)

122. Kolbitsch, C., Camparetti, P., Kruegel, C., Kirda, E., Zhou, X., Wang, X.: Effective and efficient malware detection at the end host. In: USENIX Security Symposium, pp. 351–366 (2009)

123. Kolbitsch, C., Holz, T., Kruegel, C., Kirda, E.: Inspector gadget: automated extraction of proprietary gadgets from malware binaries. In: IEEE Symposium on Security and Privacy, pp. 29–44 (2010)

124. Park, Y., Reeves, D., Mulukutla, V., Sundaravel, M.: Fast malware classification by automated behavioral graph matching. In: Sixth Annual Workshop on Cyber Security and Information Intelligence Research, pp. 45/1–4 (2010)

125. Park, Y., Reeves, D., Stamp, M.: Deriving common malware behavior through graph clustering. Comput. Secur. **39**, 419–430 (2013)

126. Martignoni, L., Stinson, E., Fredrikson, M., Jha, S., Mitchell, J.: A layered architecture for detecting malicious behaviors. In: International Workshop on Recent Advances in Intrusion Detection, pp. 78–97 (2008)

127. Rieck, K., Holz, T., Willems, C., Düssel, P., Laskov, P.: Learning and classification of malware behavior. In: International Conference on Detection of Intrusions and Malware, and Vulnerability Assessment, pp. 108–125 (2008)

128. Willems, C., Holz, T., Freiling, F.: CWSandbox: towards automated dynamic binary analysis. IEEE Secur. Priv. **5**, 32–39 (2007)

129. Hunt, G., Brubacher, D.: Detours: binary interception of Win32 functions. In: Third USENIX Windows NT Symposium, pp. 145–154 (1999)

130. Stiborek, J., Pevný, T., Řehák, M.: Multiple instance learning for malware classification. Expert Syst. Appl. **93**, 346–357 (2018)

131. Anderson, B., Storlie, C., Lane, T.: Improving malware classification: bridging the static/dynamic gap. In: Fifth ACM Workshop on Security and Artificial Intelligence, pp. 3–14 (2012)

132. Gönen, M., Alpaydm, E.: Multiple kernel learning algorithms. J. Mach. Learn. Res. **12**, 2211–2268 (2011)
133. Santos, I., Devesa, J., Brezo, F., Nieves, J., Bringas, P.: OPEM: A static-dynamic approach for machine-learning-based malware detection. In: International Joint Conference CISIS'12-ICEUTE'12-SOCO'12 Special Sessions, pp. 271–280 (2013)
134. Beek, C., Dunton, T., Grobman, S., Karlton, M., Minihane, N., Palm, C., Peterson, E., Samani, R., Schmugar, C., Sims, R., Sommer, D., Sun, B.: McAfee labs threat report. Technical Report, McAfee (2017)
135. Bontchev, V.: Current status of the CARO malware naming scheme. Slides presented in Virus Bulletin (2005)
136. Kuo, J., Beck, D.: The common malware enumeration (CME) initiative. Presented at the Virus Bulletin Conference (2005)
137. Depren, O., Topallar, M., Amarim, E., Ciliz, M.: An intelligent intrusion detection system (IDS) for anomaly and misuse detection in computer networks. Expert Syst. Appl. **29**, 713–722 (2005)
138. Gong, Y., Mabu, S., Chen, C., Wang, Y., Hirasawa, K.: Intrusion detection system combining misuse detection and anomaly detection using genetic network programming. In: ICCAS-SICE, pp. 3463–3467 (2009)
139. Anderson, J.: Computer security threat monitoring and surveillance. Technical Report, James P. Anderson Company, Fort Washington (1980)
140. Denning, D.: An intrusion-detection model. IEEE Trans. Softw. Eng. **13**, 222–232 (1987)
141. Neumann, P.: Audit trail analysis and usage collection and processing. Technical Report, SRI International (1985)
142. Smaha, S.: Haystack: an intrusion detection system. In: 4th IEEE Aerospace Computer Security Applications Conference, pp. 37–44 (1988)
143. Snapp, S., Brentano, J., Dias, G.V., Goan, T., Heberlein, L., Ho, C.L., Levitt, K., Mukherjee, B., Smaha, S., Grance, T., Teal, D., Mansur, D.: DIDS (distributed intrusion detection system)—motivation, architecture, and an early prototype. In: National Computer Security Conference, pp. 167–176 (1991)
144. Heberlein, L., Dias, G., Levitt, K., Mukherjee, B., Wood, J., Wolber, D.: A network security monitor. In: IEEE Computer Society Symposium on Research in Security and Privacy, pp. 296–304 (1990)
145. Lunt, T.: IDES: an intelligent system for detecting intruders. In: Symposium on Computer Security, Threat and Countermeasures, pp. 30–45 (1990)
146. Anderson, D., Frivold, T., Valdes, A.: Next-generation intrusion detection expert system (NIDES): a summary. Technical Report, SRI International, Menlo Park (1995)
147. Roesch, M.: Snort-lightweight intrusion detection for networks. In: USENIX 13th Conference on System Administration, pp. 229–238 (1999)
148. Paxson, V.: Bro: a system for detecting network intruders in real-time. Comput. Netw. **31**, 2435–2463 (1999)
149. DasGupta, D.: An overview of artificial immune systems and their applications. In: Artificial Immune Systems and Their Applications, pp. 3–21. Springer (1993)
150. Debar, H., Becker, M., Siboni, D.: A neural network component for an intrusion detection system. In: IEEE Computer Society Symposium on Research in Security and Privacy, pp. 240–250 (1992)
151. Ryan, J., Lin, M., Miikkulainen, R.: Intrusion detection with neural networks. Adv. Neural Inf. Process. Syst. **10**, 943–949 (1998)
152. Cannady, J.: Artificial neural networks for misuse detection. In: National Information Systems Security Conference, pp. 368–381 (1998)
153. Mukkamala, S., Janoski, G., Sung, A.: Intrusion detection using neural networks and support vector machines. In: International Joint Conference on Neural Networks, pp. 1702–1707 (2002)
154. Deng, H., Zeng, Q., Agrawal, D.: SVM-based intrusion detection system for wireless ad-hoc networks. In: 58th IEEE Vehicular Technology Conference, pp. 2147–2151 (2003)

155. Sabhnani, M., Serpen, G.: Application of machine learning algorithms to KDD intrusion detection dataset within misuse detection context. In: International Conference on Machine Learning, Models, Technologies and Applications, pp. 209–215 (2003)
156. Li, W.: Using genetic algorithm for network intrusion detection. In: United States Department of Energy Cyber Security Group Training Conference, pp. 24–27 (2004)
157. Tsai, C., Hsu, Y., Lin, C., Lin, W.: Intrusion detection by machine learning: a review. Expert Syst. Appl. **36**, 11994–12000 (2009)
158. Wu, S., Banzhaf, W.: The use of computational intelligence in intrusion detection systems: a review. Appl. Soft Comput. **10**, 1–35 (2010)
159. Axelsson, S.: Intrusion detection systems: a survey and taxonomy. Technical Report, Chalmers University of Technology, Göteborg (2000)
160. Innella, P.: The evolution of intrusion detection systems. Technical Report, Symantec (2011)
161. Machlica, L., Bartoš, K., Sofka, M.: Learning detectors of malicious web requests for intrusion detection in network traffic (2017). arXiv preprint arxiv:1702.02530
162. Sindhu, S., Geetha, S., Kannan, A.: Decision tree based light weight intrusion detection using a wrapper approach. Expert Syst. Appl. **39**, 129–141 (2012)
163. Beek, C., Dinkar, D., Frost, D., Grandjean, E., Moreno, F., Peterson, E., Rao, P., Samani, R., Schmugar, C., Simon, R., Sommer, D., Sun, B., Valenzuela, I., Weafer, V.: McAfee labs threat report. Technical Report, McAfee (2017)
164. Lunt, T.: Detecting intruders in computer systems. In: Conference on Auditing and Computer Technology (1993)
165. Daniels, T., Spafford, E.: Identification of host audit data to detect attacks on low-level IP vulnerabilities. J. Comput. Secur. **7**, 3–35 (1999)
166. Lindqvist, U., Porras, P.: Detecting computer and network misuse through the production-based expert system toolset P-BEST. In: IEEE Symposium on Security and Privacy, pp. 146–161 (1999)
167. Eckmann, S., Vigna, G., Kemmerer, R.: STATL: an attack language for state-based intrusion detection. J. Comput. Secur. **10**, 71–103 (2002)
168. Ko, C., Ruschitzka, M., Levitt, K.: Execution monitoring of security-critical programs in distributed systems: a specification-based approach. In: IEEE Symposium on Security and Privacy, pp. 175–187 (1997)
169. Chari, S., Cheng, P.: BlueBox: a policy-driven, host-based intrusion detection system. ACM Trans. Inf. Syst. Secur. **6**, 173–200 (2003)
170. Goldberg, I., Wagner, D., Thomas, R., Brewer, E.: A secure environment for untrusted helper applications: confining the wily hacker. In: USENIX Security Symposium, Focusing on Applications of Cryptography, pp. 1–13 (1996)
171. Keragala, D.: Detecting malware and sandbox evasion techniques. Technical Report, SANS Institute InfoSec Reading Room (2016)
172. Forrest, S., Hofmeyr, S., Somayaji, A., Longstaff, T.: A sense of self for Unix processes. In: IEEE Symposium on Security and Privacy, pp. 120–128 (1996)
173. Feng, H., Kolesnikov, O., Fogla, P., Lee, W., Gong, W.: Anomaly detection using call stack information. In: IEEE Symposium on Security and Privacy, pp. 62–75 (2003)
174. Luotonen, A.: The common log file format. Technical Report, CERN (1995)
175. Hallam-Baker, P., Behlendorf, B.: Extended log file format: W3C working draft WD-logfile-960323. Technical Report, W3C (1996)
176. Vigna, G., Valeur, F., Kemmerer, R.: Designing and implementing a family of intrusion detection systems. ACM SIGSOFT Softw. Eng. Notes **28**, 88–97 (2003)
177. Low, W., Lee, J., Teoh, P.: DIDAFIT: detecting intrusions in databases through fingerprinting transactions. In: ICEIS, pp. 121–128 (2002)
178. Krügel, C., Vigna, G.: Anomaly detection of web-based attacks. In: Tenth ACM Conference on Computer and Communications Security, pp. 251–261 (2003)
179. Puš, V., Velan, P., Kekely, L., Kořenek, J., Minařík, P.: Hardware accelerated flow measurement of 100 GB Ethernet. In: IFIP/IEEE International Symposium on Integrated Network Management, pp. 1147–1148 (2015)

180. Fifield, D., Lan, C., Hynes, R., Wegmann, P., Paxson, V.: Blocking-resistant communication through domain fronting. Proc. Priv. Enhanc. Technol. **2**, 46–64 (2015)
181. Whatley, J.: SAS/OR user's guide: Version 5 netflow procedure. Technical Report. SAS Institute, Inc (1985)
182. Claise, B.: Cisco systems netflow services export version 9. Technical Report, Cisco Systems, Inc (2003)
183. Claise, B., Fullmer, M., Calato, P., Penno, R.: Ipfix protocol specification. Technical Report, Cisco Systems (2005)
184. Fung, C., Boutaba, R.: Design and management of collaborative intrusion detection networks. In: IFIP/IEEE International Symposium on Integrated Network Management, pp. 955–961 (2013)
185. Spitzner, L.: The honeynet project: trapping the hackers. IEEE Secur. Priv. **99**, 15–23 (2003)

Chapter 2
Basic Concepts Concerning Classification

In the previous chapter, we came several times across the concepts of classification and classifiers. In this chapter, we will present a simple mathematical framework for those concepts. Such a framework is indispensable if classification is to be implemented, and it is due to its existence that classification algorithms belonged to the first high-level algorithms for which routinely used implementations existed, already in 1960s and 1970s. However, even if we do not want to implement any new algorithms, but only want to use existing implementations, the framework will be needed to understand basic principles of different kinds of classifiers, the assumptions for their applicability, their advantages and disadvantages.

2.1 Classifiers and Classes

Formally, a *classifier* is a mapping of some feature set $\mathscr{X}$ to some collection of *classes*, aka *labels*, $c_1, \ldots, c_m$,

$$\phi : \mathscr{X} \to C = \{c_1, \ldots, c_m\}. \tag{2.1}$$

The collection C is sometimes called *classification* of $\mathscr{X}$, though more frequently, the term classification denotes the process of constructing a classifier ϕ and subsequently using it to predict the class of yet unseen inputs $x \in \mathscr{X}$. Several important aspects of that process will be discussed in the remaining sections of this chapter. Here, on the other hand, we will take a closer look at the domain and value set of the mapping (2.1).

1. The *feature set* $\mathscr{X}$ is the set from which the combinations $x = ([x]_1, \ldots, [x]_n)$ of values of input features are taken. Hence, it is the Cartesian product $V_1 \times \ldots \cdots \times V_n$ of sets $V_1, \ldots, V_n$ of feasible values of the individual features. However, it is important that not every combination $([x]_1, \ldots, [x]_n)$ from the Cartesian product of sets of feasible values is a feasible combination: imagine a

© Springer Nature Switzerland AG 2020
M. Holeňa et al., *Classification Methods for Internet Applications*,
Studies in Big Data 69, https://doi.org/10.1007/978-3-030-36962-0_2

recommender system and the combination of Client Age = 10 and Client Marital Status = divorced. Hence, the domain of the classifier ϕ is in general not the whole $V_1 \times \cdots \times V_n$, but only some subset of it,

$$\text{Dom}\phi = \mathscr{X} \subset V_1 \times \cdots \times V_n. \tag{2.2}$$

The *features* $[x]_1, \ldots, [x]_n$ are alternatively called also *attributes* or *variables*, and their number can be quite high: several thousands are not an exception. From the data types point of view, they can be very diverse, e.g.:

- *Continuous* data, represented by real numbers, such as sound energy of speech or music, intensity of light.
- *Ordinal* data, such as various preferences, lexicographically ordered parts of text.
- *Categorical* data, aka *nominal* data, such as sex, place of residence, or colour, with a finite set V of feasible values. The elements of V are called categories.
- *Binary* data, such as sex, are categorical data for which the cardinality $|V|$ of the set V fulfils $|V| = 2$. They are, of course, a specific kind of categorical data, but at the same time, any categorical data can be always represented as a vector of binary data, usually of binary data with the value set $\{0, 1\}$. Indeed, if the $|V|$ elements of V are enumerated as $v_1, \ldots, v_{|V|}$, then the element v_j can be represented by a vector $b_j \in \{0, 1\}^{|V|}$ such that

$$[b_j]_j = 1, [b_j]_k = 0 \text{ for } k \neq j. \tag{2.3}$$

2. The collection of *classes* $C = \{c_1, \ldots, c_m\}$ is always finite. Needless to say, this does not prevent the classes to have their internal structure, most usually to be finite-dimensional vectors,

$$c_i = ([c_i]_1, \ldots, [c_i]_d), i = 1, \ldots, m. \tag{2.4}$$

The finiteness of C then requires the components $[c_i]_1, \ldots, [c_i]_d$ to belong to some finite sets $C_1, \ldots, C_d$, hence

$$C \subset C_1 \times \cdots \times C_d. \tag{2.5}$$

As to the cardinality of the collection C of classes, most common is the case $m = 2$, called *binary classification*, e.g., spam and ham, products to be recommended and those not to be recommended, malware and harmless software, network intrusion and normal traffic. For binary classification, a different notation is frequently employed, e.g., $C = \{c_+, c_-\}$, $C = \{1, 0\}$, $C = \{1, -1\}$, the first of the involved cases being called *positive*, the second *negative*. The case $m = 3$ is sometimes obtained from binary classification through introducing an additional class for those cases causing difficulties to the classifier. The interpretation of such a class then means "to some degree positive, to some certain degree negative".

However, there exists also another, more general approach to express that objects are assigned to a class only to a certain degree: namely considering, instead of a particular class c_i from the collection C, a fuzzy set on that collection. A fuzzy set on C can most simply be viewed as a sequence $\mu_1, \ldots, \mu_m \in [0, 1]$ of membership degrees in the classes $c_1, \ldots, c_m$, respectively. Hence, the set of all fuzzy sets on C is the m-dimensional unit cube $[0, 1]^m$. Using this set, a *fuzzy classifier* can be defined as a counterpart to the usual classifier (2.1):

$$\phi : \mathcal{X} \to [0, 1]^m. \tag{2.6}$$

In contrast to the term fuzzy classifier, the usual classifier (2.1) is sometimes called *crisp classifier*. At the same time, however, classifiers (2.1) are a particular case of classifiers (2.6), mapping into the set $\{u \in \{0, 1\}^m \mid \sum_{i=1}^{m} u_i = 1\}$, which is a subset of $[0, 1]^m$.

2.1.1 How Many Classes for Spam Filtering?

2.1.1.1 Spam or Ham

Spam filters usually work with two classes, thus as binary classifiers. Either the considered message is classified as legitimate ("ham") or it is a spam (unsolicited, junk or bulk mail). In this case, it is also comparatively easy to decide about the relationship between error weights assigned to the classifier for spam and ham, which will be discussed later in Sect. 2.2. It can be for example assumed that users look into the spam folder only once in a while. Thus they can miss an important legitimate email in case of wrong classification (false positive) and such errors should be penalized more. On the other hand, occasional spam in user's inbox is not that harmful—more even, it can be used as another example of spam if reported by the user.

2.1.1.2 Spam, Ham and Quarantine

A classifier ϕ mapping the feature set $\mathcal{X}$ into the set $C = \{\text{spam, ham}\}$ of classes can be accompanied by another mapping, τ, assigning to each feature vector $x \in \mathcal{X}$ a number $\tau(x) \in [0, 1]$ representing the trust or confidence that can be given to the classification $\phi(x)$. We will discuss such mappings in some detail in Sect. 6.1.2, in the context of classifier teams. In spam filters, confidence can be used to introduce an artificial third class for those $x \in \mathcal{X}$ for which the confidence of $\phi(x) = \text{spam}$ or $\phi(x) = \text{ham}$ is low. This is typically implemented using the concept of a quarantine— a place for messages not proven to be spam, nor legitimate. Then usually only the messages in the quarantine are presented to the user for assessment, not spams with high confidence.

A quarantine brings the advantage of decreasing the number of legitimate messages misclassified as spam and the number of spams misclassified as legitimate messages. However, this advantage is paid for by the necessity to regularly check the messages in the quarantine, which can easily contain important legitimate messages. And if many messages are classified to this artificial class, then such a checking can be quite tedious.

2.1.1.3 More Classes

Other approaches use handful of classes to which emails are assigned. As an example of such approach, Priority email feature of some email services sorts out incoming messages into several groups: *Spam, Social networks, Promotions, Updates and Forums*. Messages that do not fit any of the categories are considered as *Priority*. Basic motivation is to filter out not only spam, but also messages that do not have to be dealt with immediately.

Another variation, studied in [1], is an autonomous filter sorting emails into folders defined by the user. However, for such classification, enormous amount of user input would be needed. Although this could be partially dealt with by co-training, the created filter will not be easily transferable between users and in some cases content-based classification would not be suitable at all. As stated there, some users sort messages into folders according to next required action (to reply, delete, call, wait for third party, etc.) In this case, the features extracted from messages have little use for the classification.

But even in the pure context of unsolicited email, multiple classes can be used. Currently used spam filters are required to filter away not only messages that are desperately trying to sell goods or promote services. Their task is to scan for potentially harmful or inappropriate content, malware, phishing attacks and other threats. Even though binary classifier can be used, sometimes we want to warn the user even more urgently to not to click on links, to block loading of remote pictures and resources and so on. In such cases, multi-class approach needs to be taken as well.

2.2 Measures of Classifier Performance

When solving a particular classification task, we typically have a large number of classifiers available. What helps to choose the most suitable one is on the one hand understanding their principles and underlying assumptions (actually, that is the main reason why this book has been written), on the other hand comparing different of them on relevant data. Each such comparison has two ingredients:

(i) A set, or more generally a sequence $x_1, \ldots, x_q$ of independent inputs from the feature set, such that for each x_k, $k = 1, \ldots, q$, we know the correct class c_k. For the comparison based on the pairs $(x_1, c_1), \ldots, (x_q, c_q)$ not to be biased, they must be selected independently of those used as the classifier was constructed. If $(x_1, c_1), \ldots, (x_q, c_q)$ have been selected in this way, then they are usually called *test data*.

(ii) A function evaluating the performance of the classifier on $(x_1, c_1), \ldots, (x_q, c_q)$. The value of that function has usually the meaning of some error that the classifier ϕ makes when classifying $x_1, \ldots, x_q$. Therefore, a generic function of that kind will be in the following denoted as ER_ϕ.

The function ER_ϕ depends both on the test data $(x_1, c_1), \ldots, (x_q, c_q)$ and on the classes $\phi(x_1), \ldots, \phi(x_q)$ predicted for $x_1, \ldots, x_q$ by ϕ. Thus if we restrict attention to crisp classifiers (2.1), then in general,

$$\mathrm{ER}_\phi : \mathscr{X} \times C \times C \to \mathbb{R}. \tag{2.7}$$

Frequently, ER_ϕ depends on $x_1, \ldots, x_q$ only through the predictions $\phi(x_1), \ldots, \phi(x_q)$, hence

$$\mathrm{ER}_\phi : C \times C \to \mathbb{R}. \tag{2.8}$$

In such a case, ER_ϕ is completely determined by the counts of data with the correct class c_i and classified to the class c_j,

$$q_{i,j} = |\{k|1 \le k \le q, c_k = c_i, \phi(x_k) = c_j\}|, i, j = 1, \ldots, m. \tag{2.9}$$

Together with the overall count of test data with the correct class c_i, and the overall count of test data classified to c_j,

$$q_{i\cdot} = \sum_{j=1}^{m} q_{i,j}, \text{ respectively } q_{\cdot j} = \sum_{j=1}^{m} q_{i,j}, i, j = 1, \ldots, m, \tag{2.10}$$

they form the following matrix, called *confusion matrix* of the classifier ϕ:

$$
\begin{array}{c|ccc}
q & q_{\cdot 1} & \cdots & q_{\cdot m} \\
\hline
q_{1\cdot} & q_{1,1} & \cdots & q_{1,m} \\
\cdots & \cdots & \cdots & \cdots \\
q_{m\cdot} & q_{m,1} & \cdots & q_{m,m}
\end{array}
\tag{2.11}
$$

The most commonly encountered function of the kind (2.8) is *classification error*—the proportion of test data for which $\phi(x_k) \ne c_k$:

$$\mathrm{ER}_\phi = \mathrm{ER}_{\mathrm{CE}} = \frac{1}{q} \sum_{i \ne j} q_{i,j}. \tag{2.12}$$

The complementary proportion of test data for which $\phi(x_k) = c_k$ is called *accuracy*, or frequently *predictive accuracy*, to emphasize that it means the prediction of correct class for the unseen test data,

$$AC = \frac{1}{q} \sum_{i=1}^{m} q_{i,i} = 1 - ER_{CE}. \tag{2.13}$$

Notice that according to (2.12) and (2.13), all erroneous classifications $\phi(x_k) \neq c_k$ contribute to ER_{CE} equally. This corresponds to an assumption that all kinds of erroneous classifications are equally undesirable. Such an assumption is certainly unrealistic: we are more disturbed by not receiving a legitimate email than by receiving an unrecognized spam, and an unrecognized network intrusion can cause more harm than a false alarm. Therefore, a *weighted error* (or *cost-weighted error*) is used as a more realistic counterpart of (2.12)

$$ER_{\phi} = ER_{W} = \frac{1}{q \sum_{j=1}^{m} w_{1,j}} \sum_{i=1}^{m} \sum_{j \neq i} w_{i,j} q_{i,j}, \tag{2.14}$$

where $w_{i,j}, i, j = 1, \ldots, m$, denotes the weights or costs of the misclassification $\phi(x_k) = c_j$ if the correct class is c_i, and those weights fulfill

$$\sum_{j=1}^{m} w_{i,j} = \sum_{j=1}^{m} w_{i',j} \text{ for } i, i' = 1, \ldots, m. \tag{2.15}$$

Formally, also a cost of correct classification can be introduced, $w_{i,i}, i = 1, \ldots, m$, normally set to $w_{i,i} = 0$. That simplifies (2.14) to

$$ER_{\phi} = ER_{W} = \frac{1}{q \sum_{j=1}^{m} w_{1,j}} \sum_{i,j=1}^{m} w_{i,j} q_{i,j}. \tag{2.16}$$

The traditional classification error then corresponds to the classification cost

$$w_{i,j} = \begin{cases} 1 & \text{if } i \neq j \\ 0 & \text{if } i = j. \end{cases} \tag{2.17}$$

Frequently, the costs $w_{i,j}$ are scaled so that $\sum_{i,j=1}^{m} w_{i,j} = 1$. This is always possible through dividing them by the original $\sum_{i,j=1}^{m} w_{i,j}$. It turns the costs to a probability distribution on the pairs $(i, j)_{i,j=1}^{m}$ and the cost-weighted error (2.16) to the mean value of classification error with respect to that distribution. For the traditional classification error (2.12), these scaled costs are $w_{i,j} = \frac{1}{m(m-1)}, i \neq j$.

2.2.1 Performance Measures in Binary Classification

In the case of a binary classifier $\phi : \mathscr{X} \to \{c_+, c_-\}$, there are only 4 possible values $q_{i,j}$, which have got their specific names, introduced below in Table 2.1. Frequently, they are used as rates with respect to the overall number $q_{+.}$ assigned to the class c_+ and the overall number $q_{-.}$ assigned to the class c_-, as is also explained in Table 2.1. By means of the values in this table, classification error (2.12) can be rewritten as

$$\text{ER}_{\text{CE}} = \frac{1}{q}(\text{FP} + \text{FN}), \tag{2.18}$$

accuracy (2.13) as

$$\text{AC} = \frac{1}{q}(\text{TP} + \text{TN}), \tag{2.19}$$

and cost-weighted error (2.16), using a notation analogous to $w_{i,j}$ for a classification into the classes c_+, c_-, as

$$\text{ER}_{\text{W}} = \frac{1}{q}(w_{++}\text{TP} + w_{+-}\text{FN} + w_{-+}\text{FP} + w_{--}\text{TN}). \tag{2.20}$$

Apart from (2.18)–(2.20), also the true positive rate TPr, false positive rate FPr, true negative rate TNr and two additional measures, precision and F-measure, are often used as performance measures in binary classification. *Precision* PR is defined as:

$$\text{PR} = \frac{\text{TP}}{q_{.+}}, \tag{2.21}$$

the definition of the *F-measure* FM is:

$$\text{FM} = 2\frac{\text{PR} \cdot \text{TPr}}{\text{PR} + \text{TPr}}. \tag{2.22}$$

Due to the ubiquity of binary classification, several of its performance measures are known also under alternative names. The most important among such synonyms are as follows:

Table 2.1 Confusion matrix in binary classification

Correct class	Classified as			Rate (r)	
	$c_+ : q_{.+}$	$c_- : q_{.-}$			
$c_+ : q_{+.}$	True positive (TP)	False negative (FN)	$\text{TPr} = \frac{\text{TP}}{q_{+.}}$	$\text{FNr} = \frac{\text{FN}}{q_{+.}}$	
$c_- : q_{-.}$	False positive (FP)	True negative (TN)	$\text{FPr} = \frac{\text{FP}}{q_{-.}}$	$\text{TNr} = \frac{\text{TN}}{q_{-.}}$	

- *predictive value* is a synonym for precision,
- *sensitivity* and *recall* are synonyms for true positive rate,
- *specificity* is a synonym for true negative rate.

In binary classification, classifier performance is very often characterized not by a single performance measure, but by two such measures simultaneously. Most common are the pairs of measures (FPr, TPr), (AC, PR) a (PR, TPr). Notice that an ideal classifier, i.e., one for which true positive rate is 1 and false positive rate is 0, has the following values of those three pairs of measures:

$$(\text{FPr}, \text{TPr}) = (0, 1), (\text{AC}, \text{PR}) = (1, 1), (\text{PR}, \text{TPr}) = (1, 1). \qquad (2.23)$$

A pair of performance measures is particularly useful in the following situations:

(i) The performance of a classifier has been measured with different test data, typically with different subsequences of the sequence $(x_1, c_1), \ldots, (x_q, c_q)$.
(ii) The performance has been measured not for a single classifier, but for a set of classifiers, typically classifiers of the same kind, differing through the values of one or several parameters.

In both situations, the resulting pairs form a set in the 2-dimensional space, which can be connected with a curve according to increasing values of one of the two involved measures. For the pair of measures (FPr, TPr), such curves are called *receiver operating characteristics (ROC)* because they were first proposed for classification tasks

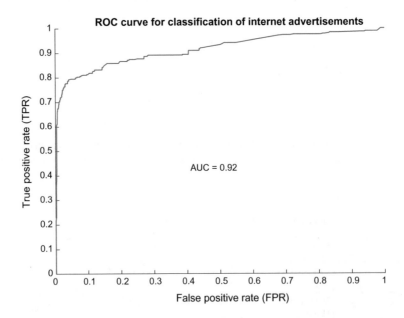

Fig. 2.1 ROC curve for the classification of graphical objects on web pages into advertisements and non-advertisements (cf. [2]). As classifier, random forests were used (Sect. 6.2)

in radar detection. As an example, Fig. 2.1 shows a ROC curve for the classification of graphical objects on web pages into advertisements and non-advertisements (cf. [2]). If the ROC curve is constructed in the situation (i), then it provides an additional performance measure of the considered classifier. The area under the ROC curve, i.e. the area delimited from above by the curve, from below by the value $\text{TPr} = 0$ and from the left and right by the values $\text{FPr} = 0$ and $\text{FPr} = 1$, has the size $\text{AUC} = \int_0^1 \text{TPr}\, d\text{FPr}$. Because the highest possible value of TPr is 1, AUC is delimited by

$$\text{AUC} = \int_0^1 \text{TPr}\, d\text{FPr} \leq \int_0^1 1\, d\text{FPr} = 1. \tag{2.24}$$

This performance measure summarizes the pairs of measures obtained for several sequences of test data (those used to construct the ROC curve) into one value.

2.3   Linear Separability

The easiness of constructing a classifier for particular classes decisively depends on the form of surfaces delimiting the individual classes, or more generally, on the form of surfaces through which the classes can be separated from each other (cf. Fig. 2.2). If $\mathcal{X} \subset \mathbb{R}^n$, then similarly to many other situations, also here the linear case is the easiest. This means that any two classes have to be separable by a *linear surface*, i.e., a *hyperplane*, in the space $\mathbb{R}^n$ spanned by the features $[x]_1, \ldots, [x]_n$. Because each hyperplane divides $\mathbb{R}^n$ into two halfspaces, classification into m classes then leads to $\frac{1}{2}m(m-1)$ hyperplanes and $m(m-1)$ halfspaces, and each class has to be a subset of some polytope resulting from the intersection of $m-1$ halfspaces.

Linear separability is a specific property and is missing in many real-world data sets. However, it can be always achieved through pre-processing based on mapping the original input data into a high-dimensional linear space. This can be easily exem-

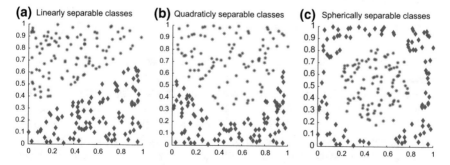

Fig. 2.2 Shapes of two-dimensional data separable by different one-dimensional surfaces in $\mathbb{R}^2$: **a** linear, **b** quadratic, **c** spherical

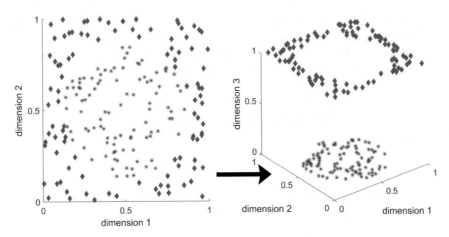

Fig. 2.3 Easy transformation of classes from Fig. 2.2 c, not separable in 2D, into classes separable in 3D

plified on the data in Fig. 2.2 c. These data were sampled from the following classes in $\mathbb{R}^2$

$$c_+ = \{x = ([x]_1, [x]_2) \in \mathbb{R}^2 | ([x]_1 - 0.5)^2 + ([x]_2 - 0.5)^2 < 0.325^2\} \quad (2.25)$$

$$\text{and } c_- = \{x = ([x]_1, [x]_2) \in \mathbb{R}^2 | ([x]_1 - 0.5)^2 + ([x]_2 - 0.5)^2 > 0.425^2\}. \quad (2.26)$$

These two classes can be separated by no line, i.e., by no hyperplane in $\mathbb{R}^2$. However, if an additional feature $[x]_3$ is added, assuming the value 0 for elements of c_+ and the value 1 for elements of c_-, then c_+ is mapped onto the set C'_+ in $\mathbb{R}^3$,

$$C'_+ = \{x = ([x]_1, [x]_2, [x]_3) \in \mathbb{R}^3 | ([x]_1 - 0.5)^2 + ([x]_2 - 0.5)^2 < 0.325^2, [x]_3 = 0\}, \quad (2.27)$$

and c_- is mapped onto C'_- in $\mathbb{R}^3$,

$$C'_- = \{x = ([x]_1, [x]_2, [x]_3) \in \mathbb{R}^3 | ([x]_1 - 0.5)^2 + ([x]_2 - 0.5)^2 > 0.425^2, [x]_3 = 1\}. \quad (2.28)$$

The sets C'_+ and C'_- are separable by any hyperplane $[x]_3 = \theta$ with $\theta \in (0, 1)$ (Fig. 2.3).

Needless to say, the choice of the space $\mathbb{R}^3$ and of the mapping from $\mathbb{R}^2$ to it in the above example relies on our a priori knowledge of what exactly the classes C'_+ and C'_- are, which is not available in real-world applications (if such a knowledge were available for the classes, then there would be no need to construct a classifier for them). Nevertheless, the principle illustrated in the example—namely, linear separability of subsets of data can be achieved through mapping them into a linear space with a sufficiently high dimension—remains valid even without such a priori knowledge. In the general case, on the other hand, the space to which the data are mapped has not only 1 dimension more than the original space containing the data, like in the example, but is has a dimension approximately equal to the number of data. More precisely, if the p inputs $x_1, \ldots, x_p \in \mathcal{X}$ occurring in the data are in a general position in an N-dimensional space, i.e., no $k \leq N$ of them, $2 \leq k \leq p$, lie on a linear manifold of a dimension lower than $k - 1$, then it is known [3] that they contain at least

$$\sum_{i=0}^{\min(p-1,N)} \binom{p-1}{i} - 1 \text{ pairs } (S_a, S_b) \text{ of nonempty subsets of } \{x_1, \ldots, x_p\} \text{ such that}$$

$$S_a \cap S_b = \emptyset, S_a \cup S_b = \{x_1, \ldots, x_p\}, S_a \text{ and } S_b \text{ are separable by a hyperplane.}$$
$$(2.29)$$

In particular, if $N \geq p - 1$, than any of the $2^{p-1} - 1$ possible pairs (S_a, S_b) of nonempty complementary subsets of $\{x_1, \ldots, x_p\}$ is separable by a hyperplane. Now, we are going to explain a method to construct an N-dimensional linear space with $N = p$.

2.3.1 Kernel-Based Mapping of Data into a High-Dimensional Space

A general method to construct, for given inputs $x_1, \ldots, x_p \in \mathcal{X}$, a linear space of dimension p is based on using kernel functions. A *kernel function* or simply *kernel*[1] is in the most general case any symmetric real function on the Cartesian product of the feature set with itself,

$$\kappa : \mathcal{X} \times \mathcal{X} \to \mathbb{R}, (\forall x, x' \in \mathcal{X}) \, \kappa(x, x') = \kappa(x', x). \qquad (2.30)$$

Typically, kernels are constructed in two steps: first, kernels for particular kinds of features are selected, and then they are combined to achieve applicability to the whole feature set $\mathcal{X}$. As to kernels for individual kinds of features, we will explain here kernels for real-valued features, for features with values in finite sets, and for features with values in subsets of given sets, which altogether cover nearly all features encountered in real-world applications.

1. If among the n considered features $[x]_1, \ldots, [x]_n$, there are $n' \leq n$ *real-valued* features, then a kernel function corresponding to all of them is a mapping

$$\kappa : \mathbb{R}^{n'} \times \mathbb{R}^{n'} \to \mathbb{R}, (\forall x, x' \in \mathbb{R}^{n'}) \, \kappa(x, x') = \kappa(x', x). \qquad (2.31)$$

---

[1] The term "kernel function" is due to the fact that such functions were first used, in the 19th century, as kernels of integral operators.

The most commonly used functions of that kind are:

- *Gaussian kernel* with a parameter $\varsigma > 0$,

$$\kappa(x, x') = \exp\left(-\frac{1}{\varsigma} d_E(x, x')^2\right) = \exp\left(-\frac{1}{\varsigma}\|x - x'\|^2\right), x, x' \in \mathbb{R}^{n'},$$
(2.32)

where d_E and $\|\cdot\|$ denote the Euclidean distance and Euclidean norm, respectively;

- *polynomial kernel*, more precisely, homogeneous polynomial kernel, with a parameter $d \in \mathbb{N}$,

$$\kappa(x, x') = (x^\top x')^d, x, x' \in \mathbb{R}^{n'}$$
(2.33)

where $x^\top$ denotes the transpose of x;

- and *inhomogeneous polynomial kernel* with parameters $d \in \mathbb{N}$ and $c > 0$,

$$\kappa(x, x') = (x^\top x' + c)^d, x, x' \in \mathbb{R}^{n'}.$$
(2.34)

All these three kernels share one additional noteworthy property: For any sequence of points from the feature set $x_1, \ldots, x_k \in \mathcal{X}, k \in \mathbb{N}$, the matrix

$$G_\kappa(x_1, \ldots, x_k) = \begin{pmatrix} \kappa(x_1, x_1) & \ldots & \kappa(x_1, x_k) \\ \ldots\ldots\ldots \\ \kappa(x_k, x_1) & \ldots & \kappa(x_k, x_k) \end{pmatrix}$$
(2.35)

which is called *Gramm matrix* of $x_1, \ldots, x_k$, is *positive semidefinite*, i.e.,

$$(\forall y \in \mathbb{R}^k) \; y^\top G_\kappa(x_1, \ldots, x_k) y \geq 0.$$
(2.36)

2. If among the n considered features $[x]_1 \ldots [x]_n$, there are n' features $[x]_{j_1} \ldots [x]_{j_{n'}}$ with *finite sets of values* $V_{j_1}, \ldots, V_{j_{n'}}$, then each value $[x]_{j_i}$ of the i-th among those features is actually a character from the finite alphabet $A = V_{j_1} \cup \cdots \cup V_{j_{n'}}$, thus combinations of values $[x]_{j_1}, \ldots, [x]_{j_{n'}} \in A^{n'}$ are strings over A. For such strings, a kernel function can be defined in the following way. Denote $\mathcal{I}_k$ the set of index combinations indexing substrings of length $k \leq n'$ of strings from $A^{n'}$,

$$\mathcal{I}_k = \{\iota = ([\iota]_1, \ldots, [\iota]_k) \in \mathbb{N}^k \mid 1 \leq [\iota]_1 < \cdots < [\iota]_k \leq n'\},$$
(2.37)

and for each string $s \in A^{n'}$ and each $\iota \in \mathcal{I}_k$ denote $s(\iota) = ([s]_{[\iota]_1}, \ldots, [s]_{[\iota]_k})$. Then a kernel κ_k on $A^{n'}$ with a *decay parameter* $\lambda \in (0, 1]$ can be defined by

$$\kappa_k(x, x') = \sum_{u \in A^k} \sum_{\substack{\iota \in \mathcal{I}_k, \\ x(\iota) = u}} \sum_{\substack{\iota' \in \mathcal{I}_k, \\ x'(\iota') = u}} \lambda^{([\iota]_k - [\iota]_1 + [\iota']_k - [\iota']_1)}, x, x' \in A^{n'}.$$
(2.38)

3. Finally, consider a feature $[x]_k$ the values of which are subsets of some set Ω, more precisely, elements of some system $\mathscr{S}$ of subsets of Ω that is closed under complement, countable intersections and countable unions. Each such system is called σ-field or σ-algebra, and its most common example is the set of all subsets of Ω, i.e., the *power set* $\mathscr{P}(\Omega)$. The set Ω itself can be quite arbitrary, in particular, it can be finite like were the sets $V_{j_1}, \ldots, V_{j_{n'}}$ in 2., countable like the natural numbers $\mathbb{N}$ or uncountable like the real numbers $\mathbb{R}$. On $\mathscr{S}$, a kernel can be defined by

$$\kappa(x, x') = P(x \cap x') - P(x)P(x'), \ x, x' \in \mathscr{S}, \tag{2.39}$$

where $P : \mathscr{S} \to [0, 1]$ is a probability on Ω. Also for this kernel κ, the Gramm matrix (2.35) is positive semidefinite.

As to the second step, combining kernels for particular kinds of features to obtain a kernel on the whole feature set, it typically relies on the fact that, given kernels κ_1 on $\mathscr{X}_1 \times \mathscr{X}_1$ and κ_2 on $\mathscr{X}_2 \times \mathscr{X}_2$, the following two functions are kernels on $(\mathscr{X}_1 \times \mathscr{X}_2) \times (\mathscr{X}_1 \times \mathscr{X}_2)$:

(i) the function $\kappa_1 \oplus \kappa_2$, called *direct sum* of κ_1 and κ_2, which is defined

$$\kappa_1 \oplus \kappa_2((x_1, x_2), (x'_1, x'_2)) =$$
$$= \kappa_1(x_1, x'_1) + \kappa_2(x_2, x'_2), \ (x'_1, x'_2) \in \mathscr{X}_1 \times \mathscr{X}_2; \tag{2.40}$$

(ii) the function $\kappa_1 \otimes \kappa_2$, called *tensor product* of κ_1 and κ_2, defined

$$\kappa_1 \otimes \kappa_2((x_1, x_2), (x'_1, x'_2)) =$$
$$= \kappa_1(x_1, x'_1)\kappa_2(x_2, x'_2), \ (x_1, x_2), (x'_1, x'_2) \in \mathscr{X}_1 \times \mathscr{X}_2); \tag{2.41}$$

Once we have a kernel κ for the whole feature set $\mathscr{X}$, it can be used to obtain p functions on $\mathscr{X}$ for the given data $x_1, \ldots, x_p \in \mathscr{X}$ through fixing one input argument of κ, i.e., the functions $\kappa(\cdot, x_1), \ldots, \kappa(\cdot, x_p)$. Since the set of all real functions on $\mathscr{X}$ is a linear space, it is possible to define its subspace $\mathscr{L}$ as the span, i.e., linear hull, of those p functions,

$$\mathscr{L} = \mathrm{span}(\{\kappa(\cdot, x_1), \ldots, \kappa(\cdot, x_p)\}) =$$
$$= \{f : \mathscr{X} \to \mathbb{R} \mid f = \sum_{k=1}^{p} a_k^f \kappa(\cdot, x_k), a_k^f \in \mathbb{R} \text{ for } k = 1, \ldots, p\}. \tag{2.42}$$

It is important to bear in mind that the elements of $\mathscr{L}$ are not individual combinations $([x]_1, \ldots, [x]_n)$ of features, but functions on the feature set $\mathscr{X}$. The linear space $\mathscr{L}$ is then the desired space to which the data $x_1, \ldots, x_p$ should be mapped:

1. Assuming the functions $\kappa(\cdot, x_1), \ldots, \kappa(\cdot, x_p)$ are linearly independent in $\mathscr{L}$, the dimension of $\mathscr{L}$ equals p. Moreover, if all features are real-valued and $x_1, \ldots, x_p$ are sampled from some n-dimensional continuous distribution, then this assumption is fulfilled almost surely, i.e., with probability 1.
2. The inputs $x_1, \ldots, x_p$ are mapped bijectively, i.e., one-to-one, onto the functions $\kappa(\cdot, x_1), \ldots, \kappa(\cdot, x_p)$.

Consequently, any of the $2^{p-1} - 1$ possible pairs of nonempty complementary subsets of $\kappa(\cdot, x_1), \ldots, \kappa(\cdot, x_p)$ is separable by a hyperplane in $\mathscr{L}$.

Due to (2.42), it is possible to define on $\mathscr{L}$ a function of 2 variables $\langle \cdot, \cdot \rangle$: $\mathscr{L} \times \mathscr{L} \to \mathbb{R}$ as

$$\langle f, g \rangle = \sum_{k,\ell=1}^{p} a_k^f a_\ell^g \kappa(x_k, x_\ell), \quad f, g \in \mathscr{L}. \tag{2.43}$$

Moreover, it can be shown [3] that if the kernel κ is continuous, e.g., if κ is one of the kernels (2.32)–(2.34), then (2.43) is a scalar product extensible to the completion $\bar{\mathscr{L}}$ of $\mathscr{L}$ with respect to the norm $\| \cdot \|_{\mathscr{L}}$ defined

$$\| f \|_{\mathscr{L}} = \sqrt{\langle f, f \rangle} = \sqrt{\sum_{k=1}^{p} (a_k^f)^2 \kappa(x_k, x_k)}, \quad f \in \mathscr{L}. \tag{2.44}$$

Any sequence $(f_n)_{n \in \mathbb{N}}$ of the elements of $\mathscr{L}$ that is a Cauchy sequence with respect to $\| \cdot \|_{\mathscr{L}}$, i.e., a sequence fulfilling

$$(\forall \varepsilon > 0)(\exists n_\varepsilon \in \mathbb{N})(\forall n, n' \in \mathbb{N}) \, n, n' \geq n_\varepsilon \Rightarrow \| f_n - f_{n'} \|_{\mathscr{L}} < \varepsilon, \tag{2.45}$$

has the property that it converges pointwise on $\mathscr{X}$, i.e., the sequence $(f_n(x))_{n \in \mathbb{N}}$ converges for each $x \in \mathscr{X}$. Then the *completion* of $\mathscr{L}$ with respect to $\| \cdot \|_{\mathscr{L}}$ is defined as

$$\bar{\mathscr{L}} = \{ f : \mathscr{X} \to \mathbb{R} | (\exists (f_n)_{n \in \mathbb{N}} - \text{a Cauchy sequence}$$
$$\text{with respect to } \| \cdot \|_{\mathscr{L}})(\forall x \in \mathscr{X}) \, f(x) = \lim_{n \in \mathbb{N}} f_n(x) \}. \tag{2.46}$$

The completion $\bar{\mathscr{L}}$ has the additional property of being already *complete*, i.e., a further completion of $\bar{\mathscr{L}}$ would not add any more elements because every Cauchy sequence $(f_n)_{n \in \mathbb{N}}$ of elements of $\bar{\mathscr{L}}$ has a limit in $\bar{\mathscr{L}}$,

$$(\exists f^* \in \bar{\mathscr{L}})(\forall \varepsilon > 0)(\exists n_\varepsilon \in \mathbb{N})(\forall n \in \mathbb{N}) \, \| f_n - f^* \|_{\mathscr{L}} < \varepsilon. \tag{2.47}$$

Finally, notice that according to (2.43), the scalar product of functions $\kappa(\cdot, x_k)$ and $\kappa(\cdot, x_\ell)$ is

$$\langle \kappa(\cdot, x_k), \kappa(\cdot, x_\ell) \rangle = \kappa(x_k, x_\ell), \tag{2.48}$$

hence, the relationship between the functions $\kappa(\cdot, x_k)$ and $(\cdot, x_\ell)$ represented by their scalar product "reproduces" the relationship between the corresponding data x_k and x_ℓ represented by $\kappa(x_k, x_\ell)$. Because of this property and because a complete space with scalar product is called Hilbert space, the space $\widetilde{\mathscr{L}}$ introduced above is usually called *reproducing kernel Hilbert space*.

2.4 Classifier Learning

In the remaining chapters, a plethora of methods for classifier construction will be presented. Each method entails a specific kind of classifiers with specific properties, but for those classifiers to be uniquely determined, some parameters always have to be tuned. Consequently, the choice of a particular classifier construction method does not lead to a single classifier (2.1), but to a whole set $\boldsymbol{\Phi}$ of classifiers of the considered kind,

$$\boldsymbol{\Phi} \subset \{\mathscr{X} \to C\}. \tag{2.49}$$

A single classifier $\phi \in \boldsymbol{\Phi}$ is then obtained through tuning its parameters to adapt them to the situation in which the classification should be performed, i.e., to the data that should be classified. That tuning is called *classifier learning* or *classifier training*, in analogy to human learning and training, which also serve such an adaptation purpose.

To compare different classifiers ϕ from the set $\boldsymbol{\Phi}$, we need—similarly to the case of comparing different kinds of classifiers, which was discussed in Sect. 2.2— a sequence of independent inputs $x_1, \ldots, x_p$, for each of which the correct class $c_1, \ldots, c_p$ is known. As a counterpart to the test data $(x_1, c_1), \ldots, (x_q, c_q)$, the pairs $(x_1, c_1), \ldots, (x_p, c_p)$ are called *training data*. It is crucial that they should possibly well represent the data that the classifier will later classify. Ideally, the training data should have the same distribution as the data to be classified. That is, however, hardly possible, alone due to the fact that the number of training data is finite.

Often, the classifier ϕ is selected from the set $\boldsymbol{\Phi}$ so as to minimize some error function $\mathrm{ER}_\phi : \mathscr{X} \times C \times C \to \mathbb{R}$, or $C \times C \to \mathbb{R}$ aggregated over all training data. In such a situation, classifier learning consists in solving the optimization task

$$\phi = \arg \min_{\phi' \in \boldsymbol{\Phi}} \mathrm{ER}_\phi((x_1, c_1, \phi'(x_1)), \ldots, (x_p, c_p, \phi'(x_p))), \tag{2.50}$$

where $\mathrm{ER}_\phi((x_1, c_1, \phi'(x_1)), \ldots, (x_p, c_p, \phi'(x_p)))$ stands for some aggregation of the values $\mathrm{ER}_\phi(x_k, c_k, \phi'(x_k))$, $k = 1, \ldots, p$, for example, sum or weighted sum. What exactly the optimization task means and what are the optimization methods suitable to solve it depends on the one hand on the choice of the error function ER_ϕ, on the other hand on the feature set $\mathscr{X}$. In the following chapters, when presenting various kinds of classifiers, also their respective learning methods will be sketched.

2.4.1 Classifier Overtraining

The fact that the minimization in (2.50) is based on the training data entails the danger that the classifier learns the precise distribution of the training data, which is in general different from the distribution of the data to be classified. Then the error it makes on the data to be classified is on average high, in spite of the error on the training data having been minimized. When computing (2.50), the resulting minimum $ER_\phi((x_1, c_1, \phi(x_1)), \ldots, (x_p, c_p, \phi(x_p)))$ is in general lower than what can be expected from the corresponding optimization task for the data to be classified, i.e., lower than

$$\mathbb{E} \min_{\phi' \in \Phi} ER_\phi(X, Y, \phi'(X)), \tag{2.51}$$

where X is a random vector with values in the feature set $\mathscr{X}$, Y is a random variables with values in the set of classes C, and $\mathbb{E}$ stands for the expectation. This phenomenon is called *overtraining* and is probably the most serious problem encountered in classifier learning.

There are two basic ways used to tackle overtraining: to *estimate the expected error* of the currently considered classifier ϕ on the data to be classified, and to restrict attention to only a subset of all considered features. Let us focus first to the estimation of the error on the classified data. To this end, we can use an additional sequence $x_1, \ldots, x_v$ of feature vectors sampled from that data, independent from $x_1, \ldots, x_p$ and such that their corresponding correct classes, $c_1, \ldots, c_v$ are known. The sequence $(x_1, c_1), \ldots, (x_v, c_v)$ is then called *validation data*, and its usefulness consists in the fact that the arithmetic mean of the error function on them

$$\frac{1}{v} \sum_{k=1}^{v} ER_\phi((x_k, c_k, \phi(x_k)) \tag{2.52}$$

is an unbiased consistent estimate of the expected error of ϕ on the data to be classified.

It is important to notice that in this way, validation data get involved into the construction of the classifier, thus in particular, they cannot be later employed as test data to measure classifier performance, according to Sect. 2.2. Hence, the sequences $x_1, \ldots, x_p, x_1, \ldots, x_q$ and $x_1, \ldots, x_v$ have to be mutually independent.

The simplest way how to obtain the validation data $(x_1, c_1), \ldots, (x_v, c_v)$ is to randomly split off a small part (e.g., 20 %) from all the data available for classifier construction. However, if the ultimate objective is not to estimate the expected error of ϕ, but to find the optimal value $\gamma \in \Gamma$ of a classifier property with a finite set of feasible values, or the optimal combination of values of several such properties, then a more sophisticated method is often used. That method makes a more effective use of the available data through dividing them into $H \geq 3$ parts (folds) of approximately equal size and using each part $(H - 1)$-times as training data and once as validation data. It is called *cross-validation* and is summarized in the following algorithm.

Algorithm 1 (Cross-validation)

Input:

- *finite set* $\Gamma = \Gamma_1 \times \cdots \times \Gamma_\ell$ *of feasible combinations of values of ℓ properties of the classifier*
- *validation data* $(x_1, c_1), \ldots, (x_v, c_v)$;
- *number H of cross-validation folds, $3 \leq H \leq v$.*

Step 1. *Generate a random permutation π of $\{1, \ldots, v\}$.*

Step 2. *Define the folds* $\Pi_h = \{\pi(\lfloor \frac{v(h-1)}{H} \rfloor + 1), \ldots, \pi(\lfloor \frac{vh}{H} \rfloor)\}, h = 1, \ldots, H$, *where $\lfloor \cdot \rfloor$ stands for the greatest smaller or equal integer.*

Step 3. *For $\gamma \in \Gamma, h = 1, \ldots, H$, train a classifier ϕ_h^γ, which has the combination of values of the considered properties equal to γ, on the data*

$$\{(x_k, c_k)|k = 1, \ldots, v, k \notin \Pi_h\}. \tag{2.53}$$

Step 4. *For $\gamma \in \Gamma$, define the error estimate based on validation data,*

$$\mathrm{ER}_\phi(\gamma) = \frac{1}{H} \sum_{h=1}^{H} \frac{1}{|\Pi_h|} \sum_{k \in \Pi_h} \mathrm{ER}_\phi(x_k, c_k, \phi_h^\gamma(x_k)). \tag{2.54}$$

Output: $\arg\min_{\gamma \in \Gamma} \mathrm{ER}_\phi(\gamma)$, *the combination of values of properties with the lowest error estimate.*

In the extreme case $H = v$, each fold contains exactly one element. Therefore, this variant of cross-validation is called *leave-one-out validation*.

The other basic remedy for overtraining—restricting learning to only a subset of all considered features, will be addressed in Sect. 2.5, in the context of feature selection.

2.4.2 Semi-supervised Learning

A serious problem pertaining to classifier learning is that obtaining the training data $(x_1, c_1), \ldots, (x_p, c_p)$, more precisely, the correct classes $c_1, \ldots, c_p$, can be time consuming, tedious and/or expensive. Think of the well known feedback forms in recommender systems, or the less known but much more complicated work of malware analysts and network operators. A common approach to that problem is to perform learning with only a small number p of such training data, but in addition with an additional set $\mathscr{U}$ of unclassified feature vectors, which is finite, but can be very large. Hence, the correct class is then known only for the small fraction $\frac{p}{p+|\mathscr{U}|}$ of all the feature vectors $\{x_1, \ldots, x_p\} \cup \mathscr{U}$ available for learning. Because the correct class must be obtained from outside the classifier, thus obtaining it must be incited by somebody who somehow supervises the process of classifier learning, learning in which the correct class is known only for a fraction of the employed feature vectors is usually called *semi-supervised*. Similarly, traditional learning in which the correct class is known for all employed feature vectors is called *supervised*. Finally, situations when no other information is known than the feature vectors themselves are called *unsupervised*. We will encounter such a situation in Sect. 2.6, in connection with clustering.

Several approaches to semi-supervised learning exist. We will recall here the three that are used most often.

1. *Generative modelling*. This approach is based on the assumption that feature vectors assigned to different classes are generated by distributions from the same family parametrized by a vector of parameters $\theta \in \Theta$, where Θ is a suitable set of parameter vectors. Hence, the distribution of the feature vectors assigned to a class $c \in C$ is uniquely determined by a parameter vector $\theta_c \in \Theta$ corresponding to that class. More precisely, let us assume the following:

   (i) The first n' among the considered n features are categorical or ordinal with values from some $\mathscr{X}_d \subset V_1 \times \cdots \times V_{n'}$, where $V_j, j = 1, \ldots, n'$ is the set of feasible values of the j-th feature, and such that their conditional distribution conditioned on the class $c \in C$ belongs for every c to the same parametrized family of discrete distributions. That distribution then for every $([x]_1, \ldots, [x]_{n'}) \in \mathscr{X}_d$ determines the conditional probability $P_{\theta_c}([x]_1, \ldots, [x]_{n'}|c)$ of $([x]_1, \ldots, [x]_{n'})$ conditioned on c. We admit also the boundary cases of no categorical and ordinal features ($n' = 0$) and only categorical features ($n' = n$).

   (ii) The remaining $n - n'$ features are real-valued, with values from some $\mathscr{X}_c \subset V_{n'+1} \times \cdots \times V_n$. Moreover, conditioned on both the class and the first n' features, they are continuous and their distribution belong for all $c \in C$ and all $([x]_1, \ldots, [x]_{n'}) \in \mathscr{X}_d$ to the same parametrized family of continuous distributions. That distribution then for every $([x]_{n'+1}, \ldots, [x]_n) \in \mathscr{X}_c$ determines the conditional density $f_{\theta_c}([x]_{n'+1}, \ldots, [x]_n|c, [x]_1, \ldots, [x]_{n'})$ conditioned on c and $([x]_1, \ldots, [x]_{n'})$. By far most frequently, the considered family of continuous distributions is the family of $n - n'$-dimensional normal distributions, for which $\theta = (\mu, \Sigma)$ with $\mu \in \mathbb{R}^{n-n'}$, $\Sigma \in \mathbb{R}^{n-n',n-n'}$, where the notation $\mathbb{R}^{a,b}$ stands for the set of $a \times b$-dimensional matrices.

   (iii) Individual feature vectors, no matter whether belonging to the training data or to the set $\mathscr{U}$, are generated independently.

Consequently, the probability that a feature vector $x \in \mathscr{X}$ assigned to the class $c \in C$ has particular values of categorical or ordinal features $([x]_1, \ldots, [x]_{n'})$ is

$$P_{\theta_c}([x]_1, \ldots, [x]_{n'}) = P_{\theta_c}([x]_1, \ldots, [x]_{n'}|c)P(c), \qquad (2.55)$$

where $P(c) = P(\{$ the correct class of x is $c\})$, aka the *a priori probability* of the class c. Similarly, the conditional probability that its continuous features assume values in a particular set $S \in \mathscr{X}_c$ conditioned on the particular values of categorical or ordinal features $([x]_1, \ldots, [x]_{n'})$ is

$$P_{\theta_c}(S|([x]_1, \ldots, [x]_{n'})) = \int_S f_{\theta_c}([x]_{n'+1}, \ldots, [x]_n | c, [x]_1, \ldots, [x]_{n'}) \mathrm{d}[x]_{n'+1} \ldots \mathrm{d}[x]_n. \tag{2.56}$$

Notice that the probability (2.55) quantifies how likely occurs a feature vector x assigned to the class c with particular values $([x]_1, \ldots, [x]_{n'})$ of categorical and ordinal features, whereas the density determining the probability (2.56) quantifies how likely a feature vector x assigned to the class c with particular values $([x]_1, \ldots, [x]_{n'})$ of categorical and ordinal features assumes particular values $([x]_{n'+1}, \ldots, [x]_n)$ of continuous features. Taking into account the assumption (iii), this allows to quantify how likely the particular data $\{(x_k, c_k)|k = 1, \ldots, p\} \cup \mathscr{U}$ available for classifier training occur, by

$$\prod_{k=1}^p P_{\theta_{c_k}}([x_k]_1, \ldots, [x_k]_{n'}) f_{\theta_c}([x_k]_{n'+1}, \ldots, [x_k]_n | c, [x_k]_1, \ldots, [x_k]_{n'})$$

$$\prod_{x \in \mathscr{U}} \sum_{c \in C} P_{\theta_c}([x]_1, \ldots, [x]_{n'}) f_{\theta_c}([x]_{n'+1}, \ldots, [x]_n | c, [x]_1, \ldots, [x]_{n'}). \tag{2.57}$$

Denote $\theta_C = (\theta_c)_{c \in C} \in \Theta^{|C|}$ the system of parameter vectors corresponding to all classes $c \in C$. Then the function L on $\Theta^{|C|}$, defined

$$L((\theta_c)_{c \in C}) = \prod_{k=1}^p P_{\theta_{c_k}}([x_k]_1, \ldots, [x_k]_{n'}) f_{\theta_c}([x_k]_{n'+1}, \ldots, [x_k]_n | c, [x_k]_1, \ldots, [x_k]_{n'})$$

$$\prod_{x \in \mathscr{U}} \sum_{c \in C} P_{\theta_c}([x]_1, \ldots, [x]_{n'}) f_{\theta_c}([x]_{n'+1}, \ldots, [x]_n | c, [x]_1, \ldots, [x]_{n'}), (\theta_c)_{c \in C} \in \Theta^{|C|} \tag{2.58}$$

is called *likelihood* or *likelihood function*. The principle of the generative modelling approach consists in choosing the values of the parameter vectors θ_c that maximize the likelihood,

$$(\theta_c)_{c \in C} = \arg \max_{\theta \in \Theta^{|C|}} L(\theta), \tag{2.59}$$

thus according to (2.58), that make the data available for classifier training most likely.

Once those parameter vectors are known, a feature vector $x \in \mathscr{U}$ can be assigned to the class c maximizing the probability

$$P_{\theta_c}(x \in \{([x]_1, \ldots, [x]_{n'})\} \times S_{([x]_{n'+1}, \ldots, [x]_n)}) =$$
$$= P_{\theta_c}([x]_1, \ldots, [x]_{n'}) P_{\theta_c}(S_{([x]_{n'+1}, \ldots, [x]_n)} | ([x]_1, \ldots, [x]_{n'})),$$
$$(2.60)$$

where $S_{([x]_{n'+1}, \ldots, [x]_n)}$ is some particular subset of $\mathscr{X}_c$ containing $([x]_{n'+1}, \ldots, [x]_n)$. Frequently, the limit $S_{([x]_{n'+1}, \ldots, [x]_n)} \to \{([x]_{n'+1}, \ldots, [x]_n)\}$ is used instead, yielding

$$c^* = \arg \max_{c \in C} P_{\theta_c}([x]_1, \ldots, [x]_{n'}) f_{\theta_c}([x]_{n'+1}, \ldots, [x]_n | c, [x]_1, \ldots, [x]_{n'}).$$
$$(2.61)$$

2. *Low-density separation.* The approach is based on the assumption that the closer feature vectors $x, x' \in \mathscr{U}$ are according to some distance d among feature vectors, the more likely they belong to the same class. It consists in attempting to divide $\mathscr{U}$ into disjoint subsets $S_1, \ldots, S_r, r \leq p$ such that:

   (i) $S_1, \ldots, S_r$ form clusters, i.e., distances between $x, x' \in S_i$ are on average smaller than distances between $x \in S_i, x' \in S_j$ (cf. Sect. 2.6). Consequently, the density around the border areas of the clusters is lower then in the density within the clusters.

   (ii) Each cluster contains at least one feature vector x_k from the training data $(x_1, c_1), \ldots, (x_p, c_p)$, and if it contains more of them, then all are assigned to the same class,

   $$x_j, x_k \in S_i \Rightarrow c_j = c_k, j, k = 1, \ldots, p, i = 1, \ldots, r. \qquad (2.62)$$

   If both conditions are fulfilled, data in a cluster are assigned the same class as the feature vector(s) from the training data that it contains.

   A key ingredient of the low-density separation approach is the considered distance d between feature vectors from the set $\mathscr{U}$. In this respect, it is very important that many different distances can be computed between the same feature vectors. This increases the chance that for some choice d, it will be possible to find a division $S_1, \ldots, S_r$ of $\mathscr{U}$ fulfilling the above conditions (i)-(ii). At the same time, however, it makes the result of the approach dependent on the choice of d: different distances can lead to different divisions, which can assign some $x \in \mathscr{U}$ to different classes. In Fig. 2.4, a 2-dimensional example illustrating the difference between supervised learning and semi-supervised learning using the low-density separation approach is shown. In this case, the Euclidean distance is used and the resulting division has only two clusters, which are linearly separable.

3. *Agglomerative clustering.* Also this approach relies on the assumption that feature vectors are more likely to belong to the same class if they are closer. This time, that assumption is used in the context of hierarchical agglomerative clustering (HAC) [4]. In principle, it combines the standard HAC algorithm with the condition that feature vectors assigned to different classes can not belong to the same cluster. The resulting algorithm is described below. Like low-density separation, also the result of this approach depends on the chosen distance between feature

vectors, but in addition also on the chosen linkage function, which evaluates the possibility to combine two clusters given a distance matrix of unclassified feature vectors.

Algorithm 2 (Semi-supervised learning through agglomerative clustering)
Input:
- *training data* $(x_1, c_1), \ldots, (x_p, c_p)$;
- *finite set* $\mathcal{U}$ *of unclassified feature vectors;*
- *distance d on* $\{x_1, \ldots, x_p\} \cup \mathcal{U}$,
$d : (\{x_1, \ldots, x_p\} \cup \mathcal{U}) \times (\{x_1, \ldots, x_p\} \cup \mathcal{U}) \to \mathbb{R}_{\geq 0}$;
- *linkage function*
$\ell : \mathscr{P}(\{x_1, \ldots, x_p\} \cup \mathcal{U}) \times \mathscr{P}(\{x_1, \ldots, x_p\} \cup \mathcal{U}) \times \mathbb{R}^{|\mathcal{U}|,|\mathcal{U}|} \to \mathbb{R}_{\geq 0}$,
where $\mathscr{P}(S)$ *denotes the set of all subsets of S.*
Step 1. *Compute the matrix* $\delta = (d(x, x'))_{x, x' \in \mathcal{U}} \in \mathbb{R}^{|\mathcal{U}|,|\mathcal{U}|}$.
Step 2. *Set* $s = 1$, $\mathscr{S}_1 = (\{x\})_{x \in \mathcal{U}}$.
Step 3. *Find* $S_{s,1}, S_{s,2} \in \mathscr{S}_s$ *such that*

$$(\forall x_k \in S_{s,1} \cap \{x_1, \ldots, x_p\})(\forall x_{k'} \in S_{s,2} \cap \{x_1, \ldots, x_p\}) \, c_k = c_{k'}, \qquad (2.63)$$

and

$$\ell(S_{s,1}, S_{s,2}, \delta) = \min\{\ell(S, S', \delta)|$$

$$S, S' \in \mathscr{S}_s, (\forall x_k \in S \cap \{x_1, \ldots, x_p\})(\forall x_{k'} \in S' \cap \{x_1, \ldots, x_p\}) \, c_k = c_{k'}\}. \quad (2.64)$$

Step 4. *Put* $S_s = S_{s,1} \cup S_{s,2}$.
Step 5. *If exists* $x_k \in S_s \cap \{x_1, \ldots, x_p\}$, *assign each* $x \in S_s$ *to the class* c_k.
Step 5. *Put* $\mathscr{S}_{s+1} = (\mathscr{S}_s \setminus \{S_{s,1}, S_{s,2}\}) \cup S_s$.
Increment $s \to s + 1$
Step 6. *If exists* $S, S \in \mathscr{S}_s$ *such that*

$$(\forall x_k \in' S \cap \{x_1, \ldots, x_p\})(\forall x_{k'} \in S' \cap \{x_1, \ldots, x_p\}) \, c_k = c_{k'}, \qquad (2.65)$$

return to Step 3.
Output: *Set of clusters* $\mathscr{S}_s$ *such that all feature vectors in each cluster* $S \in \mathscr{S}_s$ *are assigned to the same class.*

2.4.3 Spam Filter Learning

Spam filtering is a beautiful example of semi-supervised learning, as it is impossible to label a sufficient amount of incoming messages to reflect all kinds of spam. In addition, the dynamics of spam also requires the spam filter to adapt to new challenges and threats.

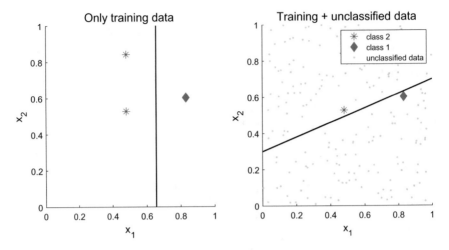

Fig. 2.4 Difference between supervised learning and semi-supervised learning using the low-density separation approach. As distance between the 2-dimensional feature vectors, the Euclidean distance is used

As discussed in Chap. 1, false positives (regular e-mail marked as spam) are much more disturbing than false negatives (spam lands in the inbox). The spam filter is, therefore, trained to underestimate the spam score of the message on purpose. If a spam message appears in the user inbox and the user marks that message as spam, the spam filter should update appropriately.

For the sake of simplicity, we will now consider only the binary spam-ham classifier training. For an extension of the following ideas to multiple classes, it is possible to use an ensemble of $\frac{1}{2}m(m-1)$ binary classifiers [5].

We will discuss feature selection more thoroughly in the next section, but let us assume that the presence of specific words in the email header and body (e.g., certain pill names or suspicious domains) is in correlation with message spam status and thus is suitable for spam filtering. The most straightforward feature set is $\mathscr{X} \subset \mathbb{N}_0^k$, where k is the size of word dictionary, and values of the vector represent the number of dictionary word occurrences in the email. It would be tempting to construct the dictionary only from the words connected to spam messages, but this would restrict the abilities of the trained filter significantly. Perhaps, the otherwise inappropriate word is a part of a regular message that should be considered ham.

Another significant selection that we need to make is which similarity measure we will use for comparison of the email messages. For text data, the cosine similarity measure is usually preferred, due to a simple reasoning: The similarity of a message to the same message concatenated several times should be maximal, i.e., should equal 1. Which holds for cosine similarity measure:

$$\text{sim}(X, Y) = \frac{X \cdot Y}{\|X\|_2 \|Y\|_2} = \frac{\sum_{i=1}^k X_i Y_i}{\sqrt{\sum_{i=1}^k X_i^2} \sqrt{\sum_{i=1}^k Y_i^2}}, \qquad (2.66)$$

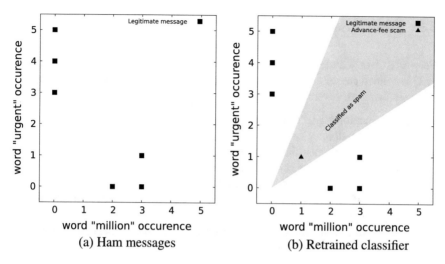

Fig. 2.5 Retraining the classifier after receiving a spam message

in which X and Y are vector representations of two distinct emails, where X_i denotes the number of occurrences of the word i.

Consider we receive a never-before-seen message from a Nigerian prince that wants to transfer millions of dollars urgently via our bank account, offering a decent share. To start the transaction, we need to send only personal information and credit card data for verification. This scam, also known as Advance-fee fraud and discussed in [6], usually contains words "million" and "urgent" in the same message. Let us assume, that we have not labelled such message as spam yet. However, we have several regular messages from our employer with urgent deadlines and several messages from our exaggerating friends that have a million problems. The decision set for the combination of these words, therefore, belongs in whole to class ham, as depicted in Fig. 2.5a.

Once we discover the message in our inbox and label it as spam, retraining of the classifier starts and (depending on the selection of boundaries, i.e. the selection of training strategy) similar-enough messages will also be classified as spam from now on. In our case, the boundary was proposed between the closest spam-ham pairs, as shown in Fig. 2.5b.

This example was oversimplified; however, the basic principles are the same in much higher dimensions and with a more sophisticated feature extraction. Other learning methods, simple to grasp but less easy to demonstrate the learning procedure, are based on error minimization using gradient descent optimization of the overall spam score including all words present in the document (as in, e.g., [7]).

2.5   Feature Selection for Classifiers

Very often, the features determining the feature set $\mathscr{X}$ on which a classifier ϕ in (2.1) is defined are selected form some larger set of available features. Such a feature selection can have several motivations:

- The optimization algorithms used for classifier learning are faster for feature sets of lower dimension.
- At the same time, classifiers can be also more accurate in such sets, in spite of the fact that less features contain less information. The reason is that some of the features may convey information irrelevant for the performed classification.
- The danger of overtraining on the space of all available features is higher than on its subspace $\mathscr{X}$ of the selected features.
- The results of classification based on less features are more easily comprehensible and interpretable.

To determine a suitable subset of the n considered features $S \subset \{1, \ldots, n\}$ requires two decisions:

1. Which subsets of features to select? The simplest approach—an exhaustive investigation of all of them—is manageable only for feature sets of moderate size because the number of subsets is exponential in the feature set size. However, various non-exhaustive heuristics are available, e.g.,

- greedy forward selection, starting with no features and adding one at a time,
- greedy backward elimination, starting with all features and removing one at a time,
- in the case of real-valued features, finding a general sufficiently suitable low-dimensional subspace of the feature space, and then choosing those features that most contribute to the base of that subspace,
- optimization of subsets with respect to suitability by evolutionary algorithms,
- a similar optimization by simulated annealing.

2. How to evaluate their suitability? Although also other approaches to the evaluation of subset suitability exist, those used by far most often are wrapping and filtering.

(i) In *wrapping*, the suitability of the subset is measured by the performance of the considered classifier if its set of features is the evaluated subset.
(ii) In *filtering*, its suitability is measured by some additional measure, independent of a particular classifier. Most important examples of such measures are:

- *Redundancy* of features in S, which is the average mutual information of those features,

$$\text{redundancy}(S) = \frac{1}{|S|^2} \sum_{j, j' \in S} I(j, j'), \qquad (2.67)$$

where $I(j, j')$ stands for the mutual information of the features j and j',

$$I(j, j') = \sum_{v_j \in V_j} \sum_{v_{j'} \in V_{j'}} P(\{x \in \mathcal{X} | [x]_j = v_j, [x]_{j'} = v_{j'}\}) \cdot$$
$$\cdot \log \left(\frac{P(\{x \in \mathcal{X} | [x]_j = v_j, [x]_{j'} = v_{j'}\})}{P(\{x \in \mathcal{X} | [x]_j = v_j\}) P(\{x \in \mathcal{X} | [x]_{j'} = v_{j'}\})} \right). \quad (2.68)$$

- *Relevance* of features in the subset S to a particular class $c \in C$,

$$\text{relevance}(S|c) =$$
$$= \frac{1}{|S|} \sum_{j \in S} \sum_{v_j \in V_j} P(\{x \in \mathcal{X} | [x]_j = v_j, \text{the correct class of } x \text{ is } c\}) \cdot$$
$$\cdot \log \left(\frac{P(\{x \in \mathcal{X} | [x]_j = v_j, \text{the correct class of } x \text{ is } c\})}{P(\{x \in \mathcal{X} | [x]_j = v_j\}) P(c)} \right), \quad (2.69)$$

where $P(c)$ is the a priori probability of the class c.
- Average *Spearman's rank correlation*,

$$\frac{2}{|S|(|S| - 1)} \sum_{j, j' \in S, j < j'} |\varrho_{j, j'}|, \quad (2.70)$$

as well as average *Kendall's rank correlation*,

$$\frac{2}{|S|(|S| - 1)} \sum_{j, j' \in S, j < j'} |\tau_{j, j'}|, \quad (2.71)$$

where $\rho_{j, j'}$ and $\tau_{j, j'}$ are the sample Spearman's rank correlation coefficient, respectively the sample Kendall's rank correlation coefficient of the j-th and j'-th feature based on the training data. For these two correlation coefficients to be available for all $j, j' \in S$, each feature has to be either real-valued or ordinal, hence, it must be linearly ordered by some ordering $\prec_j$. Examples of such an ordering are the standard ordering of real numbers if $V_j \subset \mathbb{R}$ or the lexicographic ordering if V_j contains strings. The ordering $\prec_j$ allows to assign ranks, which we will denote $\text{rank}_j(x)$, to the training feature vectors $x_1, \ldots, x_p$ according to the j-th feature in such a way that

$$\text{rank}_j(x_k) < \text{rank}_j(x_\ell) \text{ if and only if } [x_k]_j \prec_j [x_\ell]_j \text{ and } [x_k]_j = [x_\ell]_j \Rightarrow k < \ell. \quad (2.72)$$

Using those ranks, $\rho_{j, j'}$ is computed as

$$\rho_{j, j'} = 1 - \frac{6 \sum_{k=1}^{p} (\text{rank}_j(x_k) - \text{rank}_{j'}(x_k))^2}{p(p^2 - 1)}. \quad (2.73)$$

Finally, $\tau_{j,j'}$ is computed as

$$\tau_{j,j'} = \frac{2|\{(k,\ell)|([x_k]_j - [x_\ell]_j)([x_k]_{j'} - [x_\ell]_{j'}) > 0\}|}{p(p-1)}$$
$$-\frac{2|\{(k,\ell)|([x_k]_j - [x_\ell]_j)([x_k]_{j'} - [x_l]_{j'}) < 0\}}{p(p-1)} \quad (2.74)$$

The reason why $\rho_{j,j'}$ and $\tau_{j,j'}$ occur in (2.70), respectively (2.71) as absolute values is that ideally, different features should be uncorrelated, i.e., with $\rho_{j,j'} = \tau_{j,j'} = 0$. Then $|\rho_{j,j'}|$, respectively $|\tau_{j,j'}|$ are the distances of $\rho_{j,j'}$, respectively $\tau_{j,j'}$ from that ideal case.

• Average *Pearson's correlation*,

$$\frac{1}{|S|(|S|-1)} \sum_{j,j' \in S, j < j'} |r_{j,j'}|, \quad (2.75)$$

where $r_{j,j'}$ is the sample Pearson's correlation coefficient of the i-th and j-th feature based on the training data. For $r_{j,j'}$, $V_j V_{j'} \subset \mathbb{R}$ is needed, and then

$$r_{j,j'} = \frac{\sum_{k=1}^P ([x_k]_j - \frac{1}{p}\sum_{\ell=1}^P [x_\ell]_j)([x_k]_{j'} - \frac{1}{p}\sum_{\ell=1}^P [x_\ell]_{j'})}{\sqrt{\sum_{k=1}^P ([x_k]_j - \frac{1}{p}\sum_{\ell=1}^P [x_\ell]_j)^2}\sqrt{\sum_{k=1}^P ([x_k]_{j'} - \frac{1}{p}\sum_{\ell=1}^P [x_\ell]_{j'})^2}} =$$
$$= \frac{p\sum_{k=1}^P [x_k]_j [x_k]_j - \sum_{k=1}^P [x_k]_j \sum_{k=1}^P [x_k]_j}{\sqrt{p\sum_{k=1}^P [x_k]_j^2 - (\sum_{k=1}^P [x_k]_j)^2}\sqrt{p\sum_{k=1}^P [x_k]_j^2 - (\sum_{k=1}^P [x_k]_j)^2}}. \quad (2.76)$$

For the same reason as $\rho_{j,j'}$ in (2.70) and $\tau_{j,j'}$ (2.71), also $r_{j,j'}$ occurs in (2.75) as absolute value.

Redundancy and relevance are frequently combined into a measure denoted mRMR,

$$\text{mRMR} = \text{relevance}(S) - \text{redundancy}(S) =$$
$$= \frac{1}{|S|} \sum_{i \in S} \sum_{v_j \in V_j} P(\{x \in \mathscr{X} | [x]_j = v_j, \text{ the correct class of } x \text{ is } c\}) \cdot$$
$$\cdot \left(\frac{P(\{x \in \mathscr{X} | [x]_j = v_j, \text{ the correct class of } x \text{ is } c\})}{P(\{x \in \mathscr{X} | [x]_j = v_j\})P(c)} \right)$$
$$-\frac{1}{|S|^2} \sum_{j,j' \in S} I(j,j'). \quad (2.77)$$

The acronym mRMR stands for *minimum-redundancy-maximum-relevance* and originated from the fact that increasing mRMR entails decreasing redundancy and increasing relevance. However, the maximum of mRMR(S) is achieved, in general, for a different S than the minimum of redundancy(S), as well as for a different S than the maximum of relevance(S).

2.6   Classification is Neither Clustering Nor Regression

In the context of internet applications, two other statistical approaches are encountered, though less frequently than classification, which are in some extent similar to it and could be, at first glance, confused with it. These are the approaches called clustering and regression. In this section, their main principles will be explained, attention being paid primarily to the difference between each of them and classification. In Sects. 2.6.3–2.6.4, two well-known examples of clustering in internet applications are recalled, as well as one example of regression.

2.6.1   Clustering

Clustering means dividing data, more precisely feature vectors, into groups in such a way that data within a group are similar, whereas data from different groups are dissimilar. The resulting groups are called *clusters*, and once they have been found, they can be used to assign new data to the most similar group. This resembles predicting the classes of new data in classification, therefore, some authors call also clustering classification, more precisely *unsupervised classification* because there is no information about correct clusters available, as a counterpart of the information about correct classes in classification. The absence of that information is closely related to the crucial difference between classification and clustering:

- In (supervised and semi-supervised) classification, the classes exist in advance, before classification starts, although the correct assignment of feature vectors may be known only for a very small fraction of the data. And the classes are independent of the method employed for classification.
- In clustering (= unsupervised classification), the clusters emerge only as its result, and depend on the employed clustering method. Although also clustering frequently includes learning in the sense of parameter tuning, there is nothing like training data.

Each clustering method has two ingredients.

1. How the similarity of feature vectors is measured. This can be done in various ways, by far most common are:

- *Distances.* There is a plethora of possibilities how to measure a distance between feature vectors. They will be sketched in connection with k nearest neighbours classifiers in Sect. 3.2.
- *Correlation coefficients.* Although already at least 6 correlation coefficients have been proposed, nearly always is one of the three most common ones used—Spearman's, Kendall's and Pearson's, which were recalled in the previous section.

2. With a given similarity measure, how the data are grouped into resulting clusters. This has several aspects:

   (i) Whether the feature vectors may or must not belong to several clusters simultaneously, i.e., whether the clusters may be fuzzy sets or must be crisp sets.
   (ii) Whether all available data must be divided among the clusters or whether uncovered data may remain.
   (iii) Whether only one division of data into clusters is produced, or a sequence of such divisions, coarsening or refining each other. In the first case, the number of clusters in the resulting division may be given in advance, or may be found as part of the clustering itself. In a hierarchy of subsequently coarsening divisions, on the other hand, there are various possibilities how the *linkage* of two clusters at one level of the hierarchy can be performed to produce a cluster at the subsequent level. Most frequently encountered are:
      - *average linkage,* based on the average similarity of pairs of feature vectors from the Cartesian product of both clusters,
      - *single linkage,* based on the pair of most similar feature vectors from the Cartesian product of the clusters,
      - *complete linkage,* based on the pair of least similar feature vectors from the Cartesian product of the clusters.

In Fig. 2.6, a 2-dimensional visualization of the results of clustering the same audio-visual data into 4 different numbers of clusters are presented. The data records 16 000 segments of popularization lectures about science and technology [8], the clustering method was a self-organizing map (SOM), a topological kind of artificial neural networks [9].

2.6.2   *Regression*

Regression means in the simplest but most commonly encountered case establishing a functional relationship between the vector of features and some real-valued random variable depending on those features. Due to the great difference between the infinite set $\mathbb{R}$ of real numbers with a very rich structure, and finite unstructured sets of categorical data, to which also the set C of classes in classification belongs, regression and classification seem at first sight quite different. Mathematically, however,

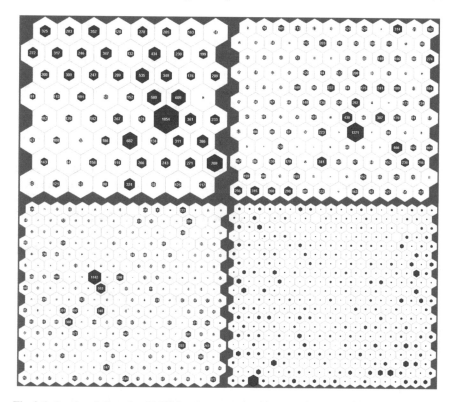

Fig. 2.6 Results of clustering 16 000 feature vectors recording segments of popularization lectures about science and technology into 64 clusters (upper left), 144 clusters (upper right), 256 clusters (lower left), and 400 clusters (lower right) [8]. The clustering method used a self-organizing map [9]

both approaches are quite similar: the result of regression is a *regression function,* a mapping of the feature set into $\mathbb{R}$,

$$\gamma : \mathscr{X} \to \mathbb{R}. \tag{2.78}$$

The definition (2.78) of a regression function is clearly analogous to the definition (2.1) of a classifier.

Like classifiers, also regression functions are obtained through learning, and also their learning consists in solving an optimization task, which is analogy to (2.50):

$$\gamma = \arg \min_{\gamma' \in \Gamma} \text{ER}_\gamma ((x_1, y_1, \gamma'(x_1)), \ldots, (x_p, y_p, \gamma'(x_p))). \tag{2.79}$$

The concepts involved in (2.79) are analogous to those involved in (2.50), namely:

1. The pairs $(x_1, y_1), \ldots, (x_p, y_p)$, called training data, in which y_k for $k = 1, \ldots, p$ is the desired value of the regression function γ in x_k.

2. The set Γ, from which the regression function γ is chosen. That set is entailed by the considered regression method, such as linear functions, polynomials up to a given degree, functions computable by a neural network with a given architecture, etc.
3. The error function ER_γ, which depends in general on the feature vector x, the prediction $\gamma(x)$ and the desired value y of γ in x,

$$ER_\gamma : \mathscr{X} \times \mathbb{R} \times \mathbb{R} \to \mathbb{R}. \qquad (2.80)$$

and an aggregation $ER_\gamma((x_1, y_1, \gamma(x_1)), \ldots, (x_p, y_p, \gamma(x_p)))$ of the values of the errors $ER_\gamma(x_k, y_k, \gamma(x_k))$, $k = 1, \ldots, p$, for example, sum or weighted sum. Similarly to classification error functions ER_ϕ, also regression error functions ER_γ frequently depend on the feature vector only through the prediction $\gamma(x)$ hence

$$ER_\gamma : \mathbb{R} \times \mathbb{R} \to \mathbb{R}. \qquad (2.81)$$

Nevertheless, the above mentioned difference of the set of real numbers $\mathbb{R}$ compared to the set of classifiers C implies that the regression error functions ER_γ differ quite a lot from the classification error functions ER_ϕ, and that for them, solving the optimization task (2.79) is more complicated than for most of the error functions ER_ϕ solving (2.50).

2.6.3 Clustering and Regression in Recommender Systems

As we have already discussed in Sect. 1.2, the recommender systems are in general based on information gathered from individual items present in the system and behaviour of the users interacting with these items. With the use of a similarity measure among the items, users can be presented with a list of similar items to consider. With a defined similarity measure among the user profiles or purchase histories, the recommender system may recommend an item that was chosen by similar users.

Both of these approaches, however, require an extensive collection of overlapping data. To make a recommendation based on user ratings, at least one different user has to give a similar rating to at least one item in common and at least one positive rating to another item. The more ratings for more items are stored in the system, and the greater their overlap on items and users, the more precise is the constructed user profile and the recommender yields better results.

2.6.3.1 Clustering for Improved Performance

Both new items and new users, therefore, pose a significant issue for the system, as the amount of available data is limited and similarities may be impossible to deduce. Consequently, this may lead to poor recommendations.

Large amounts of data, on the other hand, pose a technical problem of high demand on computational power. For example, in the approaches based on k-nearest neighbours, the recommender needs to calculate distances between all object pairs to select the nearest neighbours and base the recommendation upon them.

As a remedy, clusters of items and user profiles may be used to contain the data objects that would otherwise have an insufficient number of connections to other loose objects. To this end, the number of clusters is typically chosen to be significantly smaller than the number of original data points. On the other hand, the repetitive deduction of data point distances is carried out only within the smaller clusters, which makes recommendations feasible especially in time-critical applications, and the system can scale much better.

On the example of the MovieLens database, which contains user ratings of movies, an oversimplified item clustering can be based on the dominant genre of the movie. Such a cluster of movies can be then treated as a single item in the database and user ratings can be aggregated within these clusters. The user profile used for the recommendation itself is then much shorter and denser. Also, the profile matching would not depend on the rating of an individual movie, thus instead of a requirement to rate the same movie to enable a match of user profiles, a relaxed requirement can be used, to rate at least one movie from a genre.

On top of that, users themselves can be organised into the clusters based on their general preferred genre. The recommendation can be, therefore, based only on the nearest users in the same, pre-computed cluster, or even on the statistics of the cluster itself. For example, a user that belongs to a cluster favouring the subgenre of classic Walt Disney animated fairy-tales, with other users highly rating the movie Cinderella, may receive a recommendation of this particular movie.

The application of clustering approach in [10] shows that increasing number of clusters slightly increases the classification error of the recommender system if the recommender does not have access to data outside of the current cluster. However, the throughput increases with the number of considered clusters, as the number of members in each cluster decreases. The use of clustering in recommender system should then balance the requirements on speed and precision.

Recommendations can be also based on the presence of data points in clusters themselves. This approach, in e-commerce sometimes called market segmentation, aims at clustering of goods with the use of a pre-defined or discovered taxonomy. A user is then presented with products from clusters that match the purchase behaviour on selected level of coarseness, as the recommender presented by [11]. This approach provides a flexible system that can balance the strict taxonomy-based recommender with the ability to browse in related products.

Clustering also presents a very important approach to recommendation in the context of social networks. When the distance between two users is defined as a number of friend relationship transitions, clusters of users may correspond to groups of people with strong social interactions. Users that belong to the proposed cluster, but their linkage to other cluster members is weak, may be offered to become friends with other members of the cluster. Advertisers and other social network users may also use the proposed clusters to better target their activities and cause a higher engagement among the users in such clusters.

2.6.3.2 Regression for Improved Performance

The use of regression in recommender systems is derived from a slightly different initial question. While a cluster-based recommender searches for similar objects and users to derive a proposed set of close objects, regression attempts to predict a rating that the user will assign to an object in question. The objects with the highest predicted rating are then returned as recommendations.

This problem is, however, unrealistic to solve even for small sets of items, as the items already rated by the active user belong to an exponentially large set of possible item subsets and the rating estimation of remaining objects would require learning of exponentially many functions.

As a remedy for that, a first order rating approximation can be based on a function representing a relationship between the rated objects. Such a function can predict the rating of an item based on ratings given to other items. This reduces the number of functions to be trained.

Such an approach, presented in [12], produces a comparable error to a nearest neighbour algorithm, however with a significant speed-up in favour of the regression approach.

A benefit of the regression approach is that the estimated rating of individual objects is directly available. If the user-specific ordering of recommended items is crucial, the system can directly offer the objects with the highest estimated rating first.

2.6.4 Clustering in Malware Detection

The ever-increasing number of observed binaries is a pressing issue in the domain of malware detection and analysis since malware authors have been employing obfuscation techniques such as polymorphism and packing to hinder detection [13]. Using these techniques allows to generate an almost unlimited number of binaries with different syntactic representation, yet with the same or similar function. Grouping those binaries via a clustering became a natural choice [14].

Clustering malware into the malware families can bring several additional benefits apart from assigning a new seen binary to an already known family. Clusters can also be used to produce a more generic malware behaviour descriptions and malware signatures that have better detection capabilities [15–17]. Besides, existing clusters may be used to discover previously unknown malware families [18], or to simplify the triage of infected systems [19], significantly decreasing the number of samples that need to be analysed manually.

A good clustering should exhibit high consistency, i.e., uniform behaviour within clusters and heterogeneous behaviour between different clusters. In the case of malware behaviours that are heterogeneous by the nature of its origin, it implies that a large number of rather small clusters would be created if a high consistency is enforced. The results in [20] being compelling evidence to this phenomenon; 3698 malware samples were grouped into 403 clusters, with 100 % consistency, among which 206 (51 %) clusters contained only one sample. The desirable outcome would be a small number of large clusters containing variants of malware that belong to the same family. That leads to another important question, how to evaluate the quality of clustering in a domain as dynamic as malware detection where full ground-truth will never be available, existing labels are noisy and change in time [21]. The work of Perdisci and Man Chon [22] is an excellent starting point to delve deeper into this area of research, showing how to effectively aggregate and use contradictory results verdict of multiple anti-virus engines.

A crucial question in any clustering task is how to measure similarity; and this holds also for the clustering of executables [23]. Key approaches to malware detection using a clustering are now sketched according to the employed similarity measures and the feature types used to calculate them.

Function call graphs or control flow graphs represents relationships between subroutines in a computer program, where nodes represent functions and function calls are captured by oriented edges. In [24], the authors assume that malware binaries from the same family still have similar structure despite the polymorphism and obfuscation. So they proposed to cluster function call graphs using graph edit distance. In [25], a generic framework is presented that extracts structural information from malware programs as attributed function call graphs. Additional features are encoded as attributes at the function level. Then a similarity measure is learned from the training data to ensure that malware binaries from the same family have high similarity and binaries from different families are as dissimilar as possible.

System calls provide an interface between the operating system and system resource, e.g., accessing a file on a hard disk, creation of a new process, open a network connection. In [26], the authors used normalized histograms of n-grams from the system call logs as feature vectors. These vectors are then clustered using a hierarchical clustering algorithm. Prototypes are extracted from each cluster, employing a linear time algorithm by Gonzalez [27]. Each of these prototypes then represents a malware family and every newly observed sample is compared to them. The authors of [28] employed clustering based on Android OS system calls and demonstrated its ability to identify not only malicious, but also modified applications.

Permission-based methods present a relatively new approach how to detect and cluster malicious applications for mobile devices. Each application needs to request permission in order to access a particular resource. Therefore, the set of permissions can be used similarly to the system calls [29].

Combining multiple sources of information is a current trend in malware detection. The obvious problem is how to combine them. To this end, [30] defined similarities specifically designed to capture different aspects of individual sources of information (files, registry keys, network connections) and used clustering to solve the problem of huge dimensionality. Similarities to the resulting clusters then serve as new features.

References

1. Koprinska, I., Poon, J., Clark, J., Chan, J.: Learning to classify e-mail. Inf. Sci. **177**, 2167–2187 (2007)
2. Kushmerick, N.: Learning to remove internet advertisments. In: ACM Conference on Autonomous Agents, pp. 175–181 (1999)
3. Schölkopf, B., Smola, A.: Learning with Kernels. MIT Press, Cambridge (2002)
4. Rokach, L., Maimon, O.: Clustering Methods. Springer (2005)
5. Nasrabadi, N.: Pattern recognition and machine learning. J. Electron. Imaging **16**, 049901(2007)
6. Holt, T., Graves, D.: A qualitative analysis of advance fee fraud email schemes. Int. J. Cyber Criminol. **1**, 137–154 (2007)
7. Goodman, J., Yih, W.: Online discriminative spam filter training. In: CEAS, pp. 76–78 (2006)
8. Pulc, P., Holeňa, M.: Case study in approaches to the classification of audiovisual recordings of lectures and conferences. In: ITAT 2014. Part II., pp. 79–84 (2014)
9. Kohonen, T.: Self-Organizing Maps. Springer (1995)
10. Sarwar, B., Karypi, G., Konstan, J., Riedl, J.: Recommender systems for large-scale e-commerce: scalable neighborhood formation using clustering. In: Fifth International Conference on Computer and Information Technology, pp. 291–324 (2002)
11. Hung, L.: A personalized recommendation system based on product taxonomy for one-to-one marketing online. Expert Syst. Appl. **29**, 383–392 (2005)
12. Vucetic, S., Obradovic, Z.: A regression-based approach for scaling-up personalized recommender systems in e-commerce. In: SIGKDD Workshop on Web Mining and Web Usage Analysis, pp. 55–63 (2000)
13. Guo, F., Ferrie, P., Chiueh, T.: A study of the packer problem and its solutions. In: International Workshop on Recent Advances in Intrusion Detection, pp. 98–115 (2008)
14. Rieck, K., Holz, T., Willems, C., Düssel, P., Laskov, P.: Learning and classification of malware behavior. In: International Conference on Detection of Intrusions and Malware, and Vulnerability Assessment, pp. 108–125 (2008)
15. Park, Y., Reeves, D., Stamp, M.: Deriving common malware behavior through graph clustering. Comput. Secur. **39**, 419–430 (2013)
16. Nissim, N., Cohewn, A., Elovici, I.: ALDOCX: detection of unknown malicious microsoft office documents using designated active learning methods based on new structural feature extraction methodology. IEEE Trans. Inf. Forensics Secur. **12**, 631–646 (2017)
17. Nissim, N., Moskowitch, R., Rokach, L., Elovici, I.: Novel active learning methods for enhanced PC malware detection in windows OS. Expert Syst. Appl. **41**, 5843–5857 (2014)
18. Bayer, U., Comparetti, P., Hlauschek, C., Kruegel, C., Kirda, E.: Scalable, behavior-based malware clustering. In: NDSS'09, pp. 8–11 (2009)
19. Jang, J., Brumley, D., Venkataraman, S.: Bitshred: feature hashing malware for scalable triage and semantic analysis. In: 18th ACM Conference on Computer and Communications Security, pp. 309–320 (2011)
20. Bailey, M., Oberheide, J., Andersen, J., Mao, Z., Jahanian, F., Nazario, J.: Automated classification and analysis of internet malware. In: International Workshop on Recent Advances in Intrusion Detection, pp. 178–197 (2007)

21. Li, P., Liu, L., Gao, D., Reiter, M.: On challenges in evaluating malware clustering. In: International Workshop on Recent Advances in Intrusion Detection, pp. 238–255 (2010)
22. Perdisci, R., Man Chon, U.: VAMO: towards a fully automated malware clustering validity analysis. In: 28th Annual Computer Security Applications Conference, pp. 329–338 (2012)
23. Apel, M., Bockermann, C., Meier, M.: Measuring similarity of malware behavior. In: 34th IEEE Conference on Local Computer Networks, pp. 891–898 (2009)
24. Kinable, J., Kostakis, O.: Malware classification based on call graph clustering. J. Comput. Virol. Hacking Tech. **7**, 351–366 (2011)
25. Kong, D., Yan, G.: Discriminant malware distance learning on structural information for automated malware classification. In: 19th ACM SIGKDD International Conference on Knowledge Discovery and Data Mining, pp. 1357–1365 (2013)
26. Rieck, K., Trinius, P., Willems, C., Holz, T.: Automatic analysis of malware behavior using machine learning. J. Comput. Secur. **19**, 639–668 (2011)
27. Gonzalez, T.: Clustering to minimize the maximum intercluster distance. Theor. Comput. Sci. **38**, 293–306 (1985)
28. Burguera, I., Zurutuza, U., Nadjm-Tehrani, S.: Crowdroid: behavior-based malware detection system for Android. In: First ACM workshop on Security and Privacy in Smartphones and Mobile Devices, pp. 15–26 (2011)
29. Aung, Z., Zaw, W.: Permission-based android malware detection. Int. J. Sci. Technol. Res. **2**, 228–234 (2013)
30. Stiborek, J., Pevný, T., Řehák, M.: Multiple instance learning for malware classification. Expert Syst. Appl. **93**, 346–357 (2018)

Chapter 3
Some Frequently Used Classification Methods

This chapter, after recalling into which groups classifiers can be divided, brings a survey of the most important individual classification methods, especially those encountered in internet applications. Focus is on individual use of those methods, whereas classification by means of teams of classifiers is the topic of the last chapter of this book, Chap. 6. Excluded from the survey are key representatives of the two main trends in modern classification methods, to which the subsequent chapters are devoted:

- Support vector machines, which are the representative of a trend to possibly high predictive accuracy. They will be presented in Chap. 4.
- Classification rules, which represent the trend to the comprehensibility of classi-fication. Various kinds of classification rules will be the topic of Chap. 5.

3.1 Typology of Classification Methods

As we have seen in the previous chapter, the surfaces separating the individual classes, aka *decision boundaries*, have a crucial role for classification. Nevertheless, this does not mean that every classification method attempts to approximate the decision boundary of two classes before deciding to which of both classes a feature vector is assigned. In fact, due to high computational demands of such approximations, they are affordable only for methods developed during the last 3–4 decades. More traditional methods decide between both classes without explicitly approximating the surface that separates them. Consequently, all classification methods can be naturally divided into two large groups according to whether they explicitly approximate the surfaces separating individual classes.

© Springer Nature Switzerland AG 2020
M. Holeňa et al., *Classification Methods for Internet Applications*,
Studies in Big Data 69, https://doi.org/10.1007/978-3-030-36962-0_3

1. *Methods explicitly approximating the decision boundary.* These are comparatively recent methods, often computationally demanding. Their main representatives will be described in Sect. 3.5 and Chaps. 4, 5. Individual methods from this group differ from each other primarily through the kind of approximation they use:

   - piecewise-constant—classification rules, classification trees;
   - linear—perceptrons, linear support vector machines;
   - non-linear—multilayer perceptrons, non-linear support vector machines.

2. *Methods deciding between classes by means of alternative concepts,* computationally less demanding than the methods using an approximation of the decision boundary. These are traditional classification methods dating back to the 19th and first half of 20th century. Various methods of this group use three kinds of alternative concepts to this end:

   - *distance*, or more generally, *similarity* between the feature vector to be classified and feature vectors with known class membership—in k-nearest neighbours classifiers;
   - *class probability* of each of the classes to be the correct class of a given feature vector—in Bayes classifiers, logit method;
   - *probability distribution* of the feature vectors for each of the classes—in linear discriminant analysis, quadratic discriminant analysis.

An overview of these kinds of classification methods is given in Table 3.1.

Table 3.1 Important kinds of classification methods and their main representatives

Principle	Main examples	Sections
Methods that explicitly approximate the decision boundary		
Piecewise-constant approximation	Classification rules, classification trees	5.2–5.3
Linear approximation	Perceptrons, support vector machines	3.5, 4.2
Non-linear approximation	Multilayer perceptrons,	3.5
	Support vector machines	4.3
Methods that don't explicitly approximate the decision boundary		
Distance or similarity	k-nearest neighbours classifier	3.2
Estimated class probability	Bayes classifiers, logit method	3.3
Estimated probability distribution of feature vectors	Linear discriminant analysis, quadratic discriminant analysis	3.4

3.2   Classification Based on k-Nearest Neighbours

A very traditional way of classifying a new feature vector $x \in \mathscr{X}$ if a sequence of training data $(x_1, c_1), \ldots, (x_p, c_p)$ is available is the *nearest neighbour method*: take the x_j that is the closest to x among $x_1, \ldots, x_p$, and assign to x the class assigned to x_j, i.e., c_j.

A straightforward generalization of the nearest neighbour method is to take among $x_1, \ldots, x_p$ not one, but k feature vectors $x_{j_1}, \ldots, x_{j_k}$ closest to x. Then x is assigned the class $c \in C$ fulfilling

$$c = \arg\max_{c' \in C} |\{i, 1 \leq i \leq k | c_{j_i} = c'\}|. \tag{3.1}$$

This method is called, expectedly, *k-nearest neighbours,* or *k-nn* for short.

Hence, this is a very simple method. In particular, notice that it uses actually no learning. More precisely, its learning deserves the term *lazy learning* because in the time of learning, solely the training data is made available, whereas using the information contained in that data is deferred to the time of classifying new feature vectors.

Before the k-nn method is employed, several decisions must be made:

- How to measure the closeness of feature vectors. To this end, any from a broad spectrum of existing distance measures can be used, which will be surveyed below, in Sect. 3.2.1. However, what is important on distance measures from the point of view of the k-nn method is that they measure similarity between feature vectors. Instead of a distance, therefore, also other similarity measures can be used, such as correlation coefficients [1], the main representatives of which were recalled in Sect. 2.4.
- What number k of neighbours to consider. For different values of k, the results of classifying a new feature vector can be different (cf. Fig. 3.1).
- How to deal with ties in (3.1), e.g., choosing one of them randomly, or ordering them by means of another distance/similarity measure. In this context, it is worth realizing that ties can be avoided in binary classification, through choosing an odd k.

For its simplicity, the k-nn method pays with two serious problems:

(i) High computational demand, especially if a distance requiring intensive computation is used, such as Euclidean or Mahalanobis (cf. Sect. 3.2.1), and if really no information contained in the training data is used in the time of learning, because in that case, the distance of every new feature vector to all training data has to be computed in the classification time. However, if distances between training data are pre-computed in the time of learning, then the computational demand of nearest neighbour search in the classification time can be decreased.

(ii) Misleading classification by inappropriate neighbours. Such neighbours are obtained due to the use of inappropriate features in the nearest neighbours

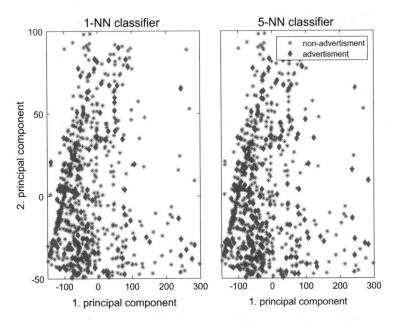

Fig. 3.1 The 1. and 2. principal component of a 1-nn (left) and 5-nn (right) classification using the Euclidean distance, of the graphical objects on web pages into advertisements and non-advertisements (cf. [2])

search, as a consequence of either inappropriate feature selection, or inappropriate feature scaling, causing misleading features to get much weight in distance computation. Countermeasures against inappropriate feature selection are feature selection methods, surveyed in Sect. 2.5, countermeasure against inappropriate feature scaling is using distances that allow reweighting individual features (cf. Sect. 3.2.1).

3.2.1 Distances Between Feature Vectors

Distance is a nonnegative function d on pairs of objects that has the following properties:

(i) symmetry: $d(x, y) = d(y, x)$,
(ii) $d(x, y) = 0$ if and only if $x = y$,
(iii) triangular inequality: $d(x, z) \leq d(x, y) + d(y, z)$.

Like kernels, which were surveyed in Sect. 2.3, also distances between the vectors of all considered features are constructed in two steps: first, distances between vectors of particular subsets of features are chosen, and then, distances for all considered subsets of features are combined together. Different distances for two subsets of vectors are needed if one subset contains real-valued features and the other contains categorical features. However, even if the features in both subsets are real-valued, or if both of them are categorical, different distances can be used for those subsets of features. On the other hand, for ordinal features with finite value sets, the same distances as for real-valued features or as for categorical features can be used.

1. *Real-valued features.* All commonly used distances between vectors of real numbers rely actually on some norm on the respective vector space,

$$\text{dist}(x, y) = \|x - y\|. \tag{3.2}$$

Therefore, we will now recall important norms on the space $\mathbb{R}^d$ of d-dimensional real vectors. Due to (3.2), the names of the norms explained below are alternatively used also as the names of the corresponding distances, e.g., Euclidean norm / Euclidean distance, Mahalanobis norm / Mahalanobis distance, etc.

   a. Most often used is definitely the *Euclidean norm*,

$$\|x\| = \sqrt{\sum_{j=1}^{d} [x]_j^2}, x \in \mathbb{R}^d. \tag{3.3}$$

     This norm has a particular importance also from a mathematical point of view because it can be actually obtained from the scalar product $x^\top y$ on $\mathbb{R}^d$,

$$\|x\| = \sqrt{x^\top x} = \sqrt{x^\top I_d x}, \tag{3.4}$$

     where I_d stands for the d-dimensional identity matrix. This norm is the default norm in the book if nothing else is said.

   b. A generalization of (3.3) is the *Minkowski norm* with a parameter $p \geq 1$,

$$\|x\|_p = \sqrt[p]{\sum_{j=1}^{d} |[x]_j|^p}, x \in \mathbb{R}^d, \tag{3.5}$$

     where $|\cdot|$ denotes the absolute value of real numbers. According to (3.5), the Euclidean norm is $\|\cdot\|_2$. Apart from it, two other special cases of (3.5) are commonly encountered, $\|\cdot\|_1$,

$$\|x\|_1 = \sum_{j=1}^{d} |[x]_j|, x \in \mathbb{R}^d, \tag{3.6}$$

     sometimes called *citi-block norm* or *manhattan norm*, and the limit case $\|\cdot\|_\infty$,

$$\|x\|_\infty = \lim_{p \to \infty} \|x\|_p = \max_{j=1,\ldots,d} |[x]_j|, \, x \in \mathbb{R}^d, \qquad (3.7)$$

called *Chebyshev norm*.

c. Another generalization of the Euclidean norm consists in generalizing the scalar product $x^\top y = x^\top I_d y$ to a scalar product $x^\top \mathrm{diag}(\lambda_1, \ldots, \lambda_d) y$, where $\lambda_1, \ldots, \lambda_d > 0$ and $\mathrm{diag}(\lambda_1, \ldots, \lambda_d)$ denotes a diagonal matrix formed by those numbers. The resulting norm

$$\|x\|_\Lambda = \sqrt{x^\top \mathrm{diag}(\lambda_1, \ldots, \lambda_d)x} = \sqrt{\sum_{j=1}^d \lambda_j [x]_j^2}, \, x \in \mathbb{R}^d, \qquad (3.8)$$

corresponds to weighting individual features with weights $\lambda_1, \ldots, \lambda_d$ and then using the Euclidean norm.

d. Similarly to using the weights $\lambda_1, \ldots, \lambda_d$ with the Euclidean norm in (3.8), they can be used also with any other Minkowski norm $\| \cdot \|_p, \, p \geq 1$:

$$\|x\|_{p,(\lambda_1,\ldots,\lambda_d)} = \sqrt[p]{\sum_{j=1}^d (\lambda_j |[x]_j|)^p}, \, x \in \mathbb{R}^d. \qquad (3.9)$$

Most frequently, such a weighted version is encountered with the citi-block norm:

$$\|x\|_{1,(\lambda_1,\ldots,\lambda_d)} = \sum_{j=1}^d \lambda_j |[x]_j|, \, x \in \mathbb{R}^d. \qquad (3.10)$$

e. A final possible generalization of the Euclidean norm is to generalize the scalar product $x^\top \mathrm{diag}(\lambda_1, \ldots, \lambda_d) y$ in Sect. 3.8 to $x^\top Q \mathrm{diag}(\lambda_1, \ldots, \lambda_d) Q^\top y$, where Q is an orthogonal matrix, i.e., a matrix fulfilling $Q^{-1} = Q^\top$. Notice that the columns of each such matrix Q form an orthonormal base of the space $\mathbb{R}^d$, and that Q then describes the transformation of coordinates of $x \in \mathbb{R}^d$ from that new base to the standard base $(1, \ldots, 0), \ldots, (0, \ldots, 1)$, whereas Q^{-1} describes the transformation of coordinates from the standard to the new base. Typically, the new base is chosen to be formed by the eigenvectors $v_1, \ldots, v_d$ of the empirical variance matrix of the training feature vectors

$$\mathrm{Var}(x_1, \ldots, x_p) = (v_{j,j'})_{j,j'=1}^d,$$

with $v_{j,j'} = \dfrac{1}{p-1} \left(\sum_{k=1}^p [x_k]_j [x_k]_{j'} - \dfrac{1}{p} \sum_{k=1}^p [x_k]_j \sum_{k=1}^p [x_k]_{j'} \right), \, j, j' = 1, \ldots, d, \quad (3.11)$

whereas the numbers $\lambda_1, \ldots, \lambda_d$ in the diagonal matrix are chosen to be reciprocal to the corresponding eigenvalues of $\mathrm{Var}(x_1, \ldots, x_p)$, or equivalently, to be the eigenvalues of $\mathrm{Var}(x_1, \ldots, x_p)^{-1}$,

$$\mathrm{Var}(x_1, \ldots, x_p)v_j = \frac{1}{\lambda_j}v_j, \text{ or equivalently, } \mathrm{Var}(x_1, \ldots, x_p)^{-1}v_j = \lambda_j v_j.$$
(3.12)

That choice is called *Mahalanobis norm*, and using the eigenvalue decompositions of $\mathrm{Var}(x_1, \ldots, x_p)$ and $\mathrm{Var}(x_1, \ldots, x_p)^{-1}$, it can be expressed as

$$\|x\|_{\mathrm{Mahalanobis}} = \sqrt{x^\top \mathrm{Var}(x_1, \ldots, c_p)^{-1}x} = \sqrt{\tilde{x}^\top \mathrm{diag}(\lambda_1, \ldots, \lambda_d)\tilde{x}}, x \in \mathbb{R}^d, \quad (3.13)$$

where $\tilde{x}$ denotes the coordinates of x in the base formed by $v_1, \ldots, v_d$, aka *principal components* of x.

2. *Categorical features.* For them, three distances are commonly encountered, but only the first of them can be used universally for any categorical features.

   a. *Edit distance* of two sequences x and y of categorical data is the number of elementary operations needed to transform x to y. Elementary operations with sequences are:
   - changing the value at a particular position of a sequence;
   - removing the value at a particular position of a sequence;
   - inserting a value between particular subsequent positions of the sequence.

   b. *Hamming distance* can be used only for categorical data sequences of the same length, and counts the number of positions at which their values differ:

   $$\text{for } x = ([x]_1, \ldots, [x]_d), y = ([y]_1, \ldots, [y]_d) \text{ is } \mathrm{dist}(x, y) = |\{j\,|\,[x]_j \neq [y]_j\}|.$$
   (3.14)

   c. *Jaccard distance* can be used only for binary vectors of the same length or for categorical data representable by such vectors. It counts the proportion of components in which both vectors differ among those for which their disjunction holds:

   $$\text{for } x, y \in \{0, 1\}^d, x = ([x]_1, \ldots, [x]_d), y = ([y]_1, \ldots, [y]_d) \text{ is}$$
   $$\mathrm{dist}(x, y) = \begin{cases} \dfrac{|\{i\,|\,[x]_i = [y]_i = 1\}|}{|\{i\,|\,\max([x]_i,[y]_i)=1\}|} & \text{if } \sum_{i=1}^d ([x]_i + [y]_i) > 0, \\ 0 & \text{if } [x]_i = [y]_i = 0, i = 1, \ldots, d. \end{cases}$$
   (3.15)

   Jaccard distance is suitable for set-valued features, more precisely, for features the values of which are subsets of some finite set S. Coding the membership of elements of S in subsets $A, B \subset S$ of S with binary vectors turns (3.15) into

$$A, B \subset S \Rightarrow \text{dist}(A, B) = \begin{cases} \frac{|A \triangle B|}{|A \cup B|} = 1 - \frac{|A \cap B|}{|A \cup B|} & \text{if } A \cup B \neq \emptyset, \\ 0 & \text{if } A \cup B = \emptyset, \end{cases} \quad (3.16)$$

where $A \triangle B$ denotes the symmetric difference of A and B, i.e., the union of the differences $A \setminus B$ and $B \setminus A$.

3. *Ordinal features* can always make use of the fact that the set V_j of feasible values of a ordinal feature, with its respective ordering $\prec_j$, is isomorphic to some subset, finite or infinite, of natural numbers $\mathbb{N}$, with one of the usual orderings $\leq$ or $\geq$ of real numbers. Then it is possible to calculate any distance pertaining to real-valued features using the respective subset of $\mathbb{N}$ instead of V_j. If in addition V_j is finite, then we have also the alternative possibility to ignore $\prec_j$ and to use a distance pertaining to categorical features.

After having recalled the most important distances used with different kinds of features, suppose that the full feature set $\mathscr{X}$ is actually a subset of Cartesian products of feature sets with subsets of the considered features,

$$\mathscr{X} \subset \mathscr{X}_1 \times \cdots \times \mathscr{X}_s \quad (3.17)$$

such that in each $\mathscr{X}_j$, $j = 1, \ldots, s$, a particular distance dist_j is used. Notice that this implies that either all features in $\mathscr{X}_j$ are real-valued and/or ordinal, or all of them are categorical and/or ordinal. Then for any feature vectors $x = (x^{(1)}, \ldots, x^{(s)})$, $y = (y^{(1)}, \ldots, y^{(s)}) \in \mathscr{X}$ such that $x^{(j)}, y^{(j)} \in \mathscr{X}_j$ for $j = 1, \ldots, s$, a vector of values of distances $(\text{dist}_1(x^{(1)}, y^{(1)}), \ldots, \text{dist}_s(x^{(s)}, y^{(s)}))$ is obtained. Since this is an s-dimensional vector of non-negative real numbers, the overall distance of x and y can be assessed using some norm of that vector, such as one of the norms 1.a.-1.e. above. Typically, some of the weighted variants (3.8)–(3.10) is used to this end, most frequently the weighted citi-block norm (3.10), which is simply the weighted mean of the distances on $\mathscr{X}_1, \ldots, \mathscr{X}_s$,

$$\text{dist}(x, y) = \|(\text{dist}_1(x^{(1)}, y^{(1)}), \ldots, \text{dist}_s(x^{(s)}, y^{(s)}))\|_{1,(\lambda_1,\ldots,\lambda_s)} = \sum_{j=1}^{s} \lambda_j \, \text{dist}_j(x^{(j)}, y^{(j)}).$$

$$(3.18)$$

3.2.2   Using k-nn Based Classifiers in Recommender Systems

For the sake of simplicity, we will stay in the most straightforward example of e-commerce recommenders, where the user is described by a vector of real-valued item ratings, and the items with the highest rating estimated from other user profiles are recommended. Such a recommender system has to fulfil the following major tasks.

3.2.2.1 Find Similarity Among Users

A straightforward application of approaches based directly on distances among feature vectors is, however, difficult.

Most importantly, not all users provide a rating to all items. A straightforward remedy is to consider only the items rated by both users for the respective distance measurement. If we consider the most common distance, based on the *Euclidean norm*, this means:

$$\text{dist}(x, y) = \sqrt{\sum_{j \in S} ([x]_j - [y]_j)^2}, \tag{3.19}$$

where x and y are the two users we determine the distance for, and S is the subset of features (dimensions, items) evaluated by both users.

This approach would be suitable if all users have the same objective interpretation and perception of the rating – for example 0 being "very bad", 0.5 being "average - nor good, nor bad", and 1 being "excellent, top-notch". This assumption is, however, hardly satisfied as people tend to have different rating scales, mostly based on the expectations that might also be very variable from person to person. The absolute value is, therefore, not as decisive for user similarity assessment as the overall correlation between the users.

The most commonly used similarity measure in recommenders is, therefore, the *Pearson correlation coefficient* on the samples from the common subset S:

$$r_{x,y} = \frac{\sum_{j \in S} ([x]_j - \overline{[x]})([y]_j - \overline{[y]})}{\sqrt{\sum_{j \in S} ([x]_j - \overline{[x]})^2} \sqrt{\sum_{j \in S} ([y]_j - \overline{[y]})^2}}, \tag{3.20}$$

where $\overline{[x]}$ denotes the average rating given by user x.

Similarity derived purely from the Pearson correlation coefficient does, however, overestimate the significance of similar rating on commonly reviewed items. Whereas, the agreement on a rating of not very common items or large variance of the rating is deemed to be more significant. For example, we cannot distinguish the individual users based on the rating given to bacon, because it presents a prevalent item and everyone has a strong opinion about it. The Pearson correlation coefficient is thus overstimulated by this feature. But rating on, for example, tiger prawns is not so common and more diverse and may, therefore, provide much better disambiguation among the individual consumers.

This effect may be mitigated by incorporating a penalty function ψ into the similarity:

$$\text{sim}_{x,y} = \frac{\sum_{j \in S} ([x]_j - \overline{[x]})([y]_j - \overline{[y]}) \psi(j, x, y)}{\sqrt{\sum_{j \in S} ([x]_j - \overline{[x]})^2} \sqrt{\sum_{j \in S} ([y]_j - \overline{[y]})^2}}. \tag{3.21}$$

This penalty function ψ may depend on the voting habits of the currently considered users (e.g. how many items had the user voted on) as well as the number of item ratings. Such functions are then reminiscent of a TF-IDF function used in natural language processing. However, in the most straightforward extensions, ψ depends only on the considered item ratings (later is denoted only $\psi(j)$). The most basic approach, mentioned in [3], uses an inverse user frequency to enhance the effect of ratings given to not widely rated items:

$$\psi(j) = \log(p/p_j),\tag{3.22}$$

where p is the number of users and p_j denotes the number of users with a stored rating to item j.

With a different assumption that the significance of rating is given by its variance, the variance weighting proposed in [4] uses

$$\psi(j) = \frac{s_j^2 - \min(s^2)}{\max(s^2)},\tag{3.23}$$

where s_j^2 is the unbiased sample variance of ratings given to item j:

$$s_j^2 = \frac{1}{p-1}\sum_{k=1}^{p}([k]_j - \overline{[\cdot]_j})^2,\tag{3.24}$$

and $\min(s^2)$ and $\max(s^2)$ is the minimal (respectively, maximal) rating variances among all items. $\overline{[\cdot]_j}$ denotes the average rating to item j.

For simplicity of our example on the k-nn based recommender system, we will use the unweighted Pearson correlation coefficient to determine the similarities among users in Table 3.2. The results are presented in Table 3.3.

3.2.2.2   Select the k-Nearest Neighbours

Next step is to select the size of a neighbourhood to consider in the estimation of unknown ratings. In theory, k has to be big enough to cover the variance in available data. On the other hand, k approaching the number of available samples (n) diminishes the benefits of neighbourhood approach and leads to results based on global statistics or noisy data. Therefore, selection of k is a challenge on its own.

In many cases, the value of $k = 1$ is used with the underlying assumption that the most similar users would give the same rating to the items of concern. For a more sophisticated, but still rough, approximation of k, a simulated study on k-nn classification [5] concluded that k should be selected in the range of $n^{2/8}$ to $n^{3/8}$ for sufficiently large n.

Table 3.2 Rating of the ten most rated movies by ten most active users from the MovieLens dataset [6] and a global rating average of considered users

x	$[x]_1$	$[x]_{50}$	$[x]_{100}$	$[x]_{121}$	$[x]_{181}$	$[x]_{258}$	$[x]_{286}$	$[x]_{288}$	$[x]_{294}$	$[x]_{300}$	$[\bar{x}]$
13	3	5	5	5	5	4	3	1	2	1	3.097
234	3	4	4		3	2	3	3	3	3	3.123
276	5	5	5	4	5	5		4	4	4	3.465
303	5	5	5	3	5	4	5	4	4	1	3.366
393	3	5	1	4	4	4		3	4		3.337
405		5			5			5			1.834
416	5	5	5	5	5	5	5	5	4	4	3.846
450	4	5	4	3	4	4	4	3	4	4	3.865
537	2	4	4	1	2	4	3	2	1	1	2.865
655	2	4	3	3	3	2	3	3	3	3	2.908

Table 3.3 Similarities measured by the Pearson correlation coefficient among the ten most active users and the global rating average, based on the ten most rated movies from the MovieLens dataset [6]

r	234	276	303	393	405	416	450	537	655
13	0.349	0.453	0.522	0.127	0.289	0.406	0.292	0.39	0.161
234		0.065	0.221	−0.268	0.408	0.024	0.394	0.24	0.731
276			0.678	0.089	0.931	0.932	0.293	−0.061	−0.103
303				−0.017	0.94	0.693	0.307	0.283	0.005
393					0.63	0.086	0.279	−0.214	0.268
405						1.0	0.163	−0.206	0.67
416							0.024	−0.123	−0.05
450								0.466	0.309
537									0.063

However, in real applications, the best value of k depends on many aspects of the data, selected similarity measure and the prediction method. Therefore, a more widespread approach to selecting k is to use a cross-validation with the available data.

3.2.2.3 Pick a Prediction Method

Because the selected neighbourhood consists of k values, we need to devise a method of combining them into a single rating prediction. Let us assume that we pick a fixed $k = 4$ for the following comparison of prediction methods and try to predict rating of movie 1 for user 13 ($[13]_1 = 3$). Therefore, we select the users 303, 276, 416 and 537 with ratings on movie 1 of 5, 5, 5 and 2.

If the number of allowed rating values is limited (in our case only five distinct rating values are possible) and we consider the ratings as categorical data, the most straightforward mechanism is to select the most common value from the set. In case of a tie, either of the most common values can be chosen. In our example, we acquire a faulty prediction of $[\hat{13}]_1 = 5$. However, this method is very simplistic and gives no assumption on the rating system; the individual rating values are assumed to be completely unrelated—even not ordered or equidistant.

Because the movie rating system is based on assigning a certain number of stars, we may safely make an assumption that the individual rating values are perceived as strictly ordered and equidistant. The rating prediction is then, in fact, a k-nn regression deducted as an average of ratings in the set. In our case, the predicted rating is 4.25, which is slightly better.

Another improvement of the prediction can be made with weighting of the neighbour ratings. To this end, we may advantageously use the already computed correlation among users as discussed in [7]:

$$[\hat{x}]_j = \overline{[x]} + \frac{\sum_{y \in N}([y]_j - \overline{[y]})r_{x,y}}{\sum_{y \in N}|r_{x,y}|}, \tag{3.25}$$

where N denotes the set of neighbours of x that contain known rating for j. In our case, the weighted prediction is equal to:

$$[\hat{13}]_1 = 3.097 + \frac{0.852 + 0.694 + 0.469 - 0.338}{0.522 + 0.453 + 0.406 + 0.39} = 4.045 \tag{3.26}$$

As the error of classification (measured as $\sum_x \sum_j |[x]_j - [\hat{x}]_j|$) is the lowest with the last mentioned method, we will select it for further predictions.

3.2.2.4 Predict Rating and Provide Recommendation

The last step is to predict the rating of items the user has not rated so far, order them in descending order and present the top portion of the list to the user.

Given that our example includes a user 405 with only a few ratings among the top ten most rated movies, this presents an ideal user to provide recommendations. To compute the predictions, Eq. (3.25) is used with all available data from Tables 3.2 and 3.3, enabling to follow the computation. The resulting predictions are presented in Table 3.4. The second row of the same table then presents the predictions based on ratings and correlations from the whole dataset.

Based on the full dataset, movies 300, 286 and 121 can be now presented to the user in this order as recommendations.

Table 3.4 Predicted ratings of the ten most rated movies that are not yet rated by user 405 from the MovieLens dataset

	$[\hat{405}]_1$	$[\hat{405}]_{100}$	$[\hat{405}]_{121}$	$[\hat{405}]_{258}$	$[\hat{405}]_{286}$	$[\hat{405}]_{294}$	$[\hat{405}]_{300}$
From Tables 3.2 and 3.3	2.488	2.484	2.369	2.327	2.56	2.172	1.452
Full dataset	1.379	1.669	2.246	2.186	2.335	1.473	2.455

3.2.3 Using k-nn Based Classifiers in Malware and Network Intrusion Detection

The usage of k-nn in the field of malware detection dates back to the early 1990s. At that time, researchers realised that malicious computer viruses spread in a similar manner as their biological counterparts [8]. For that reason, early research in malware detection field was inspired by biology, e.g., [9–11]. In [11], the authors tried to apply the nearest neighbour search directly to the source code, treating it as a text and using Hamming distance as a similarity measure. In that way, however, nearest neighbour classification performed poorly because most of the code of malicious samples and legitimate samples are very similar. This shows that using the right features and similarity measure is crucial, for a k-nn classifier.

In [12], binary executables were modelled using statistical properties of their network traffic payload. During a training phase, a profile of byte frequency distribution and standard deviation of the application payload is computed for each application. In the detection phase, a Mahalanobis distance is used to calculate the similarity of new data to the pre-computed profiles.

System calls can be used in both discussed domains, malware detection and intrusion detection. In [13], the authors used system calls together with the TF-IDF (term frequency - inverse document frequency) weighting scheme and cosine similarity.

System calls together with network connection records were used as features also in [14]. The authors defined a specific kernel for each data source and employed k-nn classifier together with clustering based anomaly detection [15].

Another great challenge is that with an increasing number of features, the differences between samples are diminishing; this effect is a consequence of the curse of dimensionality. To overcome it, the authors of [16] employed principal component analysis (PCA) on system call chains and unix commands to reduce the dimensionality. A similar approach is used in [17]: PCA followed by a k-nn classifier, but for network data.

However, even if the dimension of the problem is reasonable, the main obstacle in using k-nn on current datasets is the quadratic computational complexity. There are many approaches how to reduce this computational burden, e.g. using advanced data structures such as kd-trees [18] or using various approximative algorithms [19].

An approach of a different kind is to perform a clustering of the dataset and compare the classified instances only on a small number of datapoints that represent the resulting clusters. In [20], authors used hierarchical clustering on system call logs to identify malware families. Then they extracted prototypes that would represent a malware family. A new observed malicious sample can be assigned to the corresponding malware family using the nearest neighbour search. In [21], clustering is used to split the dataset into smaller chunks and then perform the nearest neighbour search only within the cluster that is most similar to the sample in question.

3.3 Classifiers Estimating Class Probability

Two other commonly used classification methods—Bayes classifiers and logit method—rely on estimating the probability that each available class is the correct class for a given feature vector $x \in \mathscr{X}$. In other words, they estimate for x the probability distribution on the set $C = \{c_1, \ldots, c_m\}$ of classes

$$(P(c_1|x), \ldots, P(c_m|x)) \in [0, 1]^m. \tag{3.27}$$

Such a classifier then usually classifies x to the class with the highest probability. Alternatively, the classifier can output the whole probability distribution (3.27), in which case it is a fuzzy classifier (cf. Sect. 2.1).

3.3.1 Bayes Classifiers

Bayes classifiers are classifiers based on a principle discovered in the 18th century by Thomas Bayes and known as *Bayes' theorem*. Its formulation in the context of classification depends on the considered features. For categorical and ordinal features, more precisely for a feature set $\mathscr{X}$ of n-dimensional vectors of categorical or ordinal features, the Bayes' theorem reads

$$P(c|x) = \frac{P(c)P(x|c)}{P(x)}, x \in \mathscr{X}, c \in C. \tag{3.28}$$

Here, $P(x|c)$ denotes the conditional probability that the n-dimensional vector of features has the particular value x on condition that its correct class is c. Further, $P(x)$ is the unconditional probability that the vector of features has the value x, aka the a priori probability of x, and from Sect. 2.4, we already know that $P(c)$ is the a priori probability of the event that the correct class of $x \in \mathscr{X}$ is c. The probability $P(c|x)$, resulting from (3.28), is then called *a posteriori probability* of that event.

For real-valued features, more precisely for a feature set $\mathscr{X}$ of n-dimensional features that are realizations of a continuous random vector, a counterpart of (3.28) is:

$$P(c|x) = \frac{P(c)f(x|c)}{f(x)}, x \in \mathcal{X}, c \in C, \tag{3.29}$$

in which $f(\cdot|c)$ is the conditional density of that n-dimensional random vector on condition that its correct class is c, whereas f is its unconditional density.

Bayes classifier learning includes three tasks:

1. Estimating or setting the a priori probability that the correct class of the considered vector of features is c. The unbiased estimate of that probability based on the training data $(x_1, c_1), \ldots, (x_p, c_p)$ is

$$P(c) = \frac{|\{k = 1, \ldots, p | c_j = c\}|}{p}. \tag{3.30}$$

Alternatively, $P(c)$ can be set independently of the training data, typically to $P(c) = \frac{1}{m}$, which corresponds to the uniform distribution on the set C of classes.

2. Estimating the conditional probability $P(x|c)$ in (3.28) or the conditional density $f(x|c)$ in (3.29). Though in the case of categorical and ordinal features, $P(x|c)$ can be estimated analogously to (3.30), a parametrized distribution is usually assumed behind $P(x|c)$ and always behind $f(x|c)$. In the case of a n-dimensional continuous random vector, a frequent choice is a n-dimensional normal distribution $N(\mu, \Sigma)$. Estimating $P(x|c)$ or $f(x|c)$ behind which a parametrized distribution is assumed reduces to estimating the parameters of that distribution, such as the mean μ and covariance matrix Σ of the n-dimensional normal distribution. An estimation following any of these ways is typically combined with the simplifying assumption of conditional independence of the individual features $[x]_1, \ldots, [x]_n$ of the feature vector $x \in \mathcal{X}$,

$$(\exists \mathcal{X}' \subset \mathcal{X}) \, P(x \in \mathcal{X}') = 1, (\forall x \in \mathcal{X}') \, P(x|c) = \prod_{j=1}^{n} P([x]_j|c),$$

$$\text{respectively } f(x|c) = \prod_{j=1}^{n} f([x]_j|c). \tag{3.31}$$

Bayes classifiers fulfilling the assumption (3.31) are called *naïve Bayes classifiers*. If the values of features are from the set $\mathbb{N}_0$ of non-negative integers (typically, if the feature vector describes some histogram), then an alternative simplifying assumption is a multinomial distribution of that vector,

$$(\forall x \in \mathbb{N}_0) \, P(x|c) = \frac{(\sum_{j=1}^{n}[x]_j)!}{\prod_{j=1}^{n}[x]_j!} \prod_{j=1}^{n} p_j^{[x]_j}, \tag{3.32}$$

where $p_1, \ldots, p_n \in [0, 1]$, $p_1 + \cdots + p_n = 1$. Bayes classifiers fulfilling the assumption (3.32) are called *multinomial Bayes classifiers*, or multinomial naïve Bayes classifiers.

3. Estimating the a priori probability of x in (3.28) or the unconditional density f in (3.29). This task is easy once the tasks 1. and 2. have been fulfilled, due to the fact that

$$P(x) = \sum_{c \in C} P(x|c)P(c), x \in \mathscr{X}, \tag{3.33}$$

and

$$f(x) = \sum_{c \in C} f(x|c)P(c), x \in \mathscr{X}. \tag{3.34}$$

Combining 1. and 2. tasks, the joint distribution of features and classes can be estimated. Methods like this, in which it is possible to estimate the joint distribution that generated the data, are called *generative*.

In the case of features with finite sets $V_1, \ldots, V_n$ of feasible values, a joint probability distribution on $V_1 \times \ldots V_n \times C$ fulfilling the condition (3.31) for naïve Bayes classifiers is actually induced by a specific Bayesian network [22, 23]. In the classification context, a *Bayesian network* can be defined as a pair (G, Θ) with the following properties.

(i) $G = (\mathscr{V}, \mathscr{E})$ is a directed acyclic graph (DAG) with the set of vertices $\mathscr{V} = \{V_1, \ldots, V_n, C\}$ and the set of edges $\mathscr{E}$.

(ii) $\Theta = (\pi_V)_{V \in \mathscr{V}}$ is a system of mappings such that:

- if the set $\Pi(V) = \{U \in \mathscr{V}|(U, V) \in \mathscr{E}\}$ of parents of $V \in \mathscr{V}$ is empty, then

$$\pi_V : V \to [0, 1] \tag{3.35}$$

is a probability distribution on V;
- if $\Pi(V) = \{U_{V,1}, \ldots, U_{V,n_V}\} \neq \emptyset$, then π_V is a conditional probability distribution on V, conditioned by the joint probability distribution on $\Pi(V)$, i.e.,

$$\pi_V : V \times U_{V,1} \times \cdots \times U_{V,n_V} \to [0, 1] \text{ such that } \pi_V(\cdot, u_1, \ldots, u_{n_V}) \text{ for}$$
$$(u_1, \ldots, u_{n_v}) \in U_{V,1} \times \cdots \times U_{V,n_V} \text{ is a probability distribution on } V. \tag{3.36}$$

(iii) The joint probability distribution P on $V_1 \times \ldots V_n \times C$ is defined

$$P(v) = \prod_{V \in \mathscr{V}} \pi_V(([v]_U)_{U \in V \cup \Pi(V)}), v \in V_1 \times \ldots V_n \times C, \tag{3.37}$$

where the notation $v = ([v]_{V_1}, \ldots, [v]_{V_n}, [v]_C)$ for $v \in V_1 \times \ldots V_n \times C$ has been used.

Fig. 3.2 The graph of a Bayesian network by which the joint probability distribution of a naïve Bayes classifier (3.31) is induced

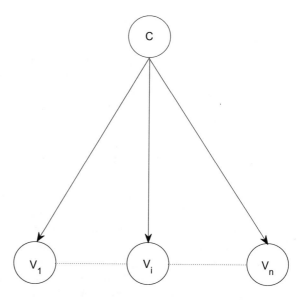

Hence, a joint probability distribution fulfilling (3.31) is indeed induced by a Bayesian network, namely by a network with the graph (Fig. 3.2)

$$G = (\mathcal{V}, \mathcal{E}_{nb}), \text{ where } \mathcal{V} \text{ is as above, and } \mathcal{E}_{nb} = \{(C, V_1), \dots, (C, V_n)\}. \quad (3.38)$$

In [23], an extension of naïve Bayes classifiers to *augmented naïve Bayes classifiers* has been proposed, which are Bayesian networks with the graph

$$G = (\mathcal{V}, \mathcal{E}_{nb} \cup \mathcal{E}_a), \text{ where } \mathcal{E}_a \subset \mathcal{V} \times \mathcal{V}, \mathcal{E}_a \cap \mathcal{E}_{nb} = \emptyset. \quad (3.39)$$

A particularly simple kind of augmented naïve Bayes classifiers are *tree-augmented naïve Bayes classifiers*, in which $(\mathcal{V}, \mathcal{E}_a)$ is a tree.

An important concept encountered when working with Bayesian networks is the *Markov blanket* MB(V) of $V \in \mathcal{V}$, which consists of parents of V, its children (i.e., vertices a parent of which is V) and vertices sharing a child with V, formally:

$$\text{MB}(V) =$$
$$= \Pi(V) \cup \{U \in \mathcal{V} | V \in \Pi(U)\} \cup \{U \in \mathcal{V} \setminus \{V\} | (\exists U' \in \mathcal{V})\{U, V\} \subset \Pi(U')\}. \quad (3.40)$$

Its importance consists in the property that the variable $[v]_V$ is conditionaly independed on other parts of the network given $MB(V)$. Formally,

$$(\forall U \in \mathcal{V} \setminus (\text{MB}(V) \cup \{V\})) \ P(V|\text{MB}(V), U) = P(V|\text{MB}(V)). \quad (3.41)$$

3.3.2 Bayes Spam Filters

Due to the simplicity of probabilistic computations, in the case of feature independence, naïve Bayes classifiers were and still are commonly used to deal with sheer
amounts of email messages and to filter spam.

First, we need to consider that presence of each dictionary word contributes to
the spam probability, according to the Bayes rule, in the following way:

$$P(c_s|[x]_j) = \frac{P(c_s)P([x]_j|c_s)}{P([x]_j)}, \tag{3.42}$$

where c_s denotes the class "spam" and $[x]_j$ the presence of word j.

The estimates of spam volume vary, the Statista portal [24] estimates that
52–61% of traffic is spam, whereas a Talos report from 2017 [25] estimates over
85% as spam. Moreover, the documentation for SpamAssassin [26] recommends
training with more ham than spam. The probability of spam class ($P(c_s)$) estimated
from the training data by (3.30) would be then inappropriate. The situation can be
simplified through assigning equal a priori probability to both classes.

Also, the calculation of the unconditional probability of individual words ($[x]_j$)
is more straightforward under this assumption. With a binary classification, the
Eq. (3.33) is now simplified to:

$$P([x]_j) = \frac{1}{2}(P([x]_j|c_s) + P([x]_j|c_h)), \tag{3.43}$$

where c_h denotes the "ham" class, and the conditional probability $P(c_s|[x]_j)$, which
can be interpreted as the spamminess of a word $[x]_j$ can be computed as:

$$P(c_s|[x]_j) = \frac{P([x]_j|c_s)}{P([x]_j|c_s) + P([x]_j|c_h)}. \tag{3.44}$$

This approach would, however, resolve only the spam probability of a single word,
whereas we want to assess the spam probability of the whole incoming message. To
this end, we need to combine the spam probability contribution of individual words.

The approach of naïve Bayes classifiers is, in accordance with (3.31):

$$P(c_s|x) = \frac{\prod_{j=1}^{n} P([x]_j|c_s)}{\prod_{j=1}^{n} P([x]_j|c_s) + \prod_{j=1}^{n} P([x]_j|c_h)}. \tag{3.45}$$

Finally, we need to assign the resulting probability to spam or ham. Commonly,
the boundary of 0.5 is used for distinguishing spam and ham messages; however, if
we want to decrease the proportion of false positives (ham marked as spam), this
threshold may be increased. Moreover, a third class can be introduced as a message
quarantine.

Now we have all ingredients for the basal decision if an incoming message is a spam or not. For word probabilities given the words from training data, based on the *UCI Spambase Data Set* [27], see Table 3.5. In that table, the probability of word presence ($P[x]_j|\cdot$) has been estimated as the fraction of documents in the class containing the considered word.

When we receive a message containing "project meeting," we compute the spam probability based on these two words as:

$$\frac{0.03 \cdot 0.01}{0.03 \cdot 0.01 + 0.10 \cdot 0.12} = 0.0241 \tag{3.46}$$

and provided we should draw a conclusion based only on those words, we may classify the message as almost definitely legitimate.

However, the message "make money over internet" with spam probability:

$$\frac{0.35 \cdot 0.38 \cdot 0.38 \cdot 0.34}{0.35 \cdot 0.38 \cdot 0.38 \cdot 0.34 + 0.15 \cdot 0.02 \cdot 0.11 \cdot 0.07} = 0.9986 \tag{3.47}$$

would be classified as spam.

There are several extensions of this naïve Bayes classification, which differ from it, e.g., through incorporating the word frequency instead of the binary presence information. Some of them are presented in [28].

3.3.3 Bayes Classification in Sentiment Analysis

The assumption of naïve Bayes classifiers that each word contained in the considered text contributes independently to the class membership is encountered in other domains as well. One of them is sentiment analysis.

This field has, however, one distinctive trait—adjectives standing in proximity to subjects can be directly assessed as positive or negative by their meaning and contribute to the score of the subject, as discussed in Sect. 1.3.2. This is a difference from spam filtering, where the spamminess of an individual word may not translate directly to the spamminess of the whole message.

For assessing the sentiment of a word, such a system needs to keep a lexicon of positive and negative words. There is a possibility to create a lexicon by hand (as in [29]), but that presents a tedious work and might fail in domains unconsidered during creation. Other authors, such as in [30], hand-edited only a small section of adjectives and proposed extending such lexicon by traversing related words in the WordNet [31]. Moreover, as this lexicon is available at https://www.cs.uic.edu/~liub/FBS/sentiment-analysis.html, it is used for many toy examples in sentiment analysis.

Table 3.5 Example training data derived from the Spambase dataset consisting of 1813 spam and 2788 ham messages (#—number of documents in the class containing the word, $P([x]_j|\cdot)$—probability that the word occurs in a document belonging to the class)

Word	Spam		Ham	
	#	$P([x]_j|\cdot)$	#	$P([x]_j|\cdot)$
you	1608	0.89	1619	0.58
your	1466	0.81	957	0.34
will	1143	0.63	1182	0.42
our	1134	0.63	614	0.22
all	1115	0.62	773	0.28
free	989	0.55	252	0.09
mail	827	0.46	475	0.17
remove	764	0.42	43	0.02
business	697	0.38	266	0.10
email	688	0.38	350	0.13
money	681	0.38	54	0.02
over	681	0.38	318	0.11
make	641	0.35	412	0.15
address	625	0.34	273	0.10
internet	619	0.34	205	0.07
receive	567	0.31	142	0.05
order	555	0.31	218	0.08
people	520	0.29	332	0.12
re	487	0.27	824	0.30
credit	377	0.21	47	0.02
addresses	287	0.16	49	0.02
report	231	0.13	126	0.05
direct	202	0.11	251	0.09
technology	112	0.06	487	0.17
font	95	0.05	22	0.01
original	85	0.05	290	0.10
edu	68	0.04	449	0.16
pm	62	0.03	322	0.12
data	61	0.03	344	0.12
hp	50	0.03	1040	0.37
project	47	0.03	280	0.10
parts	32	0.02	51	0.02
hpl	27	0.01	784	0.28
meeting	20	0.01	321	0.12
table	19	0.01	44	0.02
labs	18	0.01	451	0.16
conference	16	0.01	187	0.07
lab	12	0.01	360	0.13
george	8	0.00	772	0.28
telnet	3	0.00	290	0.10
cs	1	0.00	147	0.05

With the assumption of a single subject in the message (e.g. movie reviews, where we are interested in the overall sentiment of the movie) the sentiment may be assessed as simply as through the count of positive words minus the count of negative words.

If we take the same example as in Fig. 1.4, we detect five supposedly positive words ("spectacular", great", "magic", "beauty", "like") and five supposedly negative words ("overwhelming", "vengeance", lost", "strange", "perilous") contained in the lexicon. The review would be therefore ranked with the total score of zero as neutral. Although the Semantria algorithm used in Sect. 1.1 marked the text as negative and, upon closer inspection, the review is not advising subsequent viewing and is in fact rather negative.

The Bayesian approach can assist in the creation of more accurate classifier in a very similar manner as in the case of spam messages. To this end, we may use the pre-trained lexicon as a considered dictionary and thus reduce the dimensionality at first place. In the training phase we then deduce the probability that each such word is present in messages with the positive and the negative sentiment. The very same calculations would then lead to the probability of positive (negative) sentiment connected to the considered text.

This naive approach can be, again, extended in various ways. Another publicly available example of a simple sentiment analysis system is the system available at http://sentiment.vivekn.com/. This system is based on an enhanced naïve Bayes classifier as described in [32]. The text is processed in a bit more complicated manner, as the negations propagate to the words bearing sentiment. The process is similar to the one applied in [33]—all words between "not" or "n't" and first punctuation mark are labelled with a "not_" prefix and are treated separately. The reason is that negations of words with strong positive sentiment must not be strongly negative, and vice versa.

For example, a phrase "not bad, quite good" will be parsed into a following set of tokens: {"not", "not_bad", ",", "quite", "good"}. This way the word "bad" will no longer contribute to the sentiment of the phrase negatively. The token "not_bad" is now treated as an individual token with its separate probability of either sentiment. The complexity of training, therefore, increases slightly.

The approach of Narayanan then filters gathered features (presence of words) only to such with more than one occurrence and maximum mutual information. As the demo of this system is trained to classify only into three categories (positive, neutral and negative), the before mentioned review of movie *The Reverant* is classified as positive with 100% confidence. Although such high confidence is suspicious, it may be caused by the origin of the training dataset—movie reviews and the scores associated with them (which in this case was seven out of ten).

3.3.4   Logit Method

Logit method, more often called *logistic regression* or *logit regression*, is the classification of real-valued features by means of a specific regression, namely a regression establishing a functional relationship between those features and the probability $P(\phi(x) = C)$ that the feature vector is assigned to a particular class $c \in C$. For binary classification with $C = \{c_+, c_-\}$, the regression function (2.78) has the form

$$P(\phi(x) = c_+) = \varphi(\beta_0 + \beta_{\mathcal{X}}^\top x), x \in \mathcal{X}, \tag{3.48}$$

where $\beta_0 \in \mathbb{R}$, $\beta_{\mathcal{X}} \in \mathbb{R}^n$, and φ is a real function defined as

$$\varphi(t) = \frac{1}{1 + e^{-t}}, t \in \mathbb{R}. \tag{3.49}$$

The requirement that features have to be real-valued is hardly any restriction for the application of the method. They include not only continuous features and ordinal features with the usual ordering of real numbers, but also categorical features represented by binary vectors, as was recalled in Sect. 2. In the last case, however, the definition of the feature set (2.2) has to incorporate the constraints (2.3).

The usual names of this method are due to the fact that the function φ defined in (3.49) is called *logistic function*, whereas its inversion φ^{-1} is called *logit*. Applying φ^{-1} to (3.48) yields

$$\varphi^{-1}(P(\phi(x) = c_+)) = \beta_0 + \beta_{\mathcal{X}}^\top x, x \in \mathcal{X}, \tag{3.50}$$

thus the method is actually a linear regression of the logit of $P(\phi(x) = c_+)$ on the considered features. As usually in linear regression, the constant β_0 is called *intercept*. Notice that the definition (3.49) of φ implies for the logit of $P(\phi(x) = c_+)$

$$\varphi^{-1}(P(\phi(x) = c_+)) = \ln\left(\frac{P(\phi(x) = c_+)}{1 - P(\phi(x) = c_+)}\right) = \ln\left(\frac{P(\phi(x) = c_+)}{P(\phi(x) = c_-)}\right). \tag{3.51}$$

From (3.50) and (3.51) follows

$$P(\phi(x) = c_+) = P(\phi(x) = c_-)e^{\beta_0 + \beta_{\mathcal{X}}^\top x}, \tag{3.52}$$

which entails the distribution on C

$$P(\phi(x) = c_+) = \frac{e^{\beta_0 + \beta_{\mathcal{X}}^\top x}}{1 + e^{\beta_0 + \beta_{\mathcal{X}}^\top x}},$$
$$P(\phi(x) = c_-) = \frac{1}{1 + e^{\beta_0 + \beta_{\mathcal{X}}^\top x}}. \tag{3.53}$$

A generalization of the method for a set of m classes, $C = \{c_1, \ldots, c_m\}$ is called *multinomial logistic regression* and can be obtained through choosing an arbitrary $c_- \in C$, called *pivot class*, and employing (3.48) with c_+ replaced, in turn, by all $c \in C \setminus c_-$. For example, let the pivot class be c_m, and denote for $i = 1, \ldots, m - 1$, the intercept β_0 and vector $\beta_{\mathscr{X}} \in \mathbb{R}^n$ in (3.48) as β_{i0} and $\beta_{i\mathscr{X}}$, respectively. Then (3.50) and (3.51) yield

$$P(\phi(x) = c_i) = P(\phi(x) = c_m)e^{\beta_{i0}+\beta_{i\mathscr{X}}^\top x}, \qquad (3.54)$$

which entails the following distribution on C:

$$P(\phi(x) = c_i) = \frac{e^{\beta_{i0}+\beta_{i\mathscr{X}}^\top x}}{1 + \sum_{\ell=1}^{m-1} e^{\beta_{\ell 0}+\beta_\ell^\top x}}, i = 1, \ldots, m-1$$

$$P(\phi(x) = c_m) = \frac{1}{1 + \sum_{\ell=1}^{m-1} e^{\beta_{\ell 0}+\beta_\ell^\top x}}. \qquad (3.55)$$

For $m = 2$, (3.55) turns to (3.53). The distribution (3.55) can be directly used as the result of a fuzzy classifier $\phi : \mathscr{X} \to [0, 1]^m$, whereas a crisp classifier $\phi : \mathscr{X} \to C$ can be constructed by

$$\phi(x) = c \in C \text{ such that } P(\phi(x) = c) = \max_{i=1,\ldots,m} P(\phi(x) = c_i), x \in \mathscr{X}. \quad (3.56)$$

Needless to say, the result of (3.56) is not necessarily unique.

A serious weakness of multinomial logistic regression is that it assumes *independence of irrelevant alternatives*: according to (3.54), the relationship between $P(\phi(x) = c_i)$ and $P(\phi(x) = c_m)$ is the same if the set of classes is $C = \{c_1, \ldots, c_m\}$ as if it were only $C = \{c_i, c_m\}$. This assumption is often invalid in real world—think of a recommender system of a travel agency that should recommend, based on features characterizing the user and his/her interest, one of the destinations China, India, Indonesia, Malaysia, Singapore, Sri Lanka, Thailand, Vietnam, Egypt, and let $c_i =$ Vietnam and $c_m =$ Egypt.

Due to (3.48) and (3.55), classifier learning in logistic regression is actually simultaneous learning of the regression functions (3.48) corresponding to the choices $(\beta_0, \beta_{\mathscr{X}}) = (\beta_{i0}, \beta_{i\mathscr{X}})$ for $i = 1, \ldots, m-1$. According to Sect. 2.6, learning the regression functions corresponding to $(\beta_{i0}, \beta_{i\mathscr{X}})$ means solving the optimization task (2.79) on the set

$$\Gamma = \{\gamma : \mathscr{X} \to [0, 1] | (\forall x \in \mathscr{X}) \, \gamma(x) = \varphi(\beta_{i0} + \beta_{i\mathscr{X}}^\top x), \beta_{i0} \in \mathbb{R}, \beta_{i\mathscr{X}} \in \mathbb{R}^n\}. \qquad (3.57)$$

To this end, the sequence of classification training data $(x_1, c_1), \ldots, (x_p, c_p)$ is first transformed into $m - 1$ sequences $(x_1, y_{i1}), \ldots, (x_p, y_{ip}), i = 1, \ldots, m - 1$ for learning the $m - 1$ regression functions, as follows:

$$\text{for } i = 1, \ldots, m-1, k = 1, \ldots, p, \text{ define } y_{ik} = \begin{cases} 1 & \text{if } c_k = c_i, \\ 0 & \text{else.} \end{cases} \qquad (3.58)$$

The aggregated error function ER_γ in (2.79) is in logistic regression for crisp classifiers (3.56) usually defined to be the difference between the certainty that the training data $(x_1, c_1), \ldots, (x_p, c_p)$ have been observed, and the probability of observing them for the considered choice of $\hat{\beta}_0$ and $\hat{\beta}_{\mathcal{X}}$,

$$ER_\gamma((x_1, y_{i1}, \gamma(x_1)), \ldots, (x_p, y_{ip}, \gamma(x_p))) = 1 - P(\phi(x_1) = c_1, \ldots, \phi(x_p) = c_p). \tag{3.59}$$

Denote $(\hat{\beta}_{i0}, \hat{\beta}_{i\mathcal{X}})$ the results of learning $(\beta_{i0}, \beta_{i\mathcal{X}})$ for $i = 1, \ldots, m-1$, and put $\hat{\beta} = (\hat{\beta}_{10}, \hat{\beta}_{1\mathcal{X}}, \ldots \hat{\beta}_{(m-1)0}, \hat{\beta}_{(m-1)\mathcal{X}})$. Then taking into account (3.57) turns (2.79) into

$$\hat{\beta} = \arg \max_{\beta \in \mathbb{R}^{(m-1)(n+1)}} \lambda(\beta), \tag{3.60}$$

where $\lambda : \mathbb{R}^{(m-1)(n+1)} \to [0, 1]$ is the likelihood for the parameters β,

$$\lambda(\beta) = P(\phi(x_1) = c_1, \ldots, \phi(x_p) = c_p),$$
with $P(\phi(x_1), \ldots, \phi(x_p))$ related to β through (3.55),
$$\beta = (\beta_{10}, \beta_{1\mathcal{X}}, \ldots \beta_{(m-1)0}, \beta_{(m-1)\mathcal{X}}) \in \mathbb{R}^{(m-1)(n+1)}. \tag{3.61}$$

Hence, learning in the logit method consists in the maximization of the likelihood (3.61). Such learning is called *maximum likelihood* learning, or equivalently, maximum likelihood estimation.

If the values x of features are realizations of some random vector X, then putting the estimates $(\hat{\beta}_{i0}, \hat{\beta}_{i\mathcal{X}})$, $i = 1, \ldots, m-1$ into (3.55) yields an estimate of the conditional distribution $P(\cdot|X = x)$. Differently to Bayes classifiers, however, it is not possible to estimate the joint distribution of features and classes. Methods, in which only the conditional distribution of classes can be estimated, but not the joint probability distribution that generated the data, are called *discriminative*.

Usually, the individual training data $(x_1, c_1), \ldots, (x_p, c_p)$ are assumed mutually independent, which factorizes the likelihood to

$$\lambda(\beta) = \prod_{k=1}^{p} P(\phi(x_k) = c_k), \beta \in \mathbb{R}^{(m-1)(n+1)}. \tag{3.62}$$

A generalization of maximum likelihood learning, employed especially in the case of multinomial logistic regression, is *maximum a posteriori* learning. In this method, β is viewed as a continuous random vector on some $\Omega \subset \mathbb{R}^{(m-1)(n+1)}$ with density f, and the result of learning is obtained through maximizing the a posteriori probability density of β given the training data, which in the case of the mutual independence of the training data is

$$f(\beta|\phi(x_1) = c_1, \ldots, \phi(x_p) = c_p) = \prod_{k=1}^{p} \frac{P(\phi(x_k) = c_k)f(\beta)}{\int_{\Omega} P(\phi(x_k) = c_k)f(\beta')d\beta'}. \quad (3.63)$$

Hence $\hat{\beta}$ is obtained through the maximization

$$\hat{\beta} = \arg\max_{\beta \in \Omega} f(\beta|\phi(x_1) = c_1, \ldots, \phi(x_p) = c_p) = \arg\max_{\beta \in \Omega} \prod_{k=1}^{p} P(\phi(x_k) = c_k)f(\beta).$$
$$(3.64)$$

The density f allows to incorporate a priori knowledge about $\hat{\beta}$. Frequently, a Gaussian density on $\mathbb{R}^{(m-1)(n+1)}$ is used to this end, restricted to Ω if $\Omega \neq \mathbb{R}^{(m-1)(n+1)}$. If no a priori knowledge about β is available, but the set Ω is bounded, then the density is chosen uniform, i.e., $f(x) = \frac{1}{\int_{\Omega} d\beta'}$, which simplifies (3.64)–(3.60) with the factorized likelihood (3.62).

3.3.5 Using Bayes Classifiers and Logit Method in Recommender Systems

Inspired by the simplicity and performance of the Naïve Bayes predictions, recommender systems have also adopted the Bayesian classifier. In contrast to the memory-based methods, such as the k-nearest neighbour classifier mentioned in Sect. 3.2.2 the model-based approach of Bayesian classifier is supposed to be less vulnerable to missing data – a common problem of the recommender systems – as it uses the whole available dataset to train the model, and does not require significant feature overlap among all pairs of users.

Simple Bayesian Recommender
For a start, we will assume that all features are independent. Such assumption is seldom true (and especially in recommender systems, where we want to take into account the relations among individual features) but allows us to start with the naïve Bayes classifier:

$$P(c|[x]) = \frac{P(c) \prod_{j=1}^{n} P([x]_j|c)}{P([x])}, \quad (3.65)$$

where $P(c)$ is the a priori probability of a class, $P([x]_j|c)$ is a conditional probability of a feature given a class c and $P([x])$ is the a priori probability of the feature vector.

To simplify the equation even more, let us switch to a Simple Bayes classifier. This most straightforward case of a Bayesian network does not allow connections between individual features; the only allowed connection is from the class to each feature. Effectively, this keeps the assumption of independent features, but allows us to omit the denominator of the naïve Bayes classifier:

$$P(c, [x]) = P(c|[x]) * P([x]) = P(c) \prod_{j=1}^{n} P([x]_j|c). \tag{3.66}$$

The classifier should return the class label with the highest posterior probability, formally:

$$class([x]_j) = \arg\max_{c \in C} P(c, [x]) = \arg\max_{c \in C} P(c|x) \prod_{j=1}^{n} P([x]_j|c), \tag{3.67}$$

where C is the set of all class labels.

The last ingredient is the computation of conditional probability $P([x]_j|c)$. One possible solution is via the maximum likelihood estimator:

$$P([x]_j|c) = \frac{|\{[x]_j, c\}|}{|\{c\}|}, \tag{3.68}$$

where $|\{c\}|$ is the number of observed samples that belong to class c and $|\{[x]_j, c\}|$ is the number of samples x that also have a required class for feature j.

However, this estimator is unable to cope with very sparse data, where it gathers only zeroes and makes class selection impossible. Therefore, an additive (Laplace) smoothing is used to increase robustness:

$$P([x]_j|c) = \frac{|\{[x]_j, c\}| + 1}{|\{c\}| + |C|}, \tag{3.69}$$

where $|C|$ denotes the number of classes.

Extended Logistic Regression
The strict assumption of feature independence and methods of model estimation pose a limitation in recommendation systems. Modelling an optimal conditional probability table would be better, but is computationally unfeasible.

To this end, Greiner et al. proposed a gradient-ascent algorithm called Extended Logistic Regression [34] that is adapted to incomplete training data.

Such parameter tuning can be applied to both Simple (naïve) Bayesian and Tree Augmented Networks (Fig. 3.3) resulting in NB-ELR and TAN-ELR proposed in [35]. The TAN-ELR gives a smaller error on very sparse data, however, due to a large number of parameters that need to be optimised, NB-ELR is concluded to be more practical.

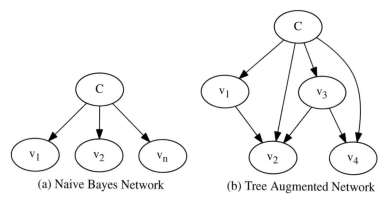

(a) Naive Bayes Network   (b) Tree Augmented Network

Fig. 3.3 Structure of Bayesian networks

3.4 Classifiers Estimating Features Distribution

The method that was considered state of the art in the classification of n-dimensional continuous feature vectors up to 1970s is based on a principle in a way opposite to that underlying the classifiers presented in the previous section: instead of estimating the probability distribution on the set $C = \{c_1, \ldots, c_m\}$ of classes, it estimates the probability distribution on $\mathscr{X}$ separately for feature vectors with the correct class equal to each of the classes $c \in C$. If all misclassifications are equally undesirable, such a classifier performs a crisp classification of x according to which class is most likely its correct class, in the sense of highest density of the corresponding probability distribution. Hence, denoting f_i the density of the probability distribution corresponding to the class $c_i, i = 1, \ldots, m$,

$$\phi(x) \in \arg \max_{i=1,\ldots,m} f_i(x), x \in \mathscr{X}. \tag{3.70}$$

In general, however, the misclassification costs $w_{i,j}$ must be taken into account, which were introduced in Sect. (2.2). Then the classification of x is performed according to the lowest density of expected costs,

$$\phi(x) \in \arg \min_{i=1,\ldots,m} \sum_{j=1}^{m} w_{i,j} f_j(x), x \in \mathscr{X}. \tag{3.71}$$

Notice that for the misclassification costs (2.17), the lowest density of expected costs (3.71) is equivalent to the highest density of the class distribution (3.70).

The distributions corresponding to the classes are in this method always parametric. Traditionally, they are n-dimensional normal distributions. With the normal distribution, the method is used in two variants, differing through whether distributions corresponding to different classes are or are not allowed to have different variances.

3.4.1 Linear and Quadratic Discriminant Analysis

Let for $i = 1, \ldots, m$, the distribution corresponding to c_i is n-dimensional normal with density f_i defined

$$f_i(x) = \frac{1}{\sqrt{(2\pi)^n \| \Sigma_i \|}} e^{(x-\mu_i)^\top \Sigma_i^{-1}(x-\mu_i)}, x \in \mathcal{X}, \qquad (3.72)$$

where $\mu_i \in \mathbb{R}^n$ is the mean of that distribution, $\Sigma_i \in \mathbb{R}^{n,n}$ is a regular covariance matrix, and $|\Sigma_i|$ denotes the determinant of Σ_i. Then (3.70) allows to discriminate between classes c_i and c_j, $j \neq i$ for all x except those fulfilling the condition $f_j(x) = f_i(x)$, which then form the border between areas in $\mathcal{X}$ classified to both classes. Taking into account (3.72), we get

$$f_j(x) = f_i(x) \Rightarrow x^\top (\Sigma_i^{-1} - \Sigma_j^{-1})x + 2x^\top (\Sigma_j^{-1}\mu_j - \Sigma_i^{-1}\mu_i)$$
$$+ \mu_i^\top (\Sigma_i^{-1})\mu_i - \mu_j^\top (\Sigma_j^{-1})\mu_j + \frac{1}{2}(\ln |\Sigma_i| - \ln |\Sigma_j|) = 0. \qquad (3.73)$$

Feature vectors fulfilling (3.73) form a quadratic surface in $\mathbb{R}^n$, due to which the method is called *quadratic discriminant analysis (QDA)*, aka Fisher discriminant analysis, as a reminder of its author.

If the condition that the distributions corresponding to all classes share the covariance matrix is added,

$$(\exists \Sigma \in \mathbb{R}^{n,n})\ \Sigma_i = \Sigma, i = 1, \ldots, m, \qquad (3.74)$$

then the Eq. (3.73) for the border between classes c_i and c_j, j simplifies to

$$2x^\top \Sigma^{-1}(\mu_j - \mu_i) + \mu_i^\top \Sigma^{-1}\mu_i - \mu_j^\top \Sigma^{-1}\mu_j = 0, \qquad (3.75)$$

which is the equation of a hyperplane, i.e., of a general linear surface. Therefore, this variant of Fisher discriminant analysis is called *linear discriminant analysis (LDA)*.

Learning a classifier based on Fisher discriminant analysis consists in solving the optimization task (2.50) with respect to $\mu_i, \Sigma_i, i = 1, \ldots, m$. To this end, the methods maximum likelihood learning and maximum a posteriori learning are used, recalled in the previous section in connection with logistic regression.

In Fig. 3.4, the application of linear and quadratic discriminant analysis to the classification of the graphical objects on web pages into advertisements and non-advertisements (cf. [2]) is shown.

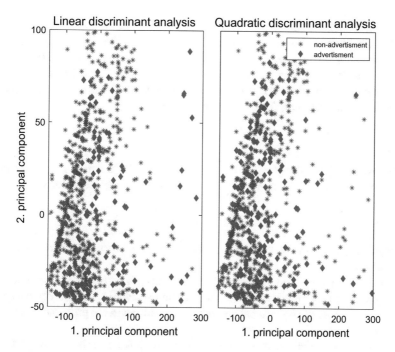

Fig. 3.4 The 1. and 2. principal component of a LDA (left) and QDA (right) classification of the graphical objects on web pages into advertisements and non-advertisements (cf. [2])

3.4.2 Discriminant Analysis in Spam Filtering

Applications of machine learning to many problems, including content-based spam filtering, suffer from the high dimensionality of input data. The number of dimensions used for processing of text content is in the simplest case, omitting n-grams, is equal to the size of the dictionary we use for the transformation of the individual documents into number vectors. Such a dictionary can be constructed with an a priori knowledge of the language, including only stems of words and disallowing terms that are too common (stop words). In many cases, the dictionary is constructed ad-hoc from the training data.

The resulting number of dimensions is typically considerable. In the *SpamAssassin datasets* `20030228_easy_ham` and `20030228_spam` [26] with 3002 documents converted to plain-text, the total size of the ad-hoc dictionary constructed with a default setup of a scikit-learn [36] vectorizer is 47655, stripping the accents results in 47088 terms, omitting terms that are unique (appear in a single document) results in 19528 terms. When considering the most restrictive bi-grams, the number of terms increases to 110886.

However, machine learning approaches require at least several data points with varying realisations in each dimension to appropriately tailor the decision boundary among the considered classes across the dimensions. Also, the size of the constructed model is proportional to the number of dimensions and, therefore, the reduction of dimensionality is convenient for both training and subsequent classification.

Such dimensionality reduction can be based upon the singular value decomposition which decomposes the original term-by-document matrix into three matrix transformations: rotation, scaling and rotation:

$$X = U \Sigma V^{\top}. \tag{3.76}$$

By using only the k highest eigenvalues (on diagonal of the matrix Σ) and replacing all others with zeroes creates an approximation of the original matrix:

$$\hat{X}_k = U_k \Sigma_k V_k^{\top}, \tag{3.77}$$

where $U_k \Sigma_k$ is commonly denoted in natural language processing as a *concept-by-document matrix*.

If we add a requirement that X is centred and normalized before the singular value decomposition (i.e. the mean is equal to zero and standard deviation equal to one), the $U_k \Sigma_k$ gives us the k principal components, resulting in the principal component analysis.

In other words, the principal component we are looking for is a result of finding a projection along which the variance of the data is maximised:

$$\max\{w^{\top} C w | \|w\| = 1\}, \tag{3.78}$$

where C is the empirical covariance matrix of the dataset computed as: $C = \frac{1}{n} X^{\top} X$.

The principal component analysis is especially useful for data visualisation and preliminary data exploration, however, as it does not utilise the information on a class assignment, it is not very suitable for discrimination tasks.

The LDA, however, uses such information and tries to find a projection that maximises the class separability:

$$\max\{w^{\top} S_b w | \|S_w^{\frac{1}{2}} w\| = 1\}, \tag{3.79}$$

where S_b is the between-class scatter matrix and S_w is the within-class scatter matrix defined as:

$$S_b = \sum_{i=1}^{N} n_i (\mu_i - \mu)(\mu_i - \mu)^{\top}$$

$$S_w = \sum_{i=1}^{N} \sum_{x \in c_i} (x - \mu_i)(x - \mu_i)^{\top}, \tag{3.80}$$

with μ_i denoting the empirical mean of data belonging to the i-th class, and n_i denoting the number of those samples.

Similarly to the principal component analysis, solutions can be obtained as eigenvectors corresponding to the eigenvalues of $S_w^{-1} S_b$ ordered in descending order.

Alternatively, to avoid computation of the covariance matrices, singular value decomposition can be utilised to gather only the scaling of covariances and subsequent singular value decomposition searches only for the projection with maximal separation of normalized class centroids. Refer to [37] for more information on an effective implementation.

To conclude our example of SpamAssassin dataset dimensionality reduction, the 19528×3002 matrix resulting from a TF-IDF vectoriser was reduced to one principal component and one linear discriminant in Fig. 3.5. While the relative distribution of both classes on the first principal component is similar, linear discriminant analysis

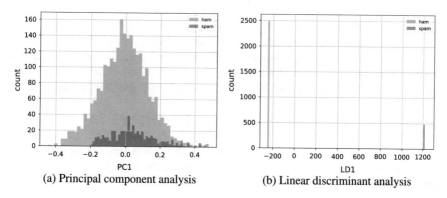

(a) Principal component analysis (b) Linear discriminant analysis

Fig. 3.5 Comparison of principal component analysis (PCA) and linear discriminant analysis (LDA) on the SpamAssassin dataset

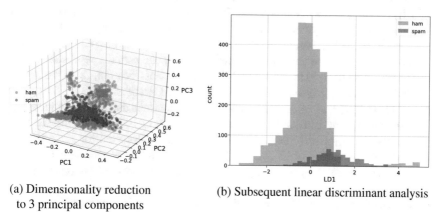

(a) Dimensionality reduction to 3 principal components (b) Subsequent linear discriminant analysis

Fig. 3.6 Result of linear discriminant analysis executed on data reduced to three principal components

performs a very distinct separation of both considered classes – the ham messages are present only in a range $[-245, -241]$, spam messages only in a range $[1190, 1229]$.

Because the spam detection problem uses only two classes and the discriminant analysis has to propose a projection maximising the distance between classes, only the most discriminant can be extracted.

To illustrate why performing low-dimensional PCA before discriminant analysis is not a good idea, refer to Fig. 3.6 that presents the result of linear discriminant analysis executed on the first three principal components from the same dataset.

3.5 Classification Based on Artificial Neural Networks

Artificial neural networks (ANNs) are mathematical models inspired by some of the functionality of the biological neural networks. The kinds of ANNs most widespread from the application point of view are, however, used not to model their biological counterparts, but because they are *universal approximators*, i.e., they can approximate with arbitrary precision any unknown function encountered in real-world applications. If the unknown function is assigning classes to feature vectors in some feature set, then such an universal approximator should be, at least theoretically, an excellent classifier.

Mathematically, an artificial neural network is a mapping $F : \mathscr{X} \subset \mathbb{R}^a \to \mathbb{R}^b$, with which a directed graph $G_F = (\mathscr{V}, \mathscr{E})$ is associated. Due to the inspiration from biological neural networks, the vertices of G_F are called *neurons* and its edges are called *connections*. In artificial neural networks, three subsets of neurons are differentiated:

1. The set $\mathscr{I} \subset \mathscr{V}$ of *input neurons* contains neurons that have only outgoing connections, hence, the in-degree of an input neuron is 0. They accept the components $[x]_1, \ldots, [x]_n$ of the inputs $x \in \mathscr{X}$. For the purpose of ANNs, the values of those components are required to be real numbers, though possibly only from some subset of $\mathbb{R}$, such as natural numbers or the set $\{0, 1\}$. They also do need not to be realizations of continuous random variables, and can actually represent ordinal or categorical data, as in (2.3). In ANNs used for classification, $\mathscr{X}$ is the feature set and $[x]_1, \ldots, [x]_n$ are the values of considered features.
2. The set $\mathscr{O} \subset \mathscr{V} \setminus \mathscr{I}$ of *output neurons* contains neurons that have only incoming connections, hence, the out-degree of an output neuron is 0. They return the components $[F(x)]_1, \ldots, [F(x)]_m$ of $F(x) \in \mathbb{R}^m$. In ANNs used for classification, $F(x)$ belongs to the set $\{b_1, \ldots, b_m\}$, where b_j is a binary representation (2.3) of the class c_j for $j = 1, \ldots, m$, or a fuzzy set on the set C of classes, therefore $|\mathscr{O}| = m$.
3. Neurons from the set $\mathscr{H} = \mathscr{V} \setminus (\mathscr{I} \cup \mathscr{O})$ are called *hidden neurons*.

The graph G_F is typically required to be acyclic. ANNs fulfilling that condition are called *feedforward networks*. Their important subclass are *layered feedforward networks*. Their set of neurons $\mathscr{V}$ has subsets $\mathscr{V}_0, \mathscr{V}_1, \ldots, \mathscr{V}_L$, called *layers*, such that:

(i) $\bigcup_{i=0}^{L} \mathcal{V}_i = \mathcal{V}$;
(ii) $\mathcal{V}_i \cap \mathcal{V}_j = \emptyset$ if $i \neq j$;
(iii) $u \in \mathcal{V}_i, i = 0, \ldots, L - 1 \Rightarrow v \in \mathcal{V}_{i+1}, (u, v) \in \mathcal{E}$;
(iv) $\mathcal{V}_0 = \mathcal{I}$;
(v) $\mathcal{V}_L = \mathcal{O}$.

If $L > 1$, then the layers $\mathcal{V}_1, \ldots, \mathcal{V}_{L-1}$ are called *hidden layers*.

Layered networks with at most one hidden layer are sometimes called *shallow*, whereas those with at least 2 hidden layers are called *deep*. In Fig. 3.7, a layered feed-forward ANN with two layers of hidden neurons for the detection and classification of network intrusion attacks is depicted.

The purpose of the graph G_F associated with the mapping F is to define a decomposition of F into simple mappings assigned to hidden and output neurons and to connections between neurons (input neurons normally only accept the components of the input, and no mappings are assigned to them). Inspired by biological terminology, mappings assigned to neurons are called *somatic*, those assigned to connections are called *synaptic*. Using somatic and synaptic mappings, the decomposition of F is defined as follows:

(i) To each hidden neuron $v \in \mathcal{H}$, a somatic mapping $f_v : \mathbb{R}^{|\text{in}(v)|} \to \mathbb{R}^{|\text{out}(v)|}$ is assigned, where $\text{in}(v)$ and $\text{out}(v)$ are, respectively, the input set and the output set of v,

$$\text{in}(v) = \{u \in \mathcal{V} \,|\, (u, v) \in \mathcal{E}\}, \tag{3.81}$$

$$\text{out}(v) = \{u \in \mathcal{V} \,|\, (v, u) \in \mathcal{E}\}. \tag{3.82}$$

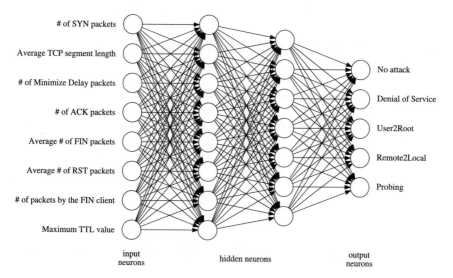

Fig. 3.7 ANN of the kind multilayer perceptron, with 2 layers of hidden neurons, for the detection and classification of network intrusion attacks

The vector of inputs to f_v is formed by outputs of the synaptic mappings $f_{(u,v)}$ assigned to the connections (u, v), $u \in \mathrm{in}(v)$, the vector of its outputs is formed by inputs to the synaptic mappings $f_{(v,u)}$ assigned to the connections (u, v), $u \in \mathrm{out}(v)$.

(ii) To each output neuron $v \in \mathcal{O}$, a somatic mapping $f_v : \mathbb{R}^{|\,\mathrm{in}(v)|} \to \mathbb{R}$ is assigned. The vector of inputs to f_v are outputs of the synaptic mappings $f_{(u,v)}$ assigned to the connections (u, v), $u \in \mathrm{in}(v)$, its output is the component $[F(x)]_j$ of $F(x)$ returned by v if the input to F is $x \in \mathcal{X}$.

(iii) To each connection $(u, v) \in \mathcal{E}$, a synaptic mapping $f_{(u,v)} : \mathbb{R} \to \mathbb{R}$ is assigned. If $u \in \mathcal{I}$, then the input of $f_{(u,v)}$ is the component of $x \in \mathcal{X}$ accepted by u, otherwise, it is a component of the output of f_u. The output of $f_{(u,v)}$ is a component of the input of f_v.

In the following three subsections, this decomposition will be explained at the detailed level of individual somatic and synaptic mappings for three particular kinds of ANNs encountered in classification.

3.5.1 Perceptrons for Linearly Separable Classes

Perceptron is one of the oldest kinds of ANNs (proposed by Rosenblatt in the late 1950s [38]), and it is also one of the simplest ones. It is its simplicity due to which it is still sometimes used in applications. Even here, however, it has only a limited importance because, as is explained below, it can be applied only to classification tasks with linearly separable classes (consequently, it is not a universal approximator).

From the point of view of the graph G_F, a perceptron has the following characteristics:

- It is the feedforward network with the smallest number of layers, $\mathcal{V} = \mathcal{I} \cup \mathcal{O}$.
- It is fully connected – $\mathcal{E} = \mathcal{I} \times \mathcal{O}$. Consequently,

$$v \in \mathcal{I} \Rightarrow \mathrm{out}(v) = \mathcal{O}, v \in \mathcal{O} \Rightarrow \mathrm{in}(v) = \mathcal{I}. \tag{3.83}$$

- The cardinalities of $\mathcal{I}$ and $\mathcal{O}$ are determined by the number n of features and the number m of classes. In particular,

$$|\mathcal{I}| = n, \tag{3.84}$$

$$|\mathcal{O}| = \begin{cases} 1 & \text{if } m = 2, \\ m & \text{if } m \geq 3. \end{cases} \tag{3.85}$$

The synaptic mapping $f_{(u,v)}$ assigned to a connection $(u, v) \in \mathcal{I} \times \mathcal{O}$ is simply the multiplication by a weight $w_{(u,v)} \in \mathbb{R}$,

$$f_{(u,v)}(\xi) = w_{(u,v)}\xi, \xi \in \mathbb{R}. \tag{3.86}$$

The somatic mapping f_v assigned to an output neuron $v \in \mathcal{O}$, in particular to the only output neuron v if $m = 2$, is a comparison of the sum of inputs with a threshold $t_v \in \mathbb{R}$,

$$f_v(\xi) = \begin{cases} 1 & \xi \in \mathbb{R}^n, [\xi]_1 + \cdots + [\xi]_n \geq t_v, \\ 0 & \xi \in \mathbb{R}^n, [\xi]_1 + \cdots + [\xi]_n < t_v, \end{cases} \tag{3.87}$$

or equivalently,

$$f_v(\xi) = \Theta\left(\sum_{j=1}^{n} [\xi]_j + b_v\right), \xi \in \mathbb{R}^n. \tag{3.88}$$

where $b_v = -t_v$ is called *bias* assigned to the neuron v, and Θ is the *Heaviside step function*, defined

$$\Theta(s) = \begin{cases} 0 & \text{if } s < 0, \\ 1 & \text{if } s \geq 0. \end{cases} \tag{3.89}$$

Due to (3.86) and (3.88), a perceptron is in the case of binary classification a mapping $F : \mathcal{X} \subset \mathbb{R}^n \to \{0, 1\}$, and in the case of classification into $m \geq 3$ classes a mapping $F : \mathcal{X} \subset \mathbb{R}^n \to \{0, 1\}^m$. Moreover, using the notation $w_v = (w_{(u,v)})_{u \in \mathcal{I}}$ for $v \in \mathcal{O}$ and $[x]_u$ for the component of $x \in \mathcal{X}$ accepted by the neuron $u \in \mathcal{I}$, the v-th component of $F(x)$ can be expressed as

$$[F(x)]_v = \Theta(x^\top w_v + b_v), \tag{3.90}$$

in particular in the case of binary classification,

$$F(x) = \Theta(x^\top w + b), \tag{3.91}$$

where the indexing of w and b by the output vector v has been omitted due to the uniqueness of that vector. Thus in that case, the set $\{x \in \mathcal{X} \,|\, F(x) = 1\}$ is a subset of a closed halfspace delimited by the hyperplane $x^\top w_v + b_v = 0$ (i.e., including the delimiting hyperplane), whereas the set $\{x \in \mathcal{X} \,|\, F(x) = 0\}$ is a subset of the complementary open halfspace (i.e., without the delimiting hyperplane). Similarly in the case of classification into $m \geq 3$ classes, the set

$$\{x \in \mathcal{X} \,|\, [F(x)]_v = 1, [F(x)]_{v'} = 0 \text{ for } v' \neq v\}, v \in \mathcal{O}, \tag{3.92}$$

is a subset of the intersection of a closed halfspace delimited by the hyperplane $x^\top w_v + b_v = 0$ and of open halfspaces delimited by the hyperplanes $x^\top w_{v'} + b_{v'} = 0$ for all $v' \neq v$. Consequently, a perceptron needs all classes to be linearly separable. As we have seen in Sect. 2.3, this could be achieved through mapping the training data into the space $\mathscr{L} = \mathrm{span}(\kappa(\cdot, x_1), \ldots, \kappa(\cdot, x_p))$, where κ is a suitable kernel. Nevertheless, this method is actually not used in connection with perceptrons. This is because, according to (2.42), the scalar products computed in (3.90) and (3.91) need to evaluate $\kappa(x, x_k)$ for all the feature vectors x_k forming the input parts of the training data $(x_1, c_1), \ldots, (x_p, c_p) \in \mathscr{X} \times \{0, 1\}$. Since mid-1990s, however, another kind of classifiers for linearly separable classes exists that usually need to evaluate $\kappa(x, x_k)$ for only a small fraction of those feature vectors. These are support vector machines, which will be explained in Chap. 4.

Notice that the vectors of weights w_v and biases b_v defining $[F(x)]_v$ are for different $v \in \mathscr{O}$ independent from each other. Hence, learning them for a perceptron (3.90) classifying into $m \geq 3$ classes is equivalent to learning each of them separately for m perceptrons (3.91). The traditional learning algorithm for a perceptron (3.91) has the following properties:

1. The training data are $(x_1, c_1), \ldots, (x_p, c_p) \in \mathscr{X} \times \{0, 1\}$, and are required to fulfil $x_1, \ldots, x_p \neq 0$.
2. The set $\boldsymbol{\Phi}$ on which the optimization (2.50) is performed is defined as:

$$\boldsymbol{\Phi} = \{F : \mathscr{X} \to \{0, 1\} | (\forall x \in \mathscr{X}) \ F(x) = \Theta(x^\top w + b), w \in \mathbb{R}^n, b \in \mathbb{R}\}. \tag{3.93}$$

3. The error function ER_ϕ simply counts the number of disagreements between $F(x_j)$ and c_j,

$$\mathrm{ER}_\phi((x_1, c_1, F(x_1)), \ldots, (x_p, c_p, F(x_p))) = |\{k | F(x_k) \neq c_k\}|. \tag{3.94}$$

4. The algorithm then proceeds as follows.

Algorithm 1 (Traditional perceptron learning algorithm)
Input:
- training data $(x_1, c_1), \ldots, (x_p, c_p) \in \mathscr{X} \times \{0, 1\}$ such that $x_1, \ldots, x_p \neq 0$,
- maximal number of iterations $i_{\max}$.
Step 1. Denote $\tilde{x}_k = (x_k, 1), k = 1, \ldots, p$.
Step 2. Initialize $k = 1, i = 0$, and $\tilde{w}_0 \in \mathbb{R}^{n+1}$ arbitrarily (e.g., randomly, or $\tilde{w}_0 = 0$).
Step 3. Set $[\tilde{w}_{i+1}]_j = [\tilde{w}_i]_j + (c_k - F(x_k))[\tilde{x}_k]_j, j = 1, \ldots, n + 1$.
Step 4. Define F_{i+1} by $F_{i+1}(x) = \Theta((x, 1)^\top \tilde{w}_{i+1}), x \in \mathscr{X}$.
Step 5. If $|\{k | F_{i+1}(x_k) \neq c_k\}| = 0$ or $i = i_{\max}$,
set $F = F_{i+1}$,
else update $i \to i + 1, k \to k \mod p + 1$, and return to Step 3.
Output: perceptron F.

5. It can be shown [39] that if the training inputs assigned to each of both classes, i.e., the sets

$$\{x_k|k = 1, \ldots, p, c_k = 1\} \text{ and } \{x_k|k = 1, \ldots, p, c_k = 0\} \qquad (3.95)$$

are linearly separable, then there exists $i_0 \in \mathbb{N}$ such that $|\{k|F_{i_0}(x_k) \neq c_k\}| = 0$. Consequently, if i_{max} is set so that it fulfils $i_{max} \geq i_0$, then the above algorithm ends after i_0 iterations with all training data correctly classified. Moreover,

$$i_0 = O\left(\frac{R^2}{\gamma_{\tilde{w}^*}^2}\right), \text{ where} \qquad (3.96)$$

$$R = \max_{k=1,\ldots,p} \|x_k\|,$$
$$\tilde{w}^* \in \mathbb{R}^{n+1}, \|([\tilde{w}^*]_1, \ldots, [\tilde{w}^*]_n)\| = 1,$$
$$c_k = 1 \Rightarrow x_k^\top \tilde{w}^* > 0, c_k = 0 \Rightarrow x_k^\top \tilde{w}^* < 0,$$
$$\gamma_{\tilde{w}^*} = \min_{k=1,\ldots,p} |x_k^\top \tilde{w}^*|.$$

3.5.2 Multilayer Perceptrons for General Classes

The term "multilayer perceptron" (MLP) evokes the idea of a chain of two or more perceptrons, in which the output layer of the 1st one is at the same time the input layer of the 2nd one, the output layer of the 2nd one is the input layer of the 3rd one, etc. The input layer of the 1st perceptron is the input layer of the whole chain, and the output layer of the last one is the output layer of the chain. All remaining layers of all chained perceptrons are from the point of view of the resulting network hidden layers (cf. Fig. 3.7). The synaptic mappings assigned to all connections of the resulting network are the multiplications (3.86), and the somatic mappings assigned to its hidden and output neurons are the mappings (3.88).

In fact, however, the term "multilayer perceptron" is used for a much broader class of feedforward networks than only the just described chain of perceptrons. This is due substantially more general somatic mappings, which will now be described.

To each hidden neuron $v \in \mathcal{H}$, a somatic mapping analogous to (3.88) is assigned, namely

$$f_v(\xi) = \varphi\left(\sum_{u \in in(v)} [\xi]_u + b_v\right), \xi \in \mathbb{R}^{|in(v)|}, \qquad (3.97)$$

where $[\xi]_u$ for $u \in in(v)$ denotes the component of ξ that is the output of the synaptic mapping $f_{u,v}$ assigned to the connection (u, v), and $\varphi : \mathbb{R} \to \mathbb{R}$ is called *activation function*. It can be shown [40] that if each somatic mappings assigned to an output neuron $v \in \mathcal{O}$ is simply a biased aggregation of its inputs,

$$f_v(\xi) = \sum_{u \in in(v)} [\xi]_u + b_v, \xi \in \mathbb{R}^{|in(v)|}. \tag{3.98}$$

then a sufficient condition for the MLP to be a universal approximator is that the activation functions assigned to all hidden neurons fulfil the following weak conditions:

(i) the value set of φ is bounded;
(ii) φ is non-constant;
(iii) φ is Borel-measurable (i.e., measurable with respect to the least σ-algebra containing all open intervals).

Activation functions encountered in applications normally have the much stronger property of being *sigmoidal*, i.e., non-decreasing, piecewise continuous, and such that

$$-\infty < \lim_{t \to -\infty} \varphi(t) < \lim_{t \to \infty} \varphi(t) < \infty. \tag{3.99}$$

Notice that a sigmoidal activation function fulfils the conditions (i)–(iii) above, and that the Heaviside step function (3.89) is sigmoidal. The activation functions most frequently encountered in MLPs are, however, two other such functions:

• the logistic function (3.49), for which $\lim_{t \to -\infty} \varphi(t) = 0$, $\lim_{t \to \infty} \varphi(t) = 1$;
• the hyperbolic tangent,

$$\varphi(t) = \tanh(t) = \frac{e^t - e^{-t}}{e^t + e^{-t}} \tag{3.100}$$

for which $\lim_{t \to -\infty} \varphi(t) = -1$, $\lim_{t \to \infty} \varphi(t) = 1$.

Moreover, the somatic mappings assigned to all hidden neurons usually include the same activation function.

The somatic mappings (3.98) need to be assigned to output neurons, to allow approximation of mappings with an unbounded value set. If the value set of the approximated mapping is known to be bounded, then also the somatic mappings (3.97) can be used. Its boundedness is in the case of classification always guaranteed—recall the binary representation (2.3) of classes. Again, the activation function included in the somatic mapping assigned to an output neuron is usually the same for all output neurons. If the Heaviside function (3.89) is used to this end, then a binary vector $b \in \{0, 1\}^m$ is obtained, which in particular can (but doesn't have to) be the binary representation (2.3) of some class $c \in C$. If the logistic function (3.49) is used, typically because it has been already included in the somatic mappings assigned to hidden neurons, then the multilayer perceptron F can either directly serve as a fuzzy classifier, with the value $F(x) \in [0, 1]^m$ for $x \in \mathcal{X}$ being the membership function of x on the set of classes $C = \{c_1, \ldots, c_m\}$, or it can be used to define a crisp classifier ϕ as follows:

$$\phi(x) = c_i \in C \text{ such that } [F(x)]_i = \max_{j=1,\ldots,m} [F(x)]_j, x \in \mathcal{X}. \tag{3.101}$$

Finally, the synaptic mapping assigned to a connection $(u, v) \in \mathcal{E}$ is in each MLP the same as in a simple chain of perceptrons, namely the multiplication by a weight $w_{(u,v)}$, which was introduced in (3.86).

Having described the somatic and synaptic mappings assigned to neurons and connections of a multilayer perceptron, we now turn to MLP learning. To this end, denote w the vector of all parameters of all somatic and synaptic mappings, i.e., of the weights $w_{(u,v)}$ of all connections $(u, v) \in \mathcal{E}$, and the biases b_v in the somatic mappings (3.97) or (3.98) assigned to neurons $v \in \mathcal{H} \cup \mathcal{O}$. Then if the MLP has the layers $L_0, L_1, \ldots, L_L$, with $L_0 = \mathcal{I}$, $L_L = \mathcal{O}$ and the remaining layers hidden, the dimension of w is

$$\dim(w) = \sum_{i=1}^{L} |L_i|(|L_{i-1}| + 1), \text{ hence } w \in \mathbb{R}^{\dim(w)}. \tag{3.102}$$

As an example, for the network in Fig. 3.7, $L = 3$, $|L_0| = |\mathcal{I}| = 8$, $|L_1| = 8$, $|L_2| = 7$, $|L_3| = |\mathcal{O}| = 5$, which yields $\dim(w) = 175$.

With respect to that parameter vector, the optimization (2.50) is now performed. The training data are $(x_1, b_1), \ldots, (x_p, b_p) \in \mathcal{X} \times \{0, 1\}^m$, where for $k = 1, \ldots, p$, the binary vector b_k is the binary representation (2.3) of some class $c \in C$. As the error function ER_ϕ, the squared error is normally used in MLP learning,

$$\text{ER}_\phi((x_1, b_1, F(x_1)), \ldots, (x_p, b_p, F(x_p))) = \sum_{k=1}^{p} \|F(x_k) - b_k\|^2. \tag{3.103}$$

From (3.97), (3.98) and (3.103) follows that the smoothness of ER_ϕ with respect to w (i.e., the degree to which the derivatives are continuous) equals the smoothness of the activation function φ. In particular, in the case of the two most commonly used activation functions, logistic function (3.49) and hyperbolic tangent (3.100), ER_ϕ is infinitely smooth, thus any possible smooth optimization method can be applied to (2.50). The simplest among such methods is the *steepest descent*, in which the vector w moves from the i-th to the $(i + 1)$-th iteration against the direction of gradient $\nabla_w \text{ER}_\phi$ of ER_ϕ with respect to w corresponding to the training data, i.e., according to

$$w_{i+1} = w_i - \alpha \nabla_w \text{ER}_\phi((x_1, b_1, F(x_1)), \ldots, (x_p, b_p, F(x_p))), \text{ with } \alpha > 0. \tag{3.104}$$

Here, α is the *step size*, obtained through a 1-demensional optimization in the direction $-\nabla_w \mathrm{ER}_\phi((x_1, b_1, F(x_1)), \ldots, (x_p, b_p, F(x_p)))$. To obtain the gradient $\nabla_w \mathrm{ER}_\phi$, the partial derivatives of ER_ϕ with respect to individual components of w have to be computed. This will now be illustrated on the example of a MLP with 1 hidden layer in which the same kind of somatic mappings (3.97) is assigned to all hidden and output neurons.

For a feature vector $x \in \mathcal{X}$, the component $[F(x)]_o$ of the MLP output $F(x)$ returned by the output neuron $o \in \mathcal{O}$ is

$$[F(x)]_o = \varphi \left(\sum_{h \in \mathcal{H}} w_{(h,o)} \varphi \left(\sum_{j \in \mathcal{I}} w_{(j,h)}[x]_j + b_h \right) + b_o \right). \tag{3.105}$$

For greater transparency of the expressions for partial derivatives, let us introduce the following shorthand symbols for $k = 1, \ldots, p, h \in \mathcal{H}, o \in \mathcal{O}$:

$$E_{k,h} = \sum_{j \in \mathcal{I}} w_{(j,h)}[x_k]_j + b_h, \tag{3.106}$$

$$E_{k,o} = 2([F(x_k)]_o - [b_k]_o)\varphi' \left(\sum_{h \in \mathcal{H}} w_{(h,o)} \varphi \left(\sum_{j \in \mathcal{I}} w_{(j,h)}[x_k]_j + b_h \right) + b_o \right) =$$

$$= 2([F(x_k)]_o - [b_k]_o)\varphi' \left(\sum_{h \in \mathcal{H}} w_{(h,o)} \varphi(E_{k,h}) + b_o \right), \tag{3.107}$$

where $[b_k]_o$ is the component of b_k corresponding to the output neuron o. Then combining (3.103) with (3.105) yields:

- for the partial derivative of ER_ϕ with respect to the weight $w_{(u,v)}$ of a connection between $u \in \mathcal{H}$ and $v \in \mathcal{O}$,

$$\frac{\partial \mathrm{ER}_\phi}{\partial w_{(u,v)}} = \sum_{k=1}^{p} E_{k,v} \varphi(E_{k,u}); \tag{3.108}$$

- for the partial derivative of ER_ϕ with respect to the weight $w_{(u,v)}$ of a connection between $u \in \mathcal{I}$ and $v \in \mathcal{H}$,

$$\frac{\partial \mathrm{ER}_\phi}{\partial w_{(u,v)}} = \sum_{k=1}^{p} \sum_{o \in \mathcal{O}} w_{(v,o)} E_{k,o} \varphi'(E_{k,v})[x_k]_u; \tag{3.109}$$

- for the partial derivative of ER_ϕ with respect to the bias b_v of a neuron $v \in \mathcal{O}$,

$$\frac{\partial \mathrm{ER}_\phi}{\partial b_v} = \sum_{k=1}^{p} E_{k,v}; \tag{3.110}$$

- for the partial derivative of ER_ϕ with respect to the bias b_v of a neuron $v \in \mathscr{H}$,

$$\frac{\partial \, \mathrm{ER}_\phi}{\partial b_v} = \sum_{k=1}^{p} \sum_{o \in \mathscr{O}} w_{(v,o)} E_{k,o} \varphi'(E_{k,v}). \tag{3.111}$$

Notice that once the partial derivatives (3.108) and (3.110) are computed, concerning the neurons in the output layer and the connections ending in them, and consequently also all expressions $E_{k,h}$ and $E_{k,o}$ for $k = 1, \ldots, p, h \in \mathscr{H}, o \in \mathscr{O}$, then it is rather easy to compute the partial derivatives (3.109) and (3.111), concerning the neurons in the hidden layer and the connections ending in them. A similar situation is also in MLPs with more hidden layers: the computation starts with connections ending in and the neurons belonging to the output layer, and then it proceeds to lower layers, against the direction of the connections in the associated graph G_F. Therefore, the steepest descent method applied to MLPs is in the area of artificial neural networks commonly called *back propagation*.

Nowadays, MLP learning is frequently performed using more efficient 2nd-order optimization methods for (2.50), such as the Levenberg-Marquardt method, the BFGS method or various conjugate gradient methods. For further reading about those methods, we recommend [41–43].

3.5.3 Deep Neural Networks Suitable for Classification

In theory, the term "deep network" could be used for any layered ANN with at least two hidden layers. In reality, however, it is normally used only in connection with networks that combine layers of different types and with different purposes (cf. Fig. 3.8). Moreover, in many such networks, the number of layers is high (dozens, or even hundreds), making them really deep, but this characterization is nevertheless secondary compared to the diversity of layers.

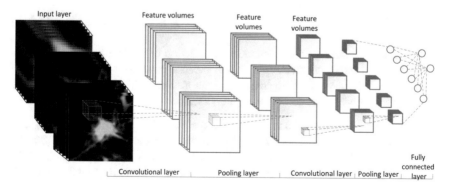

Fig. 3.8 Example deep neural network with five hidden layers of three different types (from [44])

During the last two decades, many specific types of layers serving various purposes have been proposed. In ANNs used for classification, the following are most frequently encountered:

1. *Convolutional layer—D-dimensional.* Such layers are used to classify D-dimensional connected objects, especially for $D = 2$ to classify images. The suitability of convolutional layers for the classification of low-dimensional connected objects is due to their capability to recognize matching patterns in those objects. In the most simple case, both the convolutional layer and its preceding layer need to be D-dimensional, suppose their sizes to be $c_1 \times \cdots \times c_D$ for the convolutional layer and $p_1 \times \cdots \times p_D$ for the preceding layer. The involvement of the convolutional layer in the classification is in this case controlled by the following two parameters:

- A D-dimensional array of real-valued weight vectors, i.e., $w \in \mathbb{R}^{r_1} \times \cdots \times \mathbb{R}^{r_D}$, $1 \leq r_k \leq p_k, k = 1, \ldots, D$, called *filter*, or more precisely, receptive field filter. Typically, $r_1 = \cdots = r_D$, hence, the size of the receptive field filter is a hypercube.
- A vector $(s_1, \ldots, s_D) \in \mathbb{N}^D$, the components of which are called *strides*.

Given a size $p_1 \times \cdots \times p_D$ of the preceding layer, the size $r_1 \times \cdots \times r_D$ of the receptive field filter and the vector $(s_1, \ldots, s_D)$ of values of strides imply for the size $c_1 \times \cdots \times c_D$ of the convolutional layer to fulfil the condition

$$c_k = 1 + \left\lfloor \frac{p_k - r_k}{s_k} \right\rfloor, k = 1, \ldots, D. \tag{3.112}$$

Finally, to each array $x \in \mathbb{R}^{p_1} \times \cdots \times \mathbb{R}^{p_D}$ of values in the preceding layer corresponds the array $y \in \mathbb{R}^{c_1} \times \cdots \times \mathbb{R}^{c_D}$ of values in the convolutional layer fulfilling:

$$y_{j_1, \ldots, j_D} = \sum_{i_1=1}^{r_1} \cdots \sum_{i_D=1}^{r_D} w_{i_1, \ldots, i_D} x_{(j_1-1)s_1+i_1, \ldots, (j_D-1)s_D+i_D},$$

$$\text{for } j_k \in \mathbb{N}, j_k \leq c_k, k = 1, \ldots, D. \tag{3.113}$$

In the one- and two-dimensional case, the function defined with (3.113) is sometimes called convolutional kernel, which gave the name to this type of layers. Worth mentionning are also two directions in which this most simple case of convolutinal layers are sometimes generalized:

(i) Instead of assigning a scalar $y_{j_1, \ldots, j_D} \in \mathbb{R}$ to scalars $x_{(j_1-1)s_1+i_1, \ldots, (j_D-1)s_D+i_D}$, $1 \leq r_k, k = 1, \ldots, D$, an m-dimensional vector $y_{j_1, \ldots, j_D} \in \mathbb{R}^m$ can be assigned to an n-dimensional vectors $x_{(j_1-1)s_1+i_1, \ldots, (j_D-1)s_D+i_D} \in \mathbb{R}^n$, $1 \leq r_k, k = 1, \ldots, D$. The filter w then needs to be appropriately modified.

(ii) If a larger size of the convolutional layer is desired than the one correpsonding to (3.112), then the preceding layer needs to be enlarged through appropriate padding.

2. *Max-pooling layer* and other pooling layers. A max-pooling layer computes the maxima of values assigned to subsets of its preceding layer that are such that:

   - they partition the preceding layer, i.e., that layer equals their union and they are mutually disjoint;
   - they are identically sized.

   Taking into account these two conditions, the size $p_1 \times \cdots \times p_D$ of the preceding layer and the size $r_1 \times \cdots \times r_D$ of the sets forming its partition determine the size of the max-pooling layer. For example, if the preceding layer is 3-dimensional and has the size $p_1 \times p_2 \times p_3$ with p_1, p_2 and p_3 even, and if it is partitioned into cubes $2 \times 2 \times 2$, then the max-pooling layer is 3-dimensional with the size $\lfloor \frac{p_1}{2} \times \frac{p_2}{2} \times \frac{p_3}{2} \rfloor$, whereas if the preceding layer is partitioned into sets of size $2 \times 2 \times p_3$, then the max-pooling layer is 2-dimensional with the size $\frac{p_1}{2} \times \frac{p_2}{2}$. Other commonly encountered kinds of pooling layers are *average pooling layer*, computing the averages of values assigned to sets partitioning the preceding layer, and L_2-*pooling layer*, which computes the Euclidean norm, aka L_2 norm, of those values.

3. *Batch-normalization layer.* A batch-normalization layer has the same size as its preceding layer and normalizes the values assigned to each of its neurons with respect to a subset B of training data with a cardinality $|B| \geq 2$, presented to the network simultaneously and therefore called *batch*. Typically, the standard normalization by means of the mean and standard deviation is used, i.e., the value v_j^p assigned in the preceding layer to the j-the neuron is in the batch-normalization layer transformed to

$$v_j^b = \frac{v_j^p - \bar{v}^p}{\sqrt{\frac{1}{|B|-1} \sum_{i \in B} (v_i^p - \bar{v}^p)^2}}, \quad \text{where } \bar{v}^p = \frac{1}{|B|} \sum_{i \in B} v_i^p. \tag{3.114}$$

4. *Rectified linear units (ReLU) layer.* A ReLU layer also has the same size as its preceding layer and transforms the values v assigned to the neurons of the preceding layers according to

$$\varphi(v) = \max(0, v). \tag{3.115}$$

   This function performs a nonlinear transformation, similarly to activation functions in multilayer perceptrons. Differently to them, however, (3.115) is not sigmoidal.

5. *Softmax layer.* A softmax layer has the same number of neurons as its preceding layer, say n, and the value s_j assigned to the j-the neuron of the softmax layer

is obtained from the values $p_1, \ldots, p_n$ assigned to the neurons of the preceding layer according to

$$s_j = \frac{e^{p_j}}{\sum_{i=1}^{n} e^{p_i}}, i = 1, \ldots, n. \tag{3.116}$$

Observe that (3.116) defines a probability distribution on the softmax layer. At the same time, (3.116) is also the membership function of a fuzzy set on the softmax layer. Hence, the softmax layer can be used as a fuzzy classifier. It is normally encountered at the very end of the sequence of layers.

6. *Long short-term memory (LSTM) layer.* An LSTM layer is used for classification of sequences of feature vectors, or equivalently, multidimensional time series with discrete time. Alternatively, that layer can be also employed to obtain sequences of such classifications, i.e., in situations when the neural network input is a sequence of feature vectors and its output is a sequence of classes. Differently to other commonly encountered layers, an LSTM layer consists not of simple neurons, but of units with their own inner structure. Several variants of such a structure have been proposed (e.g., [45, 46]), but all of them include at least the following four components given below. Each of them has certain properties of usual ANN neurons, in particular, the values assigned to them depend, apart from a bias, on values assigned to the unit input at the same time step and on values assigned to the unit output at the previous time step. Hence, a network using LSTM layers is a recurrent network.

   (i) *Memory cells* can store values, aka cell states, for an arbitrary time. They have no activation function, thus their output is actually a biased linear combination of unit inputs and of the values from the previous time step coming through recurrent connections.

   (ii) *Input gate* controls the extent to which values from the previous unit or from the preceding layer influence the value stored in the memory cell. It has a sigmoidal activation function, which is applied to a biased linear combination of unit inputs and of values from the previous time step, though the bias and synaptic weights of the input and recurrent connections are specific and in general different from the bias and synaptic weights of the memory cell.

   (iii) *Forget gate* controls the extent to which the memory cell state is supressed. It again has a sigmoidal activation function, which is applied to a specific biased linear combination of unit inputs and of values from the previous time step.

   (iv) *Output gate* controls the extent to which the memory cell state influences the unit output. Also this gate has a sigmoidal activation function, which is applied to a specific biased linear combination of unit inputs and of values from the previous time step, and subsequently composed either directly with the cell state or with its sigmoidal transformation, using a different sigmoid than is used by the gates.

7. *Residual layer.* A residual layer differs from all other layers in this survey through allowing connections not only from the preceding layer, but in addition also from earlier layers. Particular variants of residual layers differ as to the number of preceding layers that such additional connections may skip, as well as to whether the set of such connections is given in advance or established only during network learning, thus dependent on the training data [47, 48].

8. *Fully connected layer.* A fully connected layer, which we know already from multilayer perceptrons, i.e., a layer in which each neuron is connected with each neuron of the preceding layer. One or several fully connected layers usually occur in the last part of the network, typically succeeded only by a softmax layer.

These specific types or layers cannot be combined quite arbitrarily in a neural network. For example, it makes no sense to combine convolutional and LSTM layers because they are suitable for different kinds of input data. On the other hand, certain types of layers come nearly always together, even in a particular sequence, such as "convolutional layer—pooling layer", "convolutional layer—batch normalization layer—ReLU layer—pooling layer", or "fully connected layer/several fully connected layers—softmax layer". The whole network is typically called after the type of layers that accounts for the key computations, i.e., those that are most essential from the point of view of the task to be performed by the network. Hence, *convolutional networks* are neural networks in which the key computations are performed by convolutional layeres, in *LSTM networks* they are performed by LSTM layers, and in *residual networks* by residual layers.

Finally, several new approaches to neural network learning have been developed for deep networks. Among them, at least the following definitely should be mentioned:

- Due to a very large number of synaptic weights and other parameters of a deep network, deep learning needs also a very large amount of training data. Therefore, deep network learning is not performed with all training data at once, but rather with subsets of them, which are usually disjoint. As was already mentioned in connection with batch-normalization layer, they are called batches, actually even more frequently *mini-batches* because their size is typically much smaller than the size of the whole training dataset.
- Some new optimization methods are used to address the optimization task (2.50) for deep networks, most importantly *stochastic gradient descent*, which is a stochastic generalization of the traditional steepest descent used as back propagation for multilayer perceptron learning and recalled in the previous subsection.
- To avoid overtraining, stochastic regularization methods are most often used, primarily *dropout*, which consists in randomly leaving out individual neurons or individual connections, but also *stochastic depth*, which randomly leaves out whole layers.

For an overview of various kinds of deep neural networks, the reader is referred to the book [49] and the survey paper [50].

3.5.4 ANN-Based Classification in Spam Filtering

Similarly to the discriminant analysis approach discussed in Sect. 3.5.4, spam filters based on neural networks also commonly use the tokenized text content for classification. Moreover, as we have shown earlier, the data extracted from the SpamAssassin dataset [26] are of a large dimension and are linearly separable. A simple perceptron should, therefore, suffice for such classification.

Moreover, we need to consider that a perceptron for binary classification requires a tuning of $n + 1$ parameters, where n is the number of input dimensions (the number of neurons in the input layer), and the added one is the bias of the output neuron. Given that the considered dataset consists of only 3002 documents, it is appropriate to minimize the number of tunable parameters to avoid overfitting.

In general, the training of the perceptron differs significantly from approaches based on probability or distribution, as the parameters of the neural network are initialised randomly and tuned for lowering the loss function with each epoch.

Also, a subset of training data or separate validation set can be used to check on the progress of accuracy and loss during the training. For such purpose, we have used the 20030228_easy_ham_2 and 20030228_spam_2 datasets from Spamassassin [26].

In all examples, we use a sigmoidal activation function, Adam optimiser [51] and a binary cross-entropy loss function. We only vary the number of hidden neurons.

As the data seem to be linearly separable, we first test the simple perceptron. The accuracy of such training is presented in Fig. 3.9a.

If we cannot assume the linear separability, hidden layers must be introduced into the model. They significantly increase the number of parameters that need to be optimized. Consequently, the training phase should have even larger volumes of data and significantly more computational resources available.

Adding hidden layers for the classification of high-dimensional linearly separable data, however, hardly helps and may lead to an instability of the classifier. For the SpamAssassin data, the progress of training and validation accuracy with 10, 100 and 1 000 neurons in a single hidden layer is compared in Fig. 3.9b–d. The only significant difference is that the number of epochs required for convergence to a comparable validation accuracy is lowered to the half by increasing the number of hidden neurons 10 times. However, it also increases the number of tunable parameters 10 times and requires significantly more time to train. Another difference in the training of a neural network with 1000 neurons in the hidden layer is the presence of a swing in the validation accuracy, which may be a precursor of overtraining.

And even in more complicated scenarios of e-mail filtering, such as proposed in [52], at most one hidden layer with 20–40 neurons is typically used.

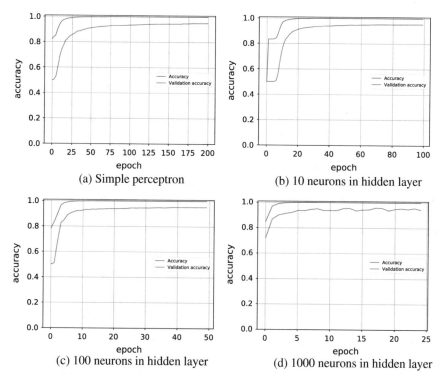

Fig. 3.9 Training and validation accuracy of ANN on a subset of SpamAssassin dataset with 3 002 training and 2 799 validation samples

3.5.5   ANN-Based Classification in Recommender Systems

One of the scenarios for the use of neural networks in recommender systems originates from the success of matrix factorization in the Netflix prize [53].

This approach represents users and items in a low-dimensional latent space and predicts the ratings with a matrix multiplication of the factor representations:

$$[\hat{x}]_i = \mu + \sum_{f=0}^{n} H_{x,f} W_{f,i}, \qquad (3.117)$$

where μ represents a normalization term that may include user and item biases, H contains the user's latent factors, W contains the latent factors of the items, and n is the number of considered factors. When only one factor is considered ($n = 1$), the prediction degrades to a *most popular recommender*, providing no personification. With an increasing number of factors, the quality of the recommendation increases, until it loses its generalization capabilities.

This matrix factorization approach (MF) can be directly transferred to a neural network that uses a separate embedding layer for users and items, followed by layer computing a dot product, such as presented in Fig. 3.10a. This transfer allows certain generalizations and is, therefore, sometimes referred to as Generalized Matrix Factorization (GMF). To adhere to the methods used in the regular matrix factorization, the mean squared error is used as a loss function. To model the embedding layer with neurons, a fully connected layer with one-hot encoding on the input can be used (see Fig. 3.10b). However, neural network frameworks usually have a separate embedding layer which uses the incoming categorical data more efficiently.

To evaluate the neural network matrix factorization, we used the MovieLens dataset [6] with 100000 movie ratings from 943 users on 1682 movies (items). The loss during the neural network training with 10% of data separated as validation set is available in Fig. 3.10c. Three latent factors were chosen for both the user and item embedding, as higher number resulted in slight overfitting of the latent space after as few as 30 epochs.

To compare the approach based on matrix factorization with the results from the k-nn model in Sect. 3.2.2, the predicted scores for user 405 after 50 epochs of training are presented in Table 3.6. Although some of the scores differ significantly in absolute values, the sets of top three recommended movies do intersect—300, 286 and 121 for k-nn and 258, 121 and 300 for matrix factorization.

The simplicity of the dot product, however, may limit the expressiveness of the neural network, as we try to model complex user-item interactions in a single low-dimensional latent space. One possible enhancement is to use element-wise multiplication of the embeddings and additional dense layer to introduce a weighting of the individual dimensions in the latent space.

Even more powerful approach is to fuse the matrix factorization with a regular multi-layer perceptron. To this end, we may use the same embeddings for both the matrix factorization pipeline and a deep neural network. However, the *Neural Matrix Factorization* (NeuMF) approach introduced in [54] proposes to train separate embeddings for GMF and MLP to provide a higher flexibility. Results of these two pipelines are then concatenated and passed through the last layer to provide the final rating prediction. The overall structure is shown in Fig. 3.11a.

To evaluate the NeuMF recommender, its authors use a hit-count among top ten recommendations and normalized discounted cumulative gain (nDCG) that takes into account also the order of the recommended items. In this case, we trained the network on the MovieLens dataset with over a million ratings, and the default settings of NeuMF—eight factors in the matrix factorization latent space and neural network with layers of 64, 32, 16 and 8 neurons. The results are shown in Fig. 3.11b.

All of the methods mentioned above used only explicit user feedback for selection of recommendations. However, an implicit feedback in recommender systems from data containing only binary information whether the user had interacted with a particular item or carried out certain action can be used in a very similar manner.

In addition to the implicit feedback from data, stored properties of the user and item can be easily added as another input to the recommender based on a neural network. One of such approaches is proposed in [55] for the selection and ranking of YouTube

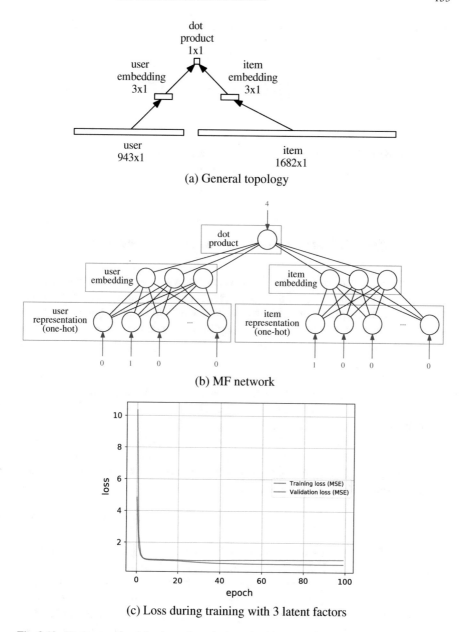

(a) General topology

(b) MF network

(c) Loss during training with 3 latent factors

Fig. 3.10 Structure and training loss of matrix factorization on MovieLens-100k dataset

Table 3.6 Predicted ratings of the ten most rated movies that are not yet rated by user 405 from the MovieLens dataset

	$[\hat{405}]_1$	$[\hat{405}]_{100}$	$[\hat{405}]_{121}$	$[\hat{405}]_{258}$	$[\hat{405}]_{286}$	$[\hat{405}]_{294}$	$[\hat{405}]_{300}$
k-nn MF	1.379	1.669	2.246	2.186	2.335	1.473	2.455
NeuMF	3.174	1.899	3.494	3.972	3.171	1.83	3.215

video recommendations. For example, the geographic embedding and gender of the user are considered, as well as the age of the particular video. The ranking network uses the embedding of video and users' language and other features.

3.5.6 Artificial Neural Networks in Sentiment Analysis

In the case of sentiment analysis, the classifier needs to work with much shorter messages than in case of spam detection. Product reviews hardly exceed one short sentence, and microblogging services (such as Twitter) limit the length of messages to a few hundred characters. On such a small set, the classifier needs to extract as much information as possible. Models based only on the presence of selected words may not suffice.

To asses the sentiment of a sentence more thoroughly, the order of individual words should also be considered. Traditional approaches based on bag-of-words may use n-grams to capture the sequence of n words. However, the high dimensionality of the input space complicates network training.

Another possible approach is to add a convolutional layer across the representation of individual words, such as presented in [56]. To this end, neural networks use an arrangement of neurons that takes words in a window of a chosen width k and combine them by a trained weight matrix W:

$$c_m = W\left(r_{m-\lfloor(k-1)/2\rfloor}, \ldots, r_m, \ldots, r_{m+\lceil(k-1)/2\rceil}\right) + b, \qquad (3.118)$$

where r_i is the representation of a word on position i and b is the trained additional bias of the convolutional layer. If the window contains indexes outside the scope of the sentence, padding the word representation is used. Note that the multiplication matrix W, as well as the bias vector b, are common for all steps of the convolution and the convolution window is also internally represented as a matrix for the multiplication.

The matrix resulting from the convolution, however, has still one of the dimensions dependent on the number of words in the sentence. This dependency presents an issue, as the follow-up processing by a deep neural network resulting in a classification requires an input of the same dimension for every input.

A max-pooling layer is used to solve this issue as it reduces the so-called temporal dimensionality, i.e., the number of members of the considered sequence, to one by

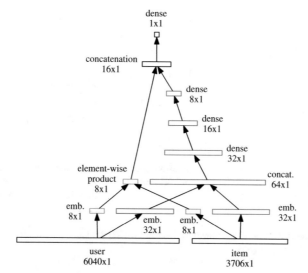

(a) General topology: blue layers represent GMF, orange ones MLP

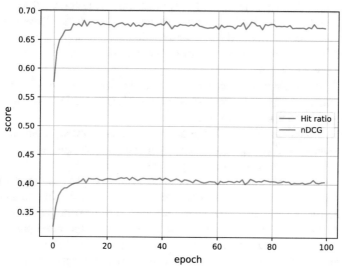

(b) Performance of NeuMF during training

Fig. 3.11 Structure and evaluation of Neural Matrix Factorization on MovieLens-1M dataset

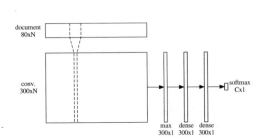
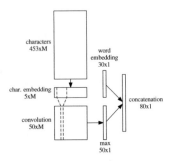

(a) Word-level convolution and classification topol-ogy; N is the number of words on input, C number of output classes

(b) Character-level convolution and word representation topology; M is the number of characters in the considered word

Fig. 3.12 Structure of a network known as *CharSCNN* [57] with metaparameters for the *Standford Twitter Sentiment corpus* [58]

taking only the maximal component across the whole convolution. The resulting vector length is then equal to the number of filters from the convolutional layer.

To classify this sentence (document) representation, additional densely connected layers are introduced, followed by either a single output neuron (in case of a binary classification) or a soft-max layer with the number of output neurons equal to the number of output classes.

The overall structure of the convolution and classification layers is presented in Fig. 3.12a on an example of a network known as *CharSCNN* [57]; this topology is, however, similar to many other systems.

The only ingredient missing is the definition of the word representation. In the simplest case, a word can be represented by a one-hot encoding from a dictionary. However, this introduces an issue of high input dimensionality.

As we have seen in the previous section, conversion to a latent factor space (embeddings) decreases the dimensionality and then reduces the number of trainable parameters in the following parts of the neural network. The embedding itself, however, needs to be trained as well. Unless we use a separate unsupervised pre-training of the word embeddings, such as word2vec [59] on a snapshot of Wikipedia corpus.

The pre-training of the embedding has a couple of advantages: we do not need to train the embedding ourselves on large corpora of our data, nor model the semantics of individual words. A disadvantage is that we are stuck with a fixed dictionary that is not specifically tailored for our application.

This issue is one of the greatest concerns in microblogging services. Words may be shortened or very specific for the individual audience, therefore, not contained in the dictionary and not having a pre-trained embedding available. Also, some of the tokens extracted from the text may contain essential keywords in the form of hashtags (starting with a symbol #).

Another issue is that the embeddings are designed not to contain much of the morphological information from the text. This approach is suitable for many natural language processing approaches, as it simplifies the means of language understanding. However, especially in short sections of text, morphology may help significantly in the assessment of the sentiment.

To tackle both of these issues, we may take into account the direct text representation. A network known as CharSCNN presented in [57] enhances the word representation given by word2vec with an additional representation created around the individual characters contained in the word. The topology of this approach is shown in Fig. 3.12b.

In a very similar manner to the word-level processing, the CharSCNN uses embeddings to represent individual characters in the considered word, convolves the embeddings with a selected window width and uses max-pooling to extract a single vector representation. This representation based on individual characters is then concatenated to the word representation from word2vec embedding and used in the word-level processing as a representation of the considered word.

3.5.7   ANN-Based Classification in Intrusion Detection

Artificial neural networks were proposed by Lunt [60] as the third component to the Intrusion detection expert system (IDES) [61, 62]. She envisioned that neural networks would be a great complement or even a future replacement of IDES statistical models. The main reasons being that neural networks don't need an accurate estimate of statistical distribution, they can adapt to different environments, due to which they are a scalable solution.

Her vision was brought to life two years later, in [63]. Audit logs were used as an input for a neural network to learn sequences of user commands and predict the next one. They proved that neural networks could learn user behaviour without knowing the statistical distribution in advance. The adaptation of the model for a different environment required retraining, which could take from several hours up to days, but in most cases, was still easier than adjusting statistical models. In this paper, authors trained a dedicated neural network for each user, which is a solution that doesn't scale. Therefore, they proposed an impressive infrastructure able to scale even with technical difficulties of 1990s. Such infrastructure should contain multiple neural networks trained for different purposes. In particular, a Kohonen's self-organizing map [64] should cluster users. Then, each cluster should have its own neural network to model the behaviour of users in that cluster.

In [65], the authors also used user commands as input, but rather than trying to learn benign and malicious command sequences, they were detecting anomalies in frequency histograms of user commands calculated for each user.

The work of Cannady [66] summarised all advantages and disadvantages of using neural networks for misuse detection. This paper outlined the path to future development in the field of intrusion detection. According to Cannady, the main advantages of the neural networks are the flexibility to analysed incomplete, distorted and noisy data and the ability to learn the characteristics of misuse attacks and generalise them. He also foresaw that neural networks would need a huge number of reliable labelled data to do that properly, but the most significant disadvantage of neural networks in the intrusion detection is their "black box" nature.

In 1999, a dataset created during 1998 DARPA Intrusion Detection Evaluation Program was released in the form of a KDD challenge. This dataset, known as KDD99 [67], became the first publicly available intrusion detection dataset and encouraged many researchers to invest into the field of intrusion detection. With millions of labelled network communication records and 38 attack types, it was almost three magnitudes larger than anything used before.

Zhang et al. [68] tested five different ANN architectures (perceptron, multi-layered perceptron, multi-layered perceptron with multi-layer connections, Fuzzy ARTMAP [69] and a network with radial basis functions, aka RBF network) for DoS attack detection. Mukkamala et al. [70] compared MLP perceptron to support vector machines. Bivens et al. [71] proposed an extension to [66] using time-window method to recognize longer multi-packet attacks. Depren et al. [72] used a hierarchical model where misuse detection based on self-organizing maps (SOMs) was coupled with random forest-based rule system to provide not only high precision, but also some sort of explanation.

The SOMs were really popular at that time; some other examples of SOM applications to intrusion detection are [73–75]. The computational complexity of SOMs limited their application in later years due to the significant growth in volume of analysed network traffic. Lei and Ghorbani [76] tried to overcome this issue by modifying SOM learning, as well as by using alternative kinds of neural networks [77].

Linda et al. [78] adapted IDS specifically for critical environments such as supervisory control of manufacturing processes or for nuclear power plants. They extracted time-window based features and used synthetically generated attack samples to learn a feed-forward MLP.

The current success of various deep learning architectures on image processing [47, 79], machine translation [80] and vision [81] was an inspiration also for the field of intrusion detection. In [82], the authors combined deep learning with spectral clustering to improve the detection rate of low frequency attacks.

The deep architectures are also well known for their ability to process raw inputs and learn their own features. Saxe et al. [83] used CNN to extract features from URLs, mutexes and changed registry keys automatically. Extracted features are then used as an input for standard MLP and used to detect malicious activity. Wang et al. [84] use CNN to learn the spatial features of network traffic and for traffic classification. Torres et al. [85] transform network traffic into a sequence of characters and then use networks with recurrent connections to learn their temporal features; those are then used to detect malware. The method proposed by [86] combines both previous approaches and uses CNN to learn the spatial features of network packets and then uses an LSTM to learn the temporal features from multiple network packets.

Pevný et al. [87] introduced stacked multiple instance learning architecture, where data are viewed not as a collection of bags but as a hierarchy of bags.

Almost every type of ANN architecture was with bigger or smaller success applied to IDS: RNN [88], LSTM [89], Deep autoencoders [90], Boltzmann machines [91] and Belief networks [92] to name a few. A comprehensive survey of machine learning methods with a focus on neural networks applied to the intrusion detection can be found in [93].

References

1. Jaskowiak, P., Campello, R.: Comparing correlation coefficients as dissimilarity measures for cancer classification in gene expression data. In: Brazilian Symposium on Bioinformatics, pp. 1–8 (2011)
2. Kushmerick, N.: Learning to remove internet advertisments. In: ACM Conference on Autonomous Agents, pp. 175–181 (1999)
3. Su, X., Khoshgoftaar, T.: A survey of collaborative filtering techniques. Adv. Artif. Intell. **2**, 14–32 (2009)
4. Herlocker, J., Konstan, J., Borchers, A., Riedl, J.: An algorithmic framework for performing collaborative filtering. In: 22nd Annual International ACM SIGIR Conference on Research and Development in Information Retrieval, pp. 230–237 (1999)
5. Enas, G., Choi, S.: Choice of the smoothing parameter and efficiency of k-nearest neighbor classification. In: Statistical Methods of Discrimination and Classification, pp. 235–244. Elsevier Science Publishers, Amsterdam (1986)
6. Harper, F., Konstan, J.: The Movielens datasets: history and context. ACM Trans. Interact. Intell. Syst. **5**, Article no. 19 (2016)
7. Resnick, P., Iacovou, N., Suchak, M., Bergstrom, P., Riedl, J.: GroupLens: an open architecture for collaborative filtering of netnews. In: Proceedings of the 1994 ACM Conference on Computer Supported Cooperative Work, pp. 175–186 (1994)
8. Kephart, J., White, S.: Directed-graph epidemiological models of computer viruses. In: IEEE Computer Society Symposium on Research in Security and Privacy, pp. 343–359 (1991)
9. Bailey, N.: The Mathematical Theory of Infectious Diseases and its Applications. Charles Griffin & Company, Bucks (1975)
10. Bellett A.J.D.: Numerical classification of some viruses, bacteria and animals according to nearest-neighbour base sequence frequency. J. Mol. Biol. **27**, 107–112 (1967)
11. Kephart, J., Sorkin, G., Arnold, W., Chess, D., G.J., T., White S.R., Watson, T.: Biologically inspired defenses against computer viruses. In: International Joint Conference on Artificial Intelligence, pp. 985–996 (1995)
12. Wang, K., Stolfo, S.: Anomalous payload-based network intrusion detection. In: International Workshop on Recent Advances in Intrusion Detection, pp. 203–222 (2004)
13. Liao, Y., Vemuri, V.: Use of k-nearest neighbor classifier for intrusion detection. Comput. Secur. **21**, 439–448 (2002)
14. Eskin, E., Arnold, A., Prerau, M., Portnoy, L., Stolfo, S.: A geometric framework for unsupervised anomaly detection. In: Applications of Data Mining in Computer Security, pp. 77–101. Springer (2002)
15. Schölkopf, B., Platt, J., Shawe-Taylor, J., Smola, A., Williamson, R.: Estimating the support of a high-dimensional distribution. Neural Comput. **13**, 1443–1471 (2001)
16. Wang, W., Guan, X., Zhang, X.: A novel intrusion detection method based on principal component analysis in computer security. In: International Symposium on Neural Networks, pp. 657–662 (2004)

17. Bouzida, Y., Cuppens, F., Cuppens-Boulahia, N., Gombault, S.: Efficient intrusion detection using principal component analysis. In: 3éme Conférence sur la Sécurité et Architectures Réseaux, pp. 381–395 (2004)

18. Silpa-Anan, C., Hartley, R.: Optimised KD-trees for fast image descriptor matching. In: IEEE Conference on Computer Vision and Patttern Recognition, pp. 1–8 (2008)

19. Dong, W., Moses, C., Li, K.: Efficient k-nearest neighbor graph construction for generic similarity measures. In: 20th International Conference on World Wide Web, pp. 577–586 (2011)

20. Rieck, K., Trinius, P., Willems, C., Holz, T.: Automatic analysis of malware behavior using machine learning. J. Comput. Secur. **19**, 639–668 (2011)

21. Deng, Z., Zhu, X., Cheng, D., Zong, M., Zhang, S.: Efficient k-nn classification algorithm for big data. Neurocomputing **195**, 143–148 (201)

22. Pearl, J.: Probabilistic Reasoning in Intelligent Systems. Morgan Kaufmann Publishers (1988)

23. Friedman, N., Geiger, D., Goldszmidt, M.: Bayesian network classifiers. Mach. Learn. **29**, 131–163 (1997)

24. Statista Inc.: Global spam volume as percentage of total e-mail traffic from January 2014 to September 2017 (2017). https://www.statista.com/statistics/420391/spam-email-traffic-share

25. Talos Intellingence Group, Cisco Inc.: Total global email and spam volume (2017). https://www.talosintelligence.com/reputation_center/email_rep

26. Apache Software Foundation, SpamAssassin Group: The Apache SpamAssassin project (2003). https://spamassassin.apache.org/old/publiccorpus/

27. University of California in Irvine: UCI repository of machine learning databases (2016). http://www.ics.uci.edu/~mlearn

28. Metsis, V., Androutsopoulos, I., Paliouras, G.: Spam filtering with Naive Bayes—which Naive Bayes? In: CEAS, pp. 28–69 (2006)

29. Das, S., Chen, M.: Yahoo! for Amazon: sentiment extraction from small talk on the web. Manag. Sci. **53**, 1375–1388 (2007)

30. Hu, M., Liu, B.: Mining and summarizing customer reviews. In: ACM SIGKDD International Conference on Knowledge Discovery and Data Mining, pp. 168–177 (2004)

31. Miller, G.: WordNet: a lexical database for English. Commun. ACM **38**, 39–41 (1995)

32. Narayanan, V., Arora, I., Bhatia, A.: Fast and accurate sentiment classification using an enhanced naive Bayes model. In: International Conference on Intelligent Data Engineering and Automated Learning, pp. 194–201 (2013)

33. Pang, B., Lee, L., Vaithyanathan, S.: Thumbs up? Sentiment classification using machine learning techniques. In: Proceedings of the ACL-02 Conference on Empirical Methods in Natural Language Processing, vol. 10, pp. 79–86 (2002)

34. Greiner, R., Su, X., Shen, B., Zhou, W.: Structural extension to logistic regression: discriminative parameter learning of belief net classifiers. Mach. Learn. **59**, 297–322 (2005)

35. Su, X., Khoshgoftaar, T.: Collaborative filtering for multi-class data using Bayesian networks. Int. J. Artif. Intell. Tools **17**, 71–85 (2008)

36. Pedregosa, F., Varoquaux, G., Gramfort, A., Michel, V., Thirion, B., Grisel, O., Blondel, M., Prettenhofer, P., Weiss, R., Dubourg, V., Vanderplas, J., Passos, A., Cournapeau, D., Brucher, M., Perrot, M., Duchesnay, E.: Scikit-learn: machine learning in python. J. Mach. Learn. Res. **12**, 2825–2830 (2011)

37. Tebbens, J., Schlesinger, P.: Improving implementation of linear discriminant analysis for the high dimension/small sample size problem. Comput. Statist. Data Anal. **52**, 423–437 (2007)

38. Rosenblatt, F.: The perceptron: a probabilistic model for information storage and organization in the brain. Psychol. Rev. **65**, 386–486 (1958)

39. Novikoff, A.: On convergence proofs on perceptrons. In: 12th Symposium on the Mathematical Theory of Automata, pp. 615–622 (1962)

40. Hornik, K.: Approximation capabilities of multilayer neural networks. Neural Netw. **4**, 251–257 (1991)

41. Dennis, J., Schnabel, R.: Numerical Methods for Unconstrained Optimization and Nonlinear Equations. Prentice Hall, Englewood Cliffs (1983)

42. Hagan, M., Demuth, H., Beale, M.: Neural Network Design. PWS Publishing, Boston (1996)

43. Scales, L.: Introduction to Non-Linear Optimization. Springer (1985)
44. Kang, G., Liu, K., Hou, B., Zhang, N.: 3D multi-view convolutional neural networks for lung nodule classification. PLoS One **12**, e0188,290 (2017)
45. Gers, F., Schmidhuber, J., Cummis, J.: Learning to forget: continual prediction with LSTM. In: 9th International Conference on Artificial Neural Networks: ICANN '99, pp. 850–855 (1999)
46. Graves, A.: Supervised Sequence Labelling with Recurrent Neural Networks. Springer (2012)
47. He, K., Zhang, X., Ren, S., Sun, J.: Deep residual learning for image recognition. In: IEEE Conference on Computer Vision and Pattern Recognition, pp. 770–778 (2016)
48. Huang, G., Liu, Z., van der Maaten, L., Weinberger, K.: Densely connected convolutional networks. In: IEEE Conference on Computer Vision and Pattern Recognition, pp. 4700–4708 (2017)
49. Goodfellow, I., Bengio, Y., Courville, A.: Deep Learning. MIT Press, Cambridge (2016)
50. Schmidhuber, J.: Deep learning in neural networks: an overview. Neural Netw. **61**, 85–117 (2015)
51. Kingma, D., Ba, J.: Adam: a method for stochastic optimization (2014). Preprint arXiv:1412.6980
52. Clark, J., Koprinska, I., Poon, J.: A neural network based approach to automated e-mail classification. In: IEEE/WIC International Conference on Web Intelligence, pp. 702–705 (2003)
53. Koren, Y., Bell, R., Volinsky, C.: Matrix factorization techniques for recommender systems. Computer **42**, 30–37 (2009)
54. He, X., Liao, L., Zhang, H., Nie, L., Hu, X., Chua, T.: Neural collaborative filtering. In: 26th International Conference on World Wide Web, pp. 173–182 (2017)
55. Covington, P., Adams, J., Sargin, E.: Deep neural networks for Youtube recommendations. In: 10th ACM Conference on Recommender Systems, pp. 191–198 (2016)
56. Kim, Y.: Convolutional neural networks for sentence classification (2014). Arxiv preprint arXiv:1408.5882
57. Dos Santos, C., Gatti, M.: Deep convolutional neural networks for sentiment analysis of short texts. In: COLING: 25th International Conference on Computational Linguistics: Technical Papers, pp. 69–78 (2014)
58. Go, A., Bhayani, R., Huang, L.: Twitter sentiment classification using distant supervision. Technical Report, Stanford University (2009)
59. Mikolov, T., Chen, K., Corrado, G., Dean, J.: Efficient estimation of word representations in vector space (2013). ArXiv preprint arxiv:1301.3781
60. Lunt, T.: IDES: An intelligent system for detecting intruders. In: Symposium on Computer Security, Threat and Countermeasures, pp. 30–45 (1990)
61. Anderson, J.: Computer security threat monitoring and surveillance. Technical Report, James P. Anderson Company, Fort Washington (1980)
62. Denning, D.: An intrusion-detection model. IEEE Trans. Softw. Eng. **13**, 222–232 (1987)
63. Debar, H., Becker, M., Siboni, D.: A neural network component for an intrusion detection system. In: IEEE Computer Society Symposium on Research in Security and Privacy, pp. 240–250 (1992)
64. Kohonen, T.: Self-Organizing Maps. Springer (1995)
65. Ryan, J., Lin, M., Miikkulainen, R.: Intrusion detection with neural networks. Adv. Neural Inf. Process. Syst. **10**, 943–949 (1998)
66. Cannady, J.: Artificial neural networks for misuse detection. In: National Information Systems Security Conference, pp. 368–381 (1998)
67. Tavallaee, M., Bagheri, E., Lu, W., Ghorbani, A.: A detailed analysis of the KDD cup 99 data set. In: IEEE Symposium on Computational Intelligence for Security and Defense Applications, pp. 288–293 (2009)
68. Zhang, Z., Li, J., Manikopoulos, C., Jorgenson, J., Ucles, J.: HIDE: a hierarchical network intrusion detection system using statistical preprocessing and neural network classification. In: IEEE Workshop on Information Assurance and Security, pp. 85–90 (2001)
69. Carpenter G.A. ad Grossberg, S., Markuzon, N., Reynolds, J., Rosen, D.: Fuzzy ARTMAP: a neural network architecture for incremental supervised learning of analog multidimensional maps. IEEE Trans. Neural Netw. **3**, 698–713 (1992)

70. Mukkamala, S., Janoski, G., Sung, A.: Intrusion detection using neural networks and support vector machines. In: International Joint Conference on Neural Networks, pp. 1702–1707 (2002)
71. Bivens, A., Palagiri, C., Smith, R., Szymanski, B., Embrechts, M.: Network-based intrusion detection using neural networks. In: Intelligent Engineering Systems through Artificial Neural Networks, pp. 579–584. ASME Press, New York (2002)
72. Depren, O., Topallar, M., Anarim, E., Ciliz, M.: An intelligent intrusion detection system (IDS) for anomaly and misuse detection in computer networks. Expert Syst. Appl. **29**, 713–722 (2005)
73. Bonifacio, J., Cansian, A., De Carvalho, A., Moreira, E.: Neural networks applied in intrusion detection systems. In: IEEE International Joint Conference on Neural Networks, pp. 205–210 (1998)
74. Ghosh, A., Wanken, J., Charron, F.: Detecting anomalous and unknown intrusions against programs. In: 14th Annual Computer Security Applications Conference, pp. 259–267 (1998)
75. Rhodes, B., Mahaffey, J., Cannady, J.: Multiple self-organizing maps for intrusion detection. In: 23rd National Information Systems Security Conference, pp. 16–19 (2000)
76. Lei, J., Ghorbani, A.: Network intrusion detection using an improved competitive learning neural network. In: Second Annual Conference on Communication Networks and Services Research, pp. 190–197 (2004)
77. Ahalt, S., Krishnamurthy, A., Chen, P., Melton, D.: Competitive learning algorithms for vector quantization. Neural Netw. **3**, 277–290 (1990)
78. Linda, O., Vollmer, T., Manic, M.: Neural network based intrusion detection system for critical infrastructures. In: International Joint Conference on Neural Networks, pp. 1827–1834 (2009)
79. Dong, C., Loy, C., He, K., Tang, X.: Learning a deep convolutional network for image super-resolution. In: European Conference on Computer Vision, pp. 184–199 (2014)
80. Wu, Y., Schuster, M., Chen, Z., Le, Q., Norouzi, M., Macherey, W., Krikun, M., Cao, Y., Gao, Q., Macherey, K., Klingner, J., Shah, A., Johnson, M., Liu, X., Kaiser, Ł., Gouws, S., Kato, Y., Kudo, T., Kazawa, H., Stevens, K., Kurian, G., Patil, N., Wang, W., Young, C., Smith, J., Riesa, J., Rudnick, A., Vinyals, O., Corrado, G., Hughes, M., Dean, J.: Google's neural machine translation system: bridging the gap between human and machine translation (2016). Arxiv preprint arXiv:1609.08144
81. Howard, A., Zhu, M., Chen, B., Kalenichenko, D., Wang, W., Weyand, T., Andreetto, M., Adam, H.: Mobilenets: efficient convolutional neural networks for mobile vision applications (2017). Arxiv preprint arXiv:1704.04861
82. Ma, T., Wang, F., Cheng, J., Yu, Y., Chen, X.: A hybrid spectral clustering and deep neural network ensemble algorithm for intrusion detection in sensor networks. Sensors **16**, Article no. 1701 (2016)
83. Saxe, J., Berlin, K.: Expose: A character-level convolutional neural network with embeddings for detecting malicious URLs, file paths and registry keys (2017). Arxiv preprint arXiv:1702.08568
84. Wang, W., Zhu, M., Zeng, X., Ye, X., Sheng, Y.: Malware traffic classification using convolutional neural network for representation learning. In: ICOIN: IEEE International Conference on Information Networking, pp. 712–717 (2017)
85. Torres, P., Catania, C., Garcia, S., Garino, C.: An analysis of recurrent neural networks for botnet detection behavior. In: ARGENCON: IEEE biennial congress of Argentina, pp. 1–6 (2016)
86. Wang, W., Sheng, Y., Wang, J., Zeng, X., Ye, X., Huang, Y., Zhu, M.: HAST-IDS: learning hierarchical spatial-temporal features using deep neural networks to improve intrusion detection. IEEE Access **6**, 1792–1806 (2017)
87. Pevný, T., Somol, P.: Discriminative models for multi-instance problems with tree structure. In: ACM Workshop on Artificial Intelligence and Security, pp. 83–91 (2016)
88. Anyanwu, L., Keengwe, J., Arome, G.: Scalable intrusion detection with recurrent neural networks. In: IEEE 7th International Conference on Information Technology: New Generations, pp. 919–923 (2010)
89. Kim, J., Kim, J., Thu, H.L.T., Kim, H.: Long short term memory recurrent neural network classifier for intrusion detection. In: PlatCon: IEEE International Conference on Platform Technology and Service, pp. 1–5 (2016)

90. Abolhasanzadeh, B.: Nonlinear dimensionality reduction for intrusion detection using auto-encoder bottleneck features. In: IKT: IEEE 7th Conference on Information and Knowledge Technology, pp. 1–5 (2015)
91. Fiore, U., Palmieri, F., Castiglione, A., De Santis, A.: Network anomaly detection with the restricted Boltzmann machine. Neurocomputing **122**, 13–23 (2013)
92. Gao, N., Gao, L., Gao, Q., Wang, H.: An intrusion detection model based on deep belief networks. In: IEEE Second International Conference on Advanced Cloud and Big Data, pp. 247–252 (2014)
93. Hodo, E., Bellekens, X., Hamilton, A., Tachtatzis, C., Atkinson, R.: Shallow and deep networks intrusion detection system: a taxonomy and survey (2017). Arxiv preprint arXiv:1701.02145

Chapter 4
Aiming at Predictive Accuracy

4.1 Relationship of Generalization to the Separating Margin

Each classifier can be trained for suitable training data in such a way that all of them are classified correctly, i.e., the classification error is 0. For example, recall from Sect. 3.5 that a perceptron has this property on assumption that in training data, any two subsets assigned to different classes are linearly separable. Nevertheless, a small classification error on training data does not necessarily generalize to unseen test data. Sometimes, the zero error on training data is a consequence of the fact that the classifier has really well learned the underlying distribution governing the data, and then also the classification error on test data will be small. In other cases, however, it might be a consequence of overtraining (cf. Sect. 2.4), and then the error on test data can be quite large.

In Sect. 2.2, we have learned that for a particular sequence of test data, the classification error is computed as the proportion of misclassified data (2.12). Typically, however, we are interested not in the error for one particular sequence, but in the expected error for all data that may occur. Therefore, the generalization ability of the classifier is better represented by the expectation of $ER_\phi(X, Y, \phi(X))$, where X is a random vector with values in the feature set $\mathcal{X}$ and Y is a random variables with values in the set of classes C (cf. (2.51)). For the classification error, this expectation turns to

$$\mathbb{E}\,ER_{CE}(X, Y, \phi(X)) = P(\phi(X) \neq Y). \tag{4.1}$$

We will now have a closer look at that probability in the simplest case of binary classification fulfilling the linear separability assumption. Recall from Sect. 2.3 that this implies for the feature set $\mathcal{X}$ that $\mathcal{X} \subset \mathbb{R}^n$.

As usual, assume a sequence of training data $(x_1, c_1), \ldots, (x_p, c_p)$. Let x_+ and x_- be points in the feature set closest to the separating hyperplane in the training data assigned each to one of the two classes, and denoted in such a way that the

© Springer Nature Switzerland AG 2020
M. Holeňa et al., *Classification Methods for Internet Applications*,
Studies in Big Data 69, https://doi.org/10.1007/978-3-030-36962-0_4

hyperplanes H_+ and H_- that are parallel to the separating hyperplane and contain the points x_+ and x_-, respectively, fulfil

$$(\exists b_+, b_- \in \mathbb{R})\; b_+ > b_-,\; x_+ \in H_+ = \{x \in \mathbb{R}^n | x^\top w + b_+ = 0\},$$
$$x_- \in H_- = \{x \in \mathbb{R}^n | x^\top w + b_- = 0\}. \tag{4.2}$$

Due to (4.2), we will denote the class to which x_+ is assigned as $+1$, and the class to which x_- is assigned as -1. Hence,

$$C = \{1, -1\}, \tag{4.3}$$

$$c_k = \begin{cases} 1 & \text{if } x^\top w + b_+ \geq 0, \\ -1 & \text{if } x^\top w + b_- \leq 0, \end{cases} k = 1, \ldots, p. \tag{4.4}$$

As a consequence of (4.2) and (4.4), the sets $\{x_k | k = 1, \ldots, p, c_k = 1\}$ and $\{x_k | k = 1, \ldots, p, c_k = -1\}$ lay always in one of the halfspaces delimited by the hyperplanes H_+ and H_-, and at the same time, they have a nonempty intersection with the respective hyperplane and the empty intersection with the other hyperplane,

$$\{x_k | k = 1, \ldots, p, c_k = 1\} \cap H_+ \neq \emptyset,\; \{x_k | k = 1, \ldots, p, c_k = -1\} \cap H_- \neq \emptyset,$$
$$\{x_k | k = 1, \ldots, p, c_k = 1\} \cap H_- = \{x_k | k = 1, \ldots, p, c_k = -1\} \cap H_+ = \emptyset. \tag{4.5}$$

Therefore, the hyperplanes H_+ and H_- are called *supporting hyperplanes* of the sets $\{x_k | k = 1, \ldots, p, c_k = 1\}$ and $\{x_k | k = 1, \ldots, p, c_k = -1\}$, respectively.

Moreover, the distance between the hyperplanes H_+ and H_- is

$$\gamma_w = d(H_+, H_-) = \frac{b_+ - b_-}{\|w\|}. \tag{4.6}$$

This distance is usually called *margin* between the sets $\{x_k | k = 1, \ldots, p, c_k = 1\}$ and $\{x_k | k = 1, \ldots, p, c_k = -1\}$. Notice that it does not depend on where exactly between H_+ and H_- the separating hyperplane lies. In particular, (4.6) holds also if the separating hyperplane is the hyperplane H_w halfway between them, i.e.,

$$H_w = \{x \in \mathbb{R}^n | x^\top w + \frac{1}{2}(b_+ + b_-) = 0\}. \tag{4.7}$$

Introducing the notation

$$b = \frac{1}{2}(b_+ + b_-),\; \rho = b_+ - b_-, \tag{4.8}$$

(4.2)–(4.7) can be summarized as follows:

$$H_w = \{x \in \mathbb{R}^n | x^\top w + b = 0\}, \tag{4.9}$$

$$c_k = \begin{cases} 1 & \text{if } x^\top w + b \geq \frac{\rho}{2}, \\ -1 & \text{if } x^\top w + b \leq -\frac{\rho}{2}, \end{cases} \quad k = 1, \ldots, p, \tag{4.10}$$

$$\gamma_w = \frac{\rho}{\|w\|}. \tag{4.11}$$

Finally, notice that (4.9)–(4.11) can be further simplified in two respects:

- According to (4.2) and (4.6), γ_w is invariant with respect to scaling of w, i.e., $\gamma_{cw} = \gamma_w$ for $c > 0$. Therefore, we can restrict attention to hyperplanes with a unit-length normal vector, $\|w\| = 1$, if it is suitable.
- Any change in b is equivalent to a suitable translation of the coordinate system. Therefore, we can restrict our investigations to the translated system corresponding to $b = 0$ if it is suitable, and only subsequently transform the results to the original coordinate system.

First, suppose that the distributions μ_+ and μ_- of feature vectors $x \in \mathscr{X}$ assigned to the classes 1 and -1 are continuous isotropic, with densities fixed with respect to the hyperplanes H_+ and H_-, respectively. For simplicity, think of normal distributions $\mu_+ = N(c_+, \sigma_+^2 I_n)$, $\mu_- = N(c_-, \sigma_-^2 I_n)$, where $c_+, c_- \in \mathbb{R}^n$, $\sigma_+, \sigma_- > 0$, I_n is the n-dimensional identity matrix, and c_+ and c_- are in fixed distances from the hyperplanes H_+ and H_-, respectively, always in the opposite halfspace than the hyperplane H_w. Then a classifier $\phi = \phi_w$ based on the hyperplane H_w, i.e.,

$$\phi(x) = \phi_w(x) = \begin{cases} 1 & x \in \mathscr{X}, x^\top w + b \geq \rho, \\ -1 & x \in \mathscr{X}, x^\top w + b < \rho, \end{cases} \tag{4.12}$$

has the expected classification error (4.1)

$$\mathbb{E}\, \mathrm{ER}_{CE}(X, Y, \phi(X)) = \mu_+(\mathscr{H}_-) + \mu_-(\mathscr{H}_+), \tag{4.13}$$

where $\mathscr{H}_+$, $\mathscr{H}_-$ are the two open halfspaces delimited by the hyperplane H_w, with $H_+ \subset \mathscr{H}_+$, $H_- \subset \mathscr{H}_-$. Due to the isotropy of μ_+ and μ_-, the error (4.1) depends only on the margin and decreases with increasing margin.

However, a similar result can be obtained even with simultaneous validity of two very weak assumptions, namely:

(i) $\boldsymbol{\Phi} = \{\phi_w : \mathscr{X} \to \{1, -1\} | w \in \mathbb{R}^n, \|w\| = 1, (\forall x \in \mathscr{X})\, \phi_w(x) = 2\Theta(x^\top w) - 1\}$ where Θ is the Heaviside step function (3.89);

(ii) $R = \sup_{x \in \mathscr{X}} \|x\| > 0$.

With these assumptions, it can be shown [1] that for each $\delta \in (0, 1)$ and each $w \in \boldsymbol{\Phi}$, a random sample $(X_1, Y_1), \ldots, (X_p, Y_p)$ with probability at least $1 - \delta$ fulfils

$$(\exists \eta > 0) \; \mathbb{E} \, \mathrm{ER}_{\mathrm{CE}}((X, Y, \phi_w(X)) | (X_1, Y_1), \ldots, (X_p, Y_p)) < \frac{1}{p} |\{k | d(X_k, H_w) <$$

$$< \frac{\gamma_w}{2}\}| + \sqrt{\frac{\eta}{p}\left(\frac{R^2}{\gamma_w^2} \ln^2 p - \ln \delta\right)}. \qquad (4.14)$$

Hence, though the error (4.14) now depends on the choice of δ, it for each δ with probability $1 - \delta$ again decreases with increasing margin.

4.2 Support Vector Machines for Linear Classification

4.2.1 Optimization Task for the Minimal Expected Error

Consider a random sample $(X_1, Y_1), \ldots, (X_p, Y_p)$ such that in the event of (4.14) fulfilled, also

$$|\{k | d(X_k, H_w) < \frac{\gamma_w}{2}\}| = 0, \qquad (4.15)$$

holds. Then in that event, minimizing $\mathbb{E} \, \mathrm{ER}_{\mathrm{CE}}((X, Y, \phi(X)) | (X_1, Y_1), \ldots, (X_p, Y_p))$ is equivalent to maximizing γ_w. Due to the invariance of $d(X_k, H_w), k = 1, \ldots, p$ with respect to the scaling of w, which entails the already recalled invariance of γ_w with respect to such scaling, this is true not only if $\|w\| = 1$, but for any $w \in \mathbb{R}^n$. Taking into account (4.10)–(4.11), then in the coordinate system corresponding to $b = 0$, the following optimization task has to be tackled:

$$\text{maximize } \frac{\rho}{\|w\|}. \qquad (4.16)$$

with constraints

$$|x_k^\top w| \geq \frac{\rho}{2}, k = 1, \ldots, p. \qquad (4.17)$$

Notice that, according to (4.10),

- if $x_k^\top w \geq \frac{\rho}{2}$, then $|x_k^\top w| = x_k^\top w = c_k x_k^\top w$,
- if $x_k^\top w \leq -\frac{\rho}{2}$, then $|x_k^\top w| = -x_k^\top w = c_k x_k^\top w$.

Therefore, the constraints (4.17) can be rewritten without absolute values, as

$$c_k x_k^\top w \geq \frac{\rho}{2}, k = 1, \ldots, p. \qquad (4.18)$$

We will solve the maximization task (4.16) in two phases. In the first, we will hold ρ fixed, and will perform the minimization of $\|w\|$, more precisely, the equivalent minimization of $\|w\|^2$, which is faster through leaving out the square root in (3.3). In the second phase, we will then add the maximization of ρ.

From the theory of non-linear optimization [2], it is known that a sufficient condition for a $w^* \in \mathbb{R}^n$ to minimize $\|w\|^2$ with the constraints (4.18) is the existence of *Lagrange multipliers* $\alpha_k^* \geq 0, k = 1, \ldots, p$ fulfilling the *Karush-Kuhn-Tucker (KKT) conditions*,

$$\alpha_k^* \left(\frac{\rho}{2} - c_k x_k^\top w^* \right) = 0, \, k = 1, \ldots, p, \tag{4.19}$$

and such that the *Lagrange function* $L : \mathbb{R}^n \times \mathbb{R}_{\geq 0}^p \to \mathbb{R}$ defined

$$L(w, \alpha) = \|w\|^2 + \sum_{k=1}^p \alpha_k \left(\frac{\rho}{2} - c_k x_k^\top w \right), \, w \in \mathbb{R}^n, \alpha = (\alpha_1, \ldots, \alpha_p) \tag{4.20}$$

has in (w^*, α^*) a saddle point, i.e.,

$$L(w^*, \alpha^*) = \min_w L(w, \alpha^*) = \max_\alpha L(w^*, \alpha). \tag{4.21}$$

Instead of the optimization task $L(w^*, \alpha^*) = \min_w L(w, \alpha^*)$, we will actually solve a family of optimization tasks

$$L(w_\alpha, \alpha) = \min_w L(w, \alpha) \tag{4.22}$$

for α from some neighbourhood of α^*. Then we obtain w^* as $w^* = w_{\alpha^*}$. Each task (4.22) implies

$$\nabla_w L(w_\alpha, \alpha) = 0, \, \text{i.e.,} \, 2w_\alpha - \sum_{k=1}^p \alpha_k c_k x_k = 0, \tag{4.23}$$

which yields

$$w_\alpha = \frac{1}{2} \sum_{k=1}^p \alpha_k c_k x_k, \tag{4.24}$$

in particular,

$$w^* = \frac{1}{2} \sum_{k=1}^p \alpha_k^* c_k x_k. \tag{4.25}$$

Putting this back to (4.20), we get $L(w_\alpha, \alpha)$ expressed using only the Lagrange multipliers,

$$L(w_\alpha, \alpha) = -\frac{1}{4} \sum_{j,k=1}^p \alpha_j \alpha_k c_j c_k x_j^\top x_k + \frac{\rho}{2} \sum_{k=1}^p \alpha_k. \tag{4.26}$$

Similarly, the KKT conditions (4.19) turn to

$$\alpha_k^*(\rho - c_k \sum_{j=1}^{p} \alpha_j^* c_j x_j^\top x_k) = 0, k = 1, \ldots, p. \tag{4.27}$$

What now remains to be done is to solve the maximization task from (4.21), i.e., $L(w^*, \alpha^*) = \max_\alpha L(w^*, \alpha)$, which is called *dual optimization task* to the original minimization of $\|w\|^2$ with the constraints (4.18). This is now possible due to $w^* = w_{\alpha^*}$ and due to (4.26), expressing L by means of α in some neighbourhood of α^*. Moreover, we make use of the fact that the dual optimization task is a maximization task, and move to the second phase of solving the maximization task (4.16), when also the maximization of ρ is added. Hence, we finally solve the maximization task

$$(\alpha^*, \rho^*) = \arg\max_{(\alpha, \rho)} -\frac{1}{4} \sum_{j,k=1}^{p} \alpha_j \alpha_k c_j c_k x_j^\top x_k + \frac{\rho}{2} \sum_{k=1}^{p} \alpha_k \tag{4.28}$$

on conditions (4.27) and $\alpha_1, \ldots, \alpha_p \geq 0, \rho > 0$. Notice that the function maximized in (4.28) is a quadratic function in α and ρ. Consequently, (4.28) is a quadratic optimization task, which is very easy to solve because every quadratic function can have only one local maximum, which is then already its global maximum.

4.2.2 Support Vectors and Their Role

Once the quadratic optimization task (4.28) is solved, we leave the coordinate system corresponding to $b = 0$ and transform the results to the original coordinate system. In particular, the KKT conditions (4.19) transform to

$$\alpha_k^*(\frac{\rho}{2} - c_k(x_k^\top w^* + b)) = 0, k = 1, \ldots, p. \tag{4.29}$$

Denote

$$\mathscr{S} = \{x_k | \alpha_k^* > 0\}. \tag{4.30}$$

Then according to the KKT conditions (4.29), the input vectors $x_1, \ldots, x_p$ from the training data fulfil

$$x_k \in \mathscr{S} \Rightarrow c_k(x_k^\top w^* + b) = \frac{\rho}{2}, \tag{4.31}$$

and taking into account (4.2), (4.7) and (4.8), they fulfil also

$$x_k \in \mathscr{S} \Rightarrow x_k \in H_+ \cup H_-. \tag{4.32}$$

Fig. 4.1 Support vectors
and their relationship to
margin between linearly
separable classes

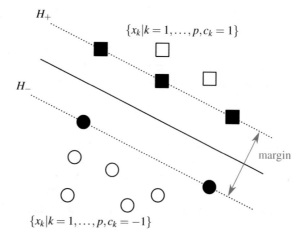

Hence, the set $\mathscr{S}$ is a subset the nonempty intersection of the training data with the supporting hyperplanes, mentioned in Sect. 4.1,

$$\mathscr{S} \subset (\{x_k | k = 1, \ldots, p, c_k = 1\} \cap H_+) \cup (\{x_k | k = 1, \ldots, p, c_k = -1\} \cap H_-).$$
(4.33)

Therefore, the elements of $\mathscr{S}$ are called *support vectors*. Usually, only a small fraction of the feature vectors x_k forming the input parts of the training data $(x_1, c_1), \ldots, (x_p, c_p) \in \mathscr{X} \times \{0, 1\}$ are support vectors (cf. Fig. 4.1).

Notice that if $x_k \notin \mathscr{S}$, then $\alpha_k^* = 0$. This turns (4.25) to

$$w^* = \frac{1}{2} \sum_{x_k \in \mathscr{S}} \alpha_k^* c_k x_k, .$$
(4.34)

Once w^* is known, then b can be obtained, due to (4.30), for any $x_k \in \mathscr{S}$ as

$$b = \frac{\rho}{2} - \frac{1}{2} \sum_{x_j \in \mathscr{S}} \alpha_j^* c_j x_k^\top x_j.$$
(4.35)

Finally, the definition (4.12) of the classifier ϕ based on the hyperplane H_w turns to

$$\phi(x) = \phi_w(x) = \begin{cases} 1 & x \in \mathscr{X}, \sum_{x_k \in \mathscr{S}} \alpha_k^* c_k x^\top x_k + b \geq 0, \\ -1 & x \in \mathscr{X}, \sum_{x_k \in \mathscr{S}} \alpha_k^* c_k x^\top x_k + b < 0, \end{cases}$$
(4.36)

i.e., ϕ is determined, among all the inputs from the training data, solely by the support vectors. It is this property due to which such classifiers are called *support vector machines (SVM)*.

4.3   Support Vector Machines for Non-linear Classification

If the sets $\{x_k | k = 1, \ldots, p, c_k = 1\}$ and $\{x_k | k = 1, \ldots, p, c_k = -1\}$ are not linearly separable, the general method of mapping them into a higher-dimensional space explained in Sect. 2.3 can be used. Recall that this space is (2.42)

$$\mathscr{L} = \text{span}(\{\kappa(\cdot, x_1), \ldots, \kappa(\cdot, x_p)\}), \tag{4.37}$$

where κ is some kernel (2.30), and recall also that for $p \geq n$ and $x_1, \ldots, x_p$ real-valued features sampled from a continuous distribution, the dimension of $\mathscr{L}$ is p with probability 1, and the sets $\{\kappa(\cdot, x_k) | k = 1, \ldots, p, c_k = 1\}$ and $\{\kappa(\cdot, x_k) | k = 1, \ldots, p, c_k = -1\}$ are linearly separable. Thus in that space, a support vector machine ϕ can be constructed in the way described in the previous section. Taking into account the definition (2.43) of the scalar product in the space $\mathscr{L}$, in particular that

$$\langle \kappa(\cdot, x_k), \kappa(\cdot, x_\ell) \rangle = \kappa(x_k, x_\ell) \text{ for } k, \ell = 1, \ldots, p, \tag{4.38}$$

the final results of that construction, (4.34)–(4.36), look in that space as follows:

$$w^* = \frac{1}{2} \sum_{x_k \in \mathscr{S}} \alpha_k^* c_k \kappa(\cdot, x_k), \tag{4.39}$$

$$b = \frac{\rho}{2} - \frac{1}{2} \sum_{x_j \in \mathscr{S}} \alpha_j^* c_j \kappa(x_k, x_j) \text{ for any } x_k \in \mathscr{S}, \tag{4.40}$$

$$\phi(x) = \begin{cases} 1 & x \in \mathscr{X}, \sum_{x_k \in \mathscr{S}} \alpha_k^* c_k \kappa(x, x_k) + b \geq 0, \\ -1 & x \in \mathscr{X}, \sum_{x_k \in \mathscr{S}} \alpha_k^* c_k \kappa(x, x_k) + b < 0. \end{cases} \tag{4.41}$$

If the number p of training data is high, then computing the prediction $\phi(x)$ of a support vector machine ϕ for a particular feature vector x according to (4.36) consists predominantly in computing the scalar products with the training feature vectors $x_1, \ldots, x_p$. In that respect, it is very important that in the non-linear case (4.41), the computation of the scalar product between the images $\kappa(\cdot, x_1), \ldots, \kappa(\cdot, x_p)$ of the training data in the space $\mathscr{L}$ actually reduces, due to (4.38), to the application of the kernel κ to the original training data $x_1, \ldots, x_p$. The reason for the importance of that property is that in such cases, the dimension n of the original training data is usually much lower than the dimension p of the space $\mathscr{L}$, $n << p$. This property is often called the *kernel trick*.

Finally, notice that the classifier ϕ is again determined, among all the inputs from the training data, solely by the support vectors. In Fig. 4.2, support vectors are indicated in the context of the SVM classification of the graphical objects on web pages into advertisements and non-advertisements (cf. [3]).

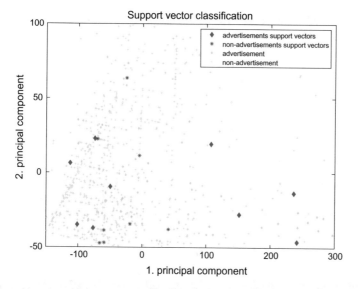

Fig. 4.2 The 1. and 2. principal component of a support vectors in the context of the SVM classification of the graphical objects on web pages into advertisements and non-advertisements (cf. [3])

4.3.1 Extension to Multiple Classes

Various extensions of support vector machines for the case that the classification is into more than two classes have been proposed. By far the most often used are the two most intuitive ones:

1. *Pairwise binary.* For each pair of classes $c_j, c_k, j \neq k$, we construct a support vector machine $\phi_{j,k} : \mathscr{X} \to \{c_j, c_k\}$ in the way described above. The resulting classifier $\phi : \mathscr{X} \to C$ is then defined to classify to the class, or more precisely one of the classes that occurred most frequently among the results of those $\frac{m(m-1)}{2}$ binary classifiers,

$$(\forall x \in \mathscr{X}) \, |\{(j,k)|j,k = 1, \ldots, p, j \neq k, \phi_{j,k}(x) = \phi(x)\}| =$$
$$= \max_{c \in C} |\{(j,k)|j,k = 1, \ldots, p, j \neq k, \phi_{j,k}(x) = c\}|.$$

$$(4.42)$$

2. *One against the rest.* For each class $c \in C$, we construct a support vector machine

$$\phi_c : \mathscr{X} \to \{c, c_\neg\}, \text{ where } c_\neg = C \setminus \{c\}.$$

$$(4.43)$$

If several such binary classifiers ϕ_c classify into their respective class c, then we choose the one among them for which its margin γ_c is maximal,

$$\gamma_{\phi(x)} = \max_{c \in C} \gamma_c, x \in \mathscr{X}. \tag{4.44}$$

4.3.2 Extension Including Noise Tolerance

Frequently, the values of features are distorted by noise. As a consequence of that distortion, the margin between the sets $\{x_k | k = 1, \ldots, p, c_k = 1\}$ and $\{x_k | k = 1, \ldots, p, c_k = -1\}$ decreases. In particular, if we assume an additive impact of the noise on the decrease of the margin, then the inequality constraints (4.18) change to

$$c_k x_k^{\top} w \geq \frac{\rho}{2} - \xi_k, \xi_k \geq 0, k = 1, \ldots, p. \tag{4.45}$$

Here, the additional variables $\xi_1, \ldots, \xi_p$ are called *slack variables*. Needless to say, it is desirable that their values are as small as possible, or equivalently, that the sum of their values is as small as possible. This is an additional objective for optimization, which has to be addressed simultaneously to the maximization of the margin (4.16).

That combined optimization task can be approached in two principally different ways: either as a multi-objective optimization task with the objectives maximal margin and minimal $\xi_1 + \cdots + \xi_p$, or as a single-objective optimization task somehow combining both objectives into a new one. We will pursue the second approach here, replacing the minimization of $\|w\|^2$ forming the first step of the maximization task (4.16) with the task

$$\text{minimize } \|w\|^2 + C \sum_{k=1}^{p} \xi_k, C > 0, \tag{4.46}$$

where the constant C must be chosen so as to express the relative importance of the sum $\xi_1 + \cdots + \xi_p$ with respect to the margin.

The minimization (4.46) with constraints (4.45) proceeds in a similar way to the minimization of $\|w\|^2$ with constraints (4.18). The Lagrange function is

$$L(w, \xi, \alpha, \beta) = \|w\|^2 + C \sum_{k=1}^{p} \xi_k + \sum_{k=1}^{p} \alpha_k (\frac{\rho}{2} - c_k x_k^{\top} w - \xi_k) - \sum_{k=1}^{p} \beta_k \xi_k,$$

$$w \in \mathbb{R}^n, \xi = (\xi_1, \ldots, \xi_p) \in \mathbb{R}^p, \alpha = (\alpha_1, \ldots, \alpha_p) \in \mathbb{R}^p_{\geq 0}, \beta = (\beta_1, \ldots, \beta_p) \in \mathbb{R}^p_{\geq 0},$$

$$\tag{4.47}$$

where α and β are the vectors of Lagrange multipliers corresponding to the conditions $c_k x_k^{\top} w \geq \frac{\rho}{2} - \xi_k$ and $\xi_k \geq 0$, respectively. The KKT conditions extend to

$$\alpha_k^* (\frac{\rho}{2} - c_k x_k^{\top} w^*) = \beta_k \xi_k = 0, k = 1, \ldots, p. \tag{4.48}$$

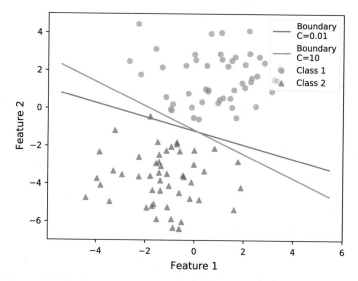

Fig. 4.3 Effect of C on the placement of the decision boundary in a linear support vector machine

The condition $\nabla_w L(w_\alpha, \alpha) = 0$ leads again to (4.24), whereas the analogous condition for the slack variables, $\nabla_\xi L(w_\alpha, \alpha) = 0$ leads to

$$\alpha = C - \beta. \tag{4.49}$$

Consequently, the set $\mathscr{S}$ of support vectors (4.30), involved in the final classifier defined by (4.36) in the linear and (4.41) in the non-linear case, fulfils the following additional implications:

$$\xi_k > 0 \Rightarrow \beta_k = 0 \Rightarrow \alpha_k = C \Rightarrow \alpha_k \in \mathscr{S}, k = 1, \dots, p. \tag{4.50}$$

4.3.3   SVM in Spam Filtering

Support vector machines seem to be an ideal choice for spam detection because the SVM decision boundary is based only on a small subset of available training data—the support vectors. The selected decision boundary should be, therefore, similar if we train an SVM classifier on different random samples from the same distribution. Assumed that the properties extracted from spam messages do not change rapidly over time, SVM should achieve the best accuracy among all classifiers.

According to comparisons mentioned in [4], SVM was able to outperform k-nn and Naïve Bayes as well as rule-based approaches that use keywords or SMTP-path.

However, support vector machines denote a family of classifiers, where the selection of the kernel changes the shape of the decision boundary significantly. Classification of spam messages based on TF-IDF (1.1) of words in message subject and body seems to be a linearly separable problem. Accordingly, almost all of the existing approaches based on message content use a linear SVM classifier.

Quite unexpectedly, SVMs have a higher testing accuracy and a lower FPr (fewer ham messages classified as spam) if we use only a binary document representation (i.e. only the information whether the word from the dictionary is present in the document or not). This behaviour is not detected only on the SpamAssassin dataset we use for our examples but was also documented in [5] on the messages collected from AT&T. However, this behaviour is not transferable to other classifiers. For example, the use of binary features in perceptron classifier leads to faster training, but the accuracy on test data is lower.

Another important decision that influences the testing accuracy is the selection of the trade-off parameter C in (4.46). Setting C to low values maximises the margin between the two classes while accepting a small training error. Selecting a high value of C allows a smaller margin but penalizes the training error. For effect of such trade-off see Fig. 4.3. In content-based spam filtering, setting a higher C leads to higher testing accuracy.

A problem in spam filtering can be long training times of SVMs because spam filters should be able both to train on large quantities of messages and simultaneously to react swiftly to changes in the spam messages or corrections made by users. To this end, either parallelization of the problem or relaxation of initial requirements is utilized.

An approach presented in [6] utilizes the MapReduce approach: It splits the training data to individual map nodes that propose a set of support vectors resulting from linear SVM and forwarding that set to a reducing node that aggregates them. As the authors conclude, this may lead to a noticeable accuracy degradation unless special care is taken during the splitting of the data to individual map nodes. However, the approach brings a substantial speed improvement.

The approach in [7] utilizes a relaxed *online training*. The transformation to online SVMs mitigates the problem of retraining the whole classifier from scratch on arrival of new training samples by using the already computed weights and bias as a seed of the optimisation task. However, this does not reduce the computational complexity increasing with the number of samples as the optimisation task is always performed on all data seen. Relaxation of the training process is proposed in three aspects: The problem size is reduced to a fixed number of 10,000 last seen samples, which also allows the classifier to better reflect a possible concept drift. The classifier retraining is not executed if the new sample is classified correctly and outside the 80% central part of the margin. And the optimisation task is limited to a single iteration. All of these changes reduce the computational complexity significantly. However, the authors warn that these relaxations are possibly allowed only because of the high linear separability on spam data which may be subject to change.

Due to the robustness of the SVM classifier, and with a significant incentive to speed up the classification of incoming messages, some of the approaches, documented in [8], limit their scope only to the message header. The involved features include binary information, such as whether SMTP server domain address exists, has a legal IP address, the sender address is legal and the same for the Return-Path and Reply-To fields, dates are valid and X-Mailer field contains an existing software identifier. Numerical features include the number of relay servers listed, number of message receivers and time span from submitting the message to receiving it on a destination message server.

As was mentioned in Sect. 1.1, some of the spam messages contain text hidden inside images, commonly containing rendered text that may or may not be possible to automatically transcribe with optical character recognition systems. However, instead of using a computationally demanding OCR, [9] proposed to use only the spam-indicative global image features, such as the fraction of image occupied by detected text regions, image saturation and colour heterogeneity (variability is the colour information from a picture quantized to a significantly smaller number of indexed colours). Based on these features, an SVM classifier recognized with a promising precision whether the image belongs to spam or one of four considered non-spam classes.

4.3.4 SVM in Recommender Systems

While even the most elemental support vector machines excel in text classification tasks, the use in recommender systems is a bit more challenging. As we presented in previous examples on recommender systems, the recommendations are commonly based on a user-item matrix (containing explicit rating or implicit records of interactions) which is usually very sparse. However, a dense matrix is favoured for the training of a model. Opposed to the tasks in text classification, where filling empty values with zeroes is meaningful (the word is not present in the document), passing zeros to the undefined elements of the user-item matrix would denote that the user provided a worst possible rating for many items, which is undesirable. On the other hand, passing an average value diminishes the power of the model and results in a very complex popularity predictor—a recommender that does not utilise the context of the individual user and gives everyone the same prediction.

To implement the context awareness to support vector machines, we need to take into account a method of neighbourhood formation. Memory-based methods, such as k-nn, utilise a distance between individual users and predict a score only from a set of most similar users. In a model based classifier, such as SVM, we might take a somewhat similar approach: train a separate model for each user, take similarity of those models and propose rating of new samples based on the most similar models of other users. For example, the model similarity defined in [10] is based on the number of training samples from one model that are identically classified also by the second model and vice versa:

$$sim(\phi_x, \phi_y) = \frac{1}{2} \left(\frac{|\phi_x(x) = \phi_y(x)|}{|x|} + \frac{|\phi_y(y) = \phi_x(y)|}{|y|} \right),$$ (4.51)

where ϕ_x is the model trained on data x.

The selection of recommendation candidates is then based on positive training samples of users with a model similarity above a defined threshold.

Apart from the model similarity approach, support vector machines can be based directly on the user-item ranking matrix. However, we need to tackle the issue of converting the score into a classification space suitable for SVM. To this end, each user rating can be represented by as many instances as the number of the possible rating values, such as proposed in [11]. In this scheme, the value of one is assigned only to the data point belonging to the combination of the assigned score for a given item and user. Rest is filled with zeros; including the items not scored by the user. The classification itself is then usually carried out by a one-against-rest scheme. This approach alone tends to yield worse results than k-nn on benchmark datasets. However, it seems to be beneficial in case of very sparse training data.

A somewhat better method for densifying the user-item matrix is presented in [12]. This heuristic approach assigns random scoring to missing elements of the matrix (the cited article uses only two classes) and repeats the steps of building a classifier for each item from ratings of other users and storing the new prediction until the accuracy on test data changes significantly. Combined with a smoothed SVM, introduced in [13], this heuristic achieved a significant accuracy boost over simpler SVM models based on either users or items.

4.3.5 SVM in Malware and Network Intrusion Detection

Support vector machines have been used for malware detection on the two platforms most frequently used for personal computers—Windows and OS X, as well as on the most common smartphone platform—Android. In [14], an SVM classifier has been used in connection with dynamic malware analysis of Windows executables. As features, n-grams of *Intel opcodes* have been used, more precisely, their empirical distributions in the analysed executable files. Hence, for each considered opcode, resp. n-gram of opcodes, the ratio of its frequency in the analysed executable to the frequency of all considered opcodes, resp. n-grams of opcodes, has been recorded. Among the 344 available Intel opcodes, 149 have been considered, including all important kinds of operations: arithmetic, logical, memory manipulations and program flow control (e.g., comparison, jump to a function and return from it). To reduce the number of obtained n-grams, principal component analysis has restricted the considered feature space to the subspace spanned by the principal components corresponding to the 8 largest eigenvalues of the empirical covariance function, which have accounted for 99.5% of the overall empirical variance. The research reported in [14] shows that most discriminative between malware and benign software are less frequent opcodes. On the other hand, the most frequent opcode *mov* is not only a poor discriminator, but if used in n-grams with less frequent opcodes, it inhibits their discriminative ability.

The empirical distribution of opcodes has been obtained in [14] through debugging the analysed exectubale for 3 minutes, ensuring that not only the loading and unpacking phases were recorded, but also that malicious activity occurred. A crucial role plays the employed debugger, named *Ollydbg*, which has two properties very important for dynamic malware analysis:

(i) It can nearly always deal with packing and encryption correctly.
(ii) For its debuggee, it is complicated to identify that it is being debugged. That property is very important from the point of view of malware detection because malware programs often attempt to detect whether they are debugged, and if they are, they switch into an evade mode, in which they change their behaviour, thus preventing a correct malware detection.

An interesting SVM-based approach to the detection of Windows malware has been presented in [15]. It has been developed specifically for *metamorhpic malware*, i.e., advanced mutating malware that changes its structure in each iteration. These changes may contain a small number of instructions for a specific functionality, or enclose multiple instructions to perform similar functionality. In each mutation, these instructions are expanded or minimized according to obfuscation techniques used. The approach proposed in [15] is based on static malware analysis. To avoid the necessity to disassemble the checked binary files, it extracts directly from them n-grams as features for classification. The feasibility of extracting features for metamorhpic malware detection directly from binaries relies on the assumption that most versions of different mutations of the same malware share a combination of several unchanged code segments. In particular, overlapping 4-grams have been extracted. Among the huge number of such n-grams, however, only 500 have been used—those that appeared most effective from the point of view of malware detection.

In [15], two important aspects of this approach to metamorhpic malware have been investigated:

1. The *kind of SVM*. Linear SVMs were considered, as well as nonlinear SVMs with Gaussian and polynomial kernels. Among all of them nonlinear SVMs with Gaussian kernels were the best in terms of accuracy, TPr, TNr, FPr and FNr.
2. The *proportion of metamorhpic malware* in the training data. Highest accuracy was achieved for the proportion of 10–50%. However, even the lower end of this range is higher than the malware incidence in real-world data. On the other hand, if the proportion of metamorhpic malware in the training data was increasing above 50%, then accuracy was decreasing.

Those investigations were performed using a set of 1020 metamorphic malware files and 2330 benign files. The malware files were obtained mostly artificially through mutations of original software, benign software were primarily system files from Windows 7 and Windows XP. The best SVM with a Gaussian kernel, trained on data containing 10% malware and 90% benign files, achieved test accuracy 99.69%. The advantage of an approach developed specifically for this kind of malware became

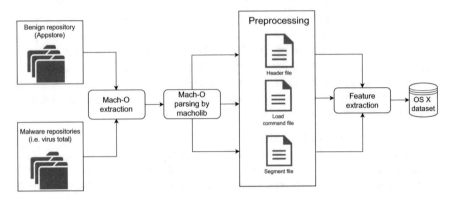

Fig. 4.4 Malware detection in the MAC OS X operating system, proposed in [16]

obvious from a comparison with several commercial anti-virus tools that the authors performed on the same test data. The best among the compared software, Kaspersky, achieved an accuracy less than 31%, and the others were even noticeably worse, e.g., AVG less than 18%, Avast less than 7%.

In [16], SVMs of several kinds have been used for malware detection in the context of the MAC OS X operating system. To this end, features obtained through static malware analysis from main parts of *Mach-O binary files* have been used (Fig. 4.4):

- *header*, e.g., CPU type, number and size of commands;
- *load commands*, e.g., imported and exported libraries, symbol table;
- *segments*, the largest part of a Mach-O file, containing application code, data and related text.

All the real-valued features used in [16] have been normalized to [0, 1]. In addition, some of them had missing values. This has been tackled using k-nearest neigbours imputation: the missing value $[x_0]_i$ of the i-th feature of a feature vector x_0 is estimated as $\frac{1}{k}\sum_{j=1}^{k}[x_j]_i$, where $x_1, \ldots, x_k$ are the k-nearest neighbours of x_0 whose $[x_j]_i$ is not missing.

The approach proposed in [16] has been validated on 152 malware Mach-O files publicly available in the internet, which have been complemented by 460 benign files from the Apple App Store, following the recommnedation in [17] that the number of benign files should be at least three times higher than the number of malware files. The benign files have been selected randomly from a set of apps obtained through putting together the top 100 apps of each of the categories Utilities, Social Network, Weather, Video and Audio, Productivity, Health and Fitness, and Network. To increase the amount of data for validation, three additional datasets with malware and benign Mach-O files were prepared from the original dataset through resampling, using the resampling method proposed in [18]. The sizes of those additional datasets were the double, triple and quintuple size of the original dataset. Apart from the considered SVM kinds, the validation included also a naïve Bayes classifier, a Bayesian network, a multilayer perceptron, and a classification tree—a kind of classifier that will be presented later in Sect. 5.3. The main results of that validation have been as follows:

1. The highest accuracy among all considered classifiers was achieved by a non-linear SVM with a Gaussian kernel: 91%. The accuracy of a non-linear SVM with a polynomial kernel was only 88%, whereas the accuracy of the linear SVM was 89%.
2. The lowest FPr among all considered classifiers was achieved by the considered SVM with a polynomial kernel: 3%. The FPr of the considered non-linear SVM with a Gaussian kernel was 3.9% and the FPr of the linear SVM was 4.1%.
3. All the considered SVMs had a lower FPr than all the considered non-SVM classifiers, and a higher accuracy than all of them except the classification tree, which had an accuracy slightly higher than the considered SVM with a polynomial kernel: 88.1%.
4. The most important part of a Mach-O binary file from the point of view of discriminating between malware and benign software were load commands. In particular, the libraries CoreGraphics, CoreLocation, CoreServices and Webkit were called a lot of more often in benign applications, whereas libc and libsqlite were called much more often by malware.

Both for Windows and for Android has been developed the SVM-based *network security management platform (NSMP)* presented in [19]. It has been developed primarily against malware infection through e-mail. It uses both static and dymamic malware analysis (Fig. 4.5). More precisely, dynamic analysis relying on behavioral modeling of sequences of events recorded in a sandbox is used if the message payload, i.e., the e-mail body is executable. Otherwise, static analysis relying mainly on decompilation and system API calls analysis is used.

Apart from malware detection, the NSMP aims also at infection prevention and at defence against the detected malware, called *digital vaccine* in [19]. Altogether, the platform has the following main components (Fig. 4.6):

(i) *Automatic system backup* of crucial system files on servers and client hosts.
(ii) *System monitoring* includes detecting changes in system files and registries, reporting to relevant networks and blacklisting infected IPs. In this way, it also contributes to infection prevention.
(iii) *System recovery* is the core of the digital vaccine, repairing the system after an infection.

On the other hand, only for Android malware detection have been developed the approaches introduced in [20–22]. In [20], an SVM with Gaussian kernel has been used in combination with static malware analysis. As features resulting from the static analysis, 14,233 function calls obtained from disassembled executables have been considered, as well as 104 requestable permissions. Among all those features, first a preselection has been made, according to the *information gain* of the feature, i.e., the entropy decrease of the probability distribution of classes caused by taking the feature into account. According to this ordering, the top 3 features were the permissions "send sms", "receive sms" and "read sms". The preselected features

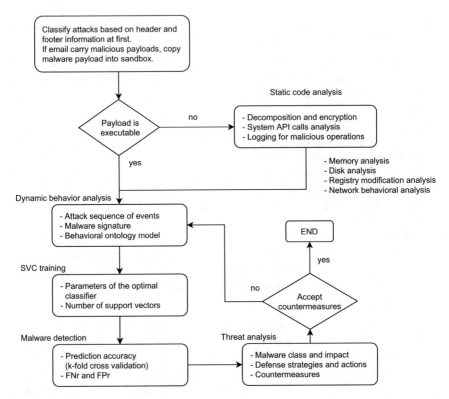

Fig. 4.5 Detection of malware infection through e-mails, according to the NSMP platform (inspired by [19])

have then been weighted with the objective to maximize the accuracy on test data of an SVM trained on the weighted features. To this end, *particle swarm optimization (PSO)* has been employed, an evolutionary optimization method inspired by the collective behaviour of swarms [23].

The proposed approach has been tested on 2,125 Android applications in [20], among which 1,016 were malware and 1,109 benign. That data has been also used to investigate the most suitable proportion of preselected features and to compare the employed SVM with with alternative classifiers:

- Preselection was tested for eight different proportions of features: 0.5, 1, 2.5,5, 10, 25, 50 and 100%, and the AUC has been measured for each of them. The highest AUC was achieved for 5% preselection and that value has subsequently been used in the comparison with other classifiers. Interestingly, the lowest AUC was achieved if 100% of features were used, i.e., without preselection.
- For the 5% preselected features with weights optimized by PSO, the testing accuracy of the employed SVM was 96.1%. For comparison, the classification into malware and benign applications was on the same conditions performed also by

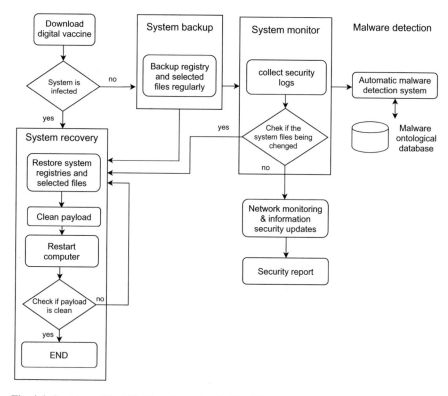

Fig. 4.6 Structure of the NSMP platform (inspired by [19])

several other classifiers, and all of them achieved a lower accuracy, for example, a logistic regression 95.8%, a *k*-nearest neighbours classifier 91.5%, and a naïve Bayes classifier only 84.3%.

The approach using linear SVMs presented in [21] relies on dynamic malware analysis. The features resulting from that analysis are *n*-grams of system calls (Fig. 4.7). In [21], an investigation of the influence of the *n*-gram size on SVM performance has been reported, for *n*=1–6. That investigation has been based on 102 Android malware application publicly available in the internet and by 100 benign applications from the Google Play. The highest accuracy on those data, 96%, and at the same time the lowest FPr, 5%, has been achieved using 3-grams. Moreover, they were not only most accurate, but the SVM accuracy was from 1-grams to 3-grams always increasing, whereas from 3-grams to 6-grams always decreasing.

The authors of [22] presented an SVM-based malware detection using a combination of static and dynamic malware analysis. An important propety of their approach is that the static analysis, relying on permissions and on constant strings from the binary file of the application, is performed between downloading the application and installing it. Hence, if a malicious application is detected already due to the

Fig. 4.7 Android malware detection using *n*-grams of system calls (inspired by [21])

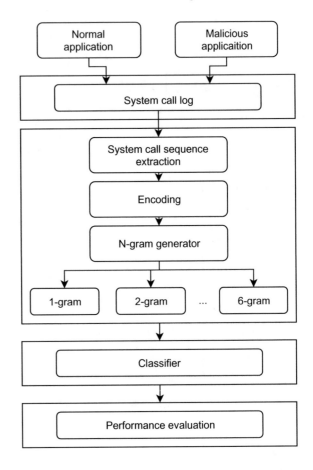

results of static analysis, its installation is blocked, avoiding the security risk for the smartphone.

Finally, an intrusion detection system based on support vector machines has been presented in [24], denoted by the authors as *DCG-IDPS (Dynamic Correlation-Based Intrusion Detection and Prevention System)*. It is intended for clouds, which are particularly attractive for attackers because of the multi-tenant environment, where cloud users keep their sensitive data and applications. This purpose motivated on the one hand a strong focus on the security of the DCG-IDPS itself, more precisely of its known vulnerabilities and of data concerning hosts configurations, on the other hand the architecture of the system, which is multitier concentric. In particular, the DCG-IDPS consists of the following six tiers:

1. The first, outer tier is formed by host-level and network-level sensors and agents that collect and check data about network traffic, memory, file systems, logs etc. to find potential intrusions. They are called *primary detectors (PDs)*. If a PD recognizes a potential malicious event, it generates a raw alert, to which some

priority is assigned according to the finding. If that priority is alarming, then the alert data is immediately passed to the next tier, else the alert generation time and the PD identifier is broadcast to all the remaining PDs.

2. The second tier consists of components called *alert aggregation units (AAUs)*. Based on a threshold, an AAU aggregates similar raw alerts and passes them to the next tier.

3. The third tier is responsible for configuring PDs, checking user data against black-listed attackers and generation of the alerts aggregated by AAUs translated to a common format denoted *intrusion detection message format*.

4. The fourth tier analyses the alerts from the outer tiers and performs SVM-based classification. As features, it uses system calls from individual virtual machines. Since the number of such features is high, a *linear SVM classifier* is employed.

5. The fifth tier is the *control centre*, which manages the information exchange in the DCG-IDPS. Its final reaction depends on the results of analysis by the fourth tier. It also updates blacklists and configurations in a global database, which are then propagated to local databases whenever necessary.

6. The sixth tier is the *innermost core* of the DCG-IDPS. It contains the most sensitive data that needs most protection from intrusion. Nobody can access the core except the control centre, i.e., the second innermost tier.

That multitier architecture combined with the correlation of alerts produced at the outer tiers is advantageous in particular in connection with alerts from multiple attack sources, no matter whether they target a single destination, like distributed denial of service, or various destinations, as in the case of a large-scale stealth scan (Fig. 4.8). In [24], the DCG-IDPS has been compared with two other intrusion detection systems for clouds—the Grid and Cloud Computing Intrusion Detection System [25] and the Heuristic Semi-Global Alignment Approach [26]. The comparison shows that the DCG-IDPS is superior with respect to false positives, alert delay, and training time.

4.4 Active Learning and Its Relevance to SVM Classifiers

At the end of this chapter, we want to show how the theoretical fundamentals under-lying support vector machines can help to deal with semi-supervised learning. We will present an approach suitable for the typical situation that after the learning starts, it is possible to obtain the correct class labels for additional unlabelled feature vectors $x \in \mathcal{U}$, chosen to this end according to some particular strategy. This approach is usually called *active learning* and is one possible solution of situations mentioned in Sect. 2.4.2, of situations when obtaining class labels of unlabelled data is time consuming or expensive, thus preferred to be performed only as little as necessary.

The relevance of active learning to support vector machines is due to their con-struction, which suggests a particular principle according to which those additional feature vectors $x \in \mathcal{U}$ can be chosen, namely, the smallest distance from the hyper-plane H_w. For simplicity, consider the linear case, when H_w is constructed in the

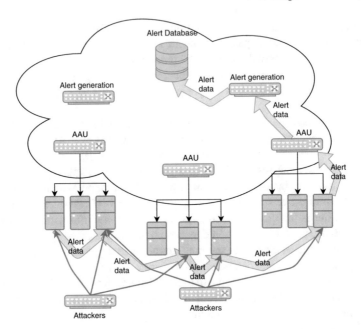

Fig. 4.8 Schema of scan attack detection by the Dynamic Correlation-Based Intrusion Detection and Prevention System (inspired by [24])

feature space $\mathscr{X}$. In the non-linear case, it is instead constructed in the space $\{\kappa(\cdot, x) | x \in \mathscr{X}\}$, a finite-dimensional subspace of which is the space $\mathscr{L}$ (2.42), recalled in Sect. 4.3. If for the feature vectors $x \in \mathscr{U}$, the unknown class is predicted by the SVM $\phi : \mathscr{X} \to \{1, -1\}$ learned using $(x_1, c_1), \ldots, (x_p, c_p)$, then ϕ is supported by all the available pairs $\{(x_1, c_1), \ldots, (x_p, c_p)\} \cup \{(x, \phi(x)) | x \in \mathscr{U}\}$.

Similarly, the margin corresponding to the expected classification error according to (4.14) is the margin between the sets $\mathscr{T}_+ \cup \mathscr{U}_+$ and $\mathscr{T}_- \cup \mathscr{U}_-$, where

$$\mathscr{T}_+ = \{x_k | k = 1, \ldots, p, c_k = 1\}, \ \mathscr{T}_- = \{x_k | k = 1, \ldots, p, c_k = -1\}, \quad (4.52)$$
$$\mathscr{U}_+ = \{x \in \mathscr{U} | \phi(x) = 1\}, \ \mathscr{U}_- = \{x \in \mathscr{U} | \phi(x) = -1\}. \quad (4.53)$$

Because the SVM ϕ is constructed to maximize the margin between $\mathscr{T}_+$ and $\mathscr{T}_-$, that margin is usually higher than the margin between $\mathscr{U}_+$ and $\mathscr{U}_-$, therefore, the margin between $\mathscr{T}_+ \cup \mathscr{U}_+$ and $\mathscr{T}_- \cup \mathscr{U}_-$ coincides with the margin between $\mathscr{U}_+$ and $\mathscr{U}_-$. This in turn, according to (4.2), (4.6) and (4.7), coincides with $\min_{x \in \mathscr{U}_+} d(x, H_w) + \min_{x \in \mathscr{U}_-} d(x, H_w)) \geq 2 \min_{x \in \mathscr{U}} d(x, H_w)$. Consequently, if a feature vector from $\arg\min_{x \in \mathscr{U}} d(x, H_w)$ is added to the training data, then the distance γ_w, which according to (4.14) determines the expected error in, is measured from that feature vector. This is indeed a justification of the above suggested principle governing the choice of additional $x \in \mathscr{U}$ for obtaining the correct class.

Taking into account (4.6)–(4.8), due to which the ordering of $x \in \mathcal{U}$ according to the distance $d(x, H_w)$ coincides with the ordering according to $|x^\top w + b|$, active learning using support vector machines can be summarized in the following algorithm.

Algorithm 4 (SVM active learning)
Input:
- training data $(x_1, c_1), \ldots, (x_p, c_p)$,
- set of unlabelled inputs $\mathcal{U}$,
- number p_a of additional feature vectors for which the correct class should be obtained,
- information whether active learning should be performed in connection with linear on non-linear SVM classification,
- in the non-linear case, in addition, a kernel κ.
Step 1. If the non-linear case is dealt with, replace:
- $x_k \leftarrow \kappa(\cdot, x_k), k = 1, \ldots, p$,
- $\mathcal{U} \leftarrow \{\kappa(\cdot, x) | x \in \mathcal{U}\}$,
- $\mathcal{X} \leftarrow \{\kappa(\cdot, x) | x \in \mathcal{X}\}$.
Step 2. Using the training data $(x_1, c_1), \ldots, (x_p, c_p)$, train a SVM ϕ based on a hyperplane with normal vector w and intercept b.
Step 3. Set $k = 1$.
Step 4. Denote $x_{p+k} = \arg\min_{x \in \mathcal{U}} |x^\top w + b|$.
Step 5. For x_{p+k}, obtain the correct class c_{p+k}.
Step 6. If $k < p_a$, increment $k \rightarrow k + 1$ and return to Step 4.
Output: Training data extended to $(x_1, c_1), \ldots, (x_{p+p_a}, c_{p+p_a})$.

4.4.1 Examples of SVM-Based Active Learning in Internet Applications

The use of active learning is beneficial in many internet applications, as a vast amount of data without a label is usually readily available compared to only a fraction of labelled samples. Training a classifier on such data also incorporates several additional challenges that favour using active learning.

Users of internet applications nowadays typically receive a high level of personalization. Therefore, each user has its model that is initially trained only on a tiny subset of data that is in some way related to the user. Alternatively, the model can be pre-trained on a dataset, if available for the given task, or on data acquired from other users. However, such a model is of rather poor quality and needs to be prepared for tuning by implicit feedback from the user on proposed instances.

The reduction of data to the most informative unlabelled sample, however, opens a possibility to ask the user explicitly to label the data. However, users should not be bothered too often. Therefore, only the samples that will provide the most information to the training should be selected for user labelling. In a sense, the user can be taken as a most valuable but expensive expert; if the user is bothered too much with irrelevant queries that will not improve the classification, the user will leave.

In general, two main approaches to active learning can be recognised: an offline approach that can select the best candidate form a possibly large pool of unlabelled instances, and an online approach that takes instances one-by-one and provides the most uncertain ones to the user.

The offline approach has all unlabelled instances available and, therefore, results in a more accurate classifier with less labels provided from the user. As presented in [27], by providing a label to only ten text documents selected by an active learning approach, the classifier has an equivalent accuracy to a classifier trained on more than a hundred of randomly chosen samples across six of the ten most frequent topic groups in Reuters-21578 dataset. On other topics, the improvement is not as high, but still noticeable. This article also discusses more sophisticated methods of sample selection for labelling.

Although the example mentioned above concerned the classification of articles into topics, the same approach can be used for document type detection, spam filtering or other text-based tasks.

Apart from that, the same author proposed in [28] a use of the same algorithm for relevance feedback on image data. In this article, the feature set is based on colour and texture features of images. The process of active SVM training presents the user with 20 images and asks him/her to select the images that are relevant to a given category. In some of their experiments, the number of iterations required to train the classifier to correctly recognise the category of a given image was reduced by up to 80%.

A very similar algorithm is used in [29] for sentiment analysis in the financial domain. In this article, the classifier is pre-trained on general tweets containing emoticons as easy-to-extract sentiment markers and consequently improved and targeted to the financial domain using several active learning approaches. Using the acquired sentiment prediction, authors were able to closely predict the changes in stock prices.

Selection of the best suitable instance from the pool of unlabelled data is, however, rather computationally expensive. Mainly due to the need to recompute the conditional probability for all unlabelled samples in each round. In large-scale systems, such as in spam filtering, this approach becomes unfeasible.

To eliminate the cost of instance selection, [30] proposed an online spam-filtering approach that takes incoming instances one-by-one, classifies them with an available classifier and picks the instance for assessment by the user with a probability that depends on the distance from the separating hyperplane. This approach, therefore, does not guarantee optimal number of user interactions, but in combination with an online SVM achieves a significant training cost reduction.

Another benefit of an online approach is that only the model has to be kept in the memory, as opposed to the whole pool of unlabelled instances. On the other hand, an updatable SVM classifier needs to be used, which may limit the flexibility of SVM by requiring to use the linear SVM classifier, possibly combined with a non-trainable kernel. Also, this approach might not be very suitable for areas with a substantial concept drift.

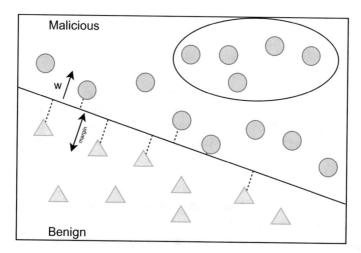

Fig. 4.9 Selecting for active learning in the halfspace corresponding to malicious software not only points close to the separating hyperplane, but also the points in the ellipse far from that hyperplane (inspired by [31])

Internet applications in which active learning is particularly useful are malware detection and network intrusion detection. The reason is that in them, assigning class labels needs time-intensive involvement of human analysts. In the remaining part of this section, several examples of combining SVM with active learning in those two application domains will be presented.

In [31], SVM in connection with active learning have been used for *malware detection* in the Windows operating system. The employed SVM has a Gaussian kernel (2.32), and active learning uses, apart the above described approach selecting the points closest to the separating hyperplane, also another approach selecting points farthest from the separating hyperplane in the halfspace corresponding to malware (Fig. 4.9). This other approach allows to acquire more malware files to extract signatures for a database of malware signatures. At the same time, if there is a cluster of similar malware files, then it is represented in the database by only one signature. Needless to say, for an SVM with a Gaussian kernel, the separating hyperplane is constructed not in the space of the data, but in the space (2.42) into which it is transformed using that kernel and in which malware and benign software are linearly separable.

The features used in [31] are based on static analysis of binary executable files. More precisely, 5-grams of their bytes are used. The authors tested also 3-, 4- and 6-grams. All n-grams are extracted from raw binary executables, without any decryption or decompression. Feature selection was with these features performed in two steps:

(i) All n-grams occurring in the training data (in the case of 5-grams, they were 1,575,804,954) were ordered according to their TF-IDF (1.1) and only the 5,500 with the highest TF-IDF were kept.

(ii) Among them, 300 features were selected using a feature selection method proposed in [32] (originally, for the area of gen-expression data). The authors tested also two other particular feature selection methods and the numbers of selected features 50, 100, 200, 1,000, 1,500 and 2,000.

An experiment reported in [31] documented for three active learning strategies the performance of the SVM classifier and the number of acquired files for the signature database during 10 trials of malware acquisition simulating the first 10 days after the classifier was trained. The experiment used 25,260 Windows executables among which exactly 10% were malware and 90% benign software. That ratio was chosen according to the suggestion in [33]. These 25,260 files were randomly divided into ten groups, each containing 2,521 executables representing those obtained each of the 10 days, and additional 50 executables for initial training. The experiment was performed in the following steps:

1. It started with training the SVM on the 50 executables for initial training. Needless to say, the correct labels (malware—-benign software) of the executables used in the experiment were known, more precisely, the result of checking them with Kapersky's anti-virus software were used to this end.
2. For each of the 10 days, a given number of executables were selected from among the 2,521 ones representing those obtained that day, according to the considered active learning strategy.
3. For those selected executables, their correct labels were recalled, representing their labelling by a human malware analyst.
4. The selected executables with their correct labels were added to the training set and the SVM was retrained.
5. At the same time, the database of malware signatures was extended with those among the selected executables that were labelled as malware, but were not sufficiently similar to any executable already present in the database
6. Unless the considered day was already the last (10th), the experiment returned to the step 2 for the next day.

As the number of files selected in the step 2, the following possibilities were tested: multiples of 10 from 10 till 350, 450, 550, 600, 650, 700, 750, 800. For their selection, the following three active learning strategies were considered:

(i) *SVM simple margin strategy*, i.e., the active learning algorithm presented in the introductory part of Sect. 4.4, which selects the points closest to the separating hyperplane. As was recalled above, the separating hyperplane is constructed in the space (2.42) into which the data is transformed using the considered Gaussian kernel.
(ii) *Exploitation strategy* selects the points farthest from the separating hyperplane in the subspace of (2.42) corresponding to malware.
(iii) *Combined strategy* starts as the SVM simple margin strategy and then it gradually moves to the exploitation strategy during the considered ten days.

The results of the reported experiment have shown that the exploitation strategy and combined strategy have similar accuracy as the SVM simple margin strategy, but

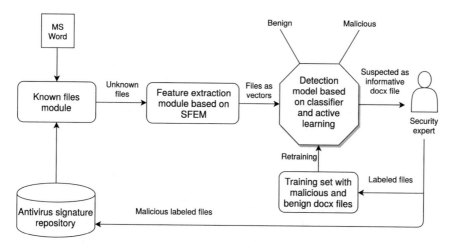

Fig. 4.10 The ALDOCX framework for malware detection in Microsoft Word ".docx" files (inspired by [34])

they achieve substantially higher numbers of malware executables acquired for the signature database. For example, if 50 files were selected in step 2 of the experiment, then the exploitation strategy acquired about 30% more malware executables than the SVM simple margin strategy, if 800 files were selected, then it achieved up to 130% more, and if 250 files were selected, then it achieved up to 160% more. It is also worth mentioning that with the simple margin strategy, the number of acquired malware executables increased only for the 1.−4. day and then it decreased, wheres for the exploitation and combined strategies, it increased for all the 10 days.

A similar group of authors extended all three above active learning strategies to malware detection in Microsoft Word ".docx" files, in a framework named *ALDOCX* [34] (Fig. 4.10). The threat pertaining to this kind of files is due to the following properties:

- Microsoft Word files are probably the kind of files that Windows users open most frequently.
- They may contain macros launching commands of the programming language Visual Basic for Applications, which can perform malicious actions.
- They may contain objects implementing the technology called Object linking and editing, used to share data between applications. In such objects, many kinds of files can be embedded, including binary files, possibly malicious, and hypertext markup language files, possibly containing malicious javascript code.

It should be mentioned that ALDOCX can detect malware only in Microsoft Word ".docx" files, not in the legacy ".doc" files, which are not XML-based and have a rather different structure.

ALDOCX took over the above active learning strategies (i)–(iii), and added another strategy:

(iv) *Comb-ploit* is the opposite of the combined strategy (iii). It starts as the exploitation strategy and then moves to SVM simple margin.

The features that ALDOCX uses to characterize a ".docx" document are extracted from the ".xml" file describing the configuration and content of that document. Among all features that can be extracted in this way, 5,000 have been selected that have evaluated as the most discriminative between malicious and benign ".docx" files, including features that concern visual basic code and object linking and embedding objects. In experiments with ALDOCX, its authors tested also subsets of those 5,000 features, sized 10, 40, 80, 100, 300, 500, 800 and 1,000.

Also in [34], a 10-trial experiment has been reported, simulating the performance of ALDOCX during the first 10 days after the classifier was trained, similar to the above described experiment reported in [31]. It used 16,811 ".docx" files, among which 327 were malicious and 16,811 benign, and those files were randomly divided into ten times 1,600 files representing the Word documents obtained each of the 10 days, and additional 811 files for initial training. Among the tested subset of features, the 100-features subsets was finally used, that choice being based on the TPr and FPr values achieved with SVMs trained on approximately 90% of all used ".docx" files. The experiment again followed the above steps 1.–6. In the step 2., the following numbers of selected files were tested: 10, 20, 50, 100 and 250. Differently to the experiment reported in [31], the TPr values of SVMs with the considered active learning strategies were compared not only with each other, but also with a SVM without active learning and with ten commonly used representatives of commercial anti-virus software. Apart from the 10-trials experiment, the authors of ALDOCX evaluated the active learning strategies also in a 280-trials experiment, though only with 13,661 ".docx" files, among which 312 were malicious.

The experiments reported in [34] have brought several important results:

- When all available data were used, then the SVM classifier achieved the highest TPr, measured using a 10-fold cross-validation, for 40 features (93.48%) and 100 features (93.34%). Based on that result, 100 features were subsequently used in the 10-trials and 280-trials experiments.
- The exploitation and combined strategies were clearly advantageous from the point of view of acquisition of new malware files. In the 10-trials of malware acquisition simulating the first 10 days after the classifier was trained, they from the 3rd day on maintained high acquisition rates of 93–94% of malicious files. For comparison, the acquisition rate of the SVM simple margin strategy was gradually decreasing, being as low as 3% on the 10th day.
- Exploitation and Combi-Ploit were the only strategies that showed an increase of acquisition rate from the 1st to the 2nd day. The combined strategy had a decrease, but then it had a very similar performance as exploitation.
- At the end of the 10-trials, the highest TPr was achieved by the SVM simple margin strategy: 93.6%, but the other strategies were only slightly worse: combi-ploit 93.2%, exploitation 93%, combined 92.3%.
- The above mentioned TPr 93.6% using the SVM simple margin strategy was achieved when 50 new files were selected for testing every day. Consequently, the

SVM was trained with 1,311 files (811 for initial training and 500 acquired during the 10 days) out of the total 16,811, which is 7.1%. For comparison, to obtain the average TPr value 93.34% using all available data, the same number of features and a 10-fold cross-validation, every SVM was trained with 9 out of the 10 folds, i.e., with 90% of the total number of files.

- For comparison, the TPr was measured also for 8 common antivirus programs. For all of them, it was lower than the above values, sometimes even substantially lower (e.g., Avast 74%, Kaspersky 68.8%, Symantec 66.1%, McAfee 63%).
- In the 280-trial experiment, exploitation and combined strategy acquired the same number of malware files after mere 12 trials, compared to all 280 trials of SVM, namely 304 files,, i.e., 97.4% of the available malware.

Let us now move from Windows to Android. The combination of SVM and active learning has been employed for malware detection in the system *RobotDroid* [35]. It uses dynamic malware analysis, which is performed, basically, in the following steps:

(i) Monitoring points are inserted in every service manager in Android, to collect information about calls of the respective service.
(ii) For each service call, the process ID and timestamp are logged, allowing to trace the behaviour of all software.
(iii) Using active learning according to Algorithm 4, some unlabelled software is labelled either as malware or as benign.

Using timestamps allows to distinguish also malware and benign software with a similar pattern of requests if their timing patterns differ.

More recently, using SVM with active learning for Android malware detection has been presented in [36] (Fig. 4.11). Also here, dynamic malware analysis is performed. Main differences compared to RobotDroid are as follows:

- The dynamic features are obtained not from service managers, but from apps' logs. To this end, the authors developed a specific tool called *DroidCat*.
- As features the occurrences of 150 predefined keywords in the apps' logs are used, as well as timestamps corresponding to 56 time-dependent features.
- The active learning is not pool-based, i.e., the set $\mathcal{U}$ of unlabelled inputs is not fixed like in Algorithm 4, but *stream-based*, i.e., a new set $\mathcal{U}$ is available at each iteration.
- The decision whether to use an unlabelled point x is based not on minimizing $|x^\top w + b|$ like in Algorithm 4, but on maximizing the expected reduction of error.

The authors of [36] validated their approach on 700 malware and 700 benign files. The malware files were obtained from the Drebin dataset [37], which is the largest public collection of Android malware. It includes more than 5,000 malware apps from 179 malware families. The 700 files were selected from Drebin in such a way that from each represented family, there are always multiple apps. The 700 benign files were selected from various categories of Android apps.

Finally in [38], the combination of SVM and active learning has been employed for *network intrusion detection*. The SVM uses a Gaussian kernel, but the active

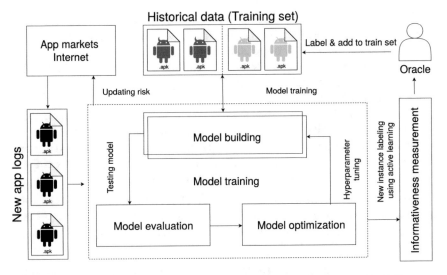

Fig. 4.11 Android malware detection using SVM and active learning as proposed in [36])

learning strategy is much more involved than the one outlined in Algorithm 4. It has the following three main ingredients:

(i) The available unlabelled inputs are clustered around the support vectors. According to the respective support vector, each cluster corresponds either to the some particular kind of intrusion or to the normal operation.

(ii) To take into account the fact that assigning the unlabelled inputs to a particular kind of intrusion or to normal operation is connected with some uncertainty, the clustering method *fuzzy c-means* [39] has been used, for which the resulting clusters are fuzzy sets. The membership of the unlabelled inputs in those fuzzy sets then allows to assess their uncertainty.

(iii) The fuzzy c-means clustering allows to obtain for unlabelled inputs an additional classification into the considered classes of intrusion and into normal operation. In [38], that additional classification is combined with the classification by the trained SVM in such a way that:

- if the label based on the fuzzy c-means clustering coincides with the label assigned by the trained SVM, the unlabelled input keeps that label;
- if the label based on the fuzzy c-means and the label assigned by the trained SVM differ, the unlabelled input receives a new label corresponding to an unknown intrusion.

The authors of [38] validated their approach on a subset of 25,252 records from the KDD99 dataset [40], which contains four kinds of intrusions: DoS, remote to local (R2L), user to root (U2R), and probe attacks.

References

1. Bartlett, P., Shawe-Taylor, J.: Generalization performance of support vector machines and other pattern classifiers. In: Schölkopf, B., Burges, C., Smola, A. (eds.) Advances in Kernel Methods—Support Vector Learning, pp. 43–54. MIT Press, Cambridge (1999)
2. Bertsekas, D.: Nonlinear Programming, 2nd edn. Athena Scientific, Nashua (2004)
3. Kushmerick, N.: Learning to remove internet advertisments. In: ACM Conference on Autonomous Agents, pp. 175–181 (1999)
4. Blanzieri, E., Bryl, A.: A survey of learning-based techniques of email spam filtering. Artif. Intell. Rev. **29**, 63–92 (2008)
5. Drucker, H., Wu, D., Vapnik, V.: Support vector machines for spam categorization. IEEE Trans. Neural Netw. **10**, 1048–1054 (1999)
6. Caruana, G., Li, M., Qi, M.: A MapReduce based parallel SVM for large scale spam filtering. In: FSKD: Eighth International Conference on Fuzzy Systems and Knowledge Discovery, pp. 2659–2662 (2011)
7. Sculley, D., Wachman, G.: Relaxed online SVMs for spam filtering. In: 30th Annual International ACM SIGIR Conference on Research and Development in Information Retrieval, pp. 415–422 (2007)
8. Al-Jarrah, O., Khater, I., Al-Duwairi, B.: Identifying potentially useful email header features for email spam filtering. In: ICDS: 6th International Conference on Digital Society, pp. 140–145 (2012)
9. Aradhye, H., Myers, G., Herson, J.: Image analysis for efficient categorization of image-based spam e-mail. In: ICDAR'05: 8th International Conference on Document Analysis and Recognition, pp. 914–918 (2005)
10. Oku, K., Nakajima, S., Miyazaki, J., Uemura, S.: Context-aware SVM for context-dependent information recommendation. In: 7th International Conference on Mobile Data Management, pp. 109/1–4 (2006)
11. Billsus, D., Pazzani, M.: Learning collaborative information filters. In: International Conference on Machine Learning, pp. 46–54 (1998)
12. Xia, Z., Dong, Y., Xing, G.: Support vector machines for collaborative filtering. In: 44th Annual Southeast Regional Conference, pp. 169–174 (2006)
13. Lee, Y., Mangasarian, O.: SSVM: a smooth support vector machine for classification. Comput. Optim. Appl. **20**, 5–22 (2001)
14. O'Kane, P., Sezer, S., McLaughlin, K., Im, E.: SVM training phase reduction using dataset feature filtering for malware detection. IEEE Trans. Inf. Forensics Secur. **8**, 500–509 (2013)
15. Khammas, B., Monemi, A., Ismail, I., Nor, S., Marsono, M.: Metamorphic malware detection based on support vector machine classification of malware sub-signatures. Telekomnika **14**, 1157–1165 (2016)
16. Pajouh, H., Dehghantanha, A., Khayami, R., Choo, K.: Intelligent OS X malware threat detection with code inspection. J. Comput. Virol. Hacking Techn. **14**, 212–223 (2018)
17. Masud, M., Khan, L., Thuraisingham, B.: A hybrid model to detect malicious executables. In: IEEE International Conference on Communications, pp. 1443–1448 (2007)
18. Chawla, N., Bowyer, K., Hall, L., Kegelmeyer, W.: SMOTE: synthetic minority over-sampling technique. J. Artif. Intell. Res. **16**, 321–357 (2002)
19. Wang, P., Wang, Y.: Malware behavioural detection and vaccine development by using a support vector model classifier. J. Comput. Syst. Sci. **81**, 1012–1026 (2015)
20. Xu, Y., Wu, C., Yheng, K., Wang, X., Niu, X., Lu, T.: Computing adaptive feature weights with PSO to improve android malware detection. Secur. Commun. Netw. **10**, article no. 3284,080 (2017)
21. Mas'ud, M., Sahib, S., Abdollah, M., Selamat, S., Yusof, R.: An evaluation of n-gram system call sequence in mobile malware detection. ARPN J. Eng. Appl. Sci. **11**, 3122–3126 (2016)
22. Rehman, Z., Khan, S., Muhammad, K., Lee, J., Lv, Z., Baik, S., Shah, P., Awan, K.: Machine learning-assisted signature and heuristic-based detection of malwares in Android devices. Comput. Electr. Eng. **69**, 828–841 (2018)

23. Kennedy, J.: Particle swarm optimization. In: Encyclopedia of Machine Learning, pp. 760–766. Springer (2011)
24. Umamaheswari, K., Sujatha, S.: Impregnable defence architecture using dynamic correlation-based graded intrusion detection system for cloud. Defence Sci. J. **67**, 645–653 (2017)
25. Vieira, K., Schulter, A., Westphall, C., Westphall, C.: Intrusion detection for grid and cloud computing. IEEE IT Professional **12**, 38–43 (2010)
26. Tupakula, U., Varadharaja, V., Akku, N.: Intrusion detection techniques for infrastructure as a service cloud. In: 9th (IEEE) International Conference on Dependable, Autonomic and Secure Computing, pp. 744–751 (2011)
27. Tong, S., Koller, D.: Support vector machine active learning with applications to text classification. J. Mach. Learn. Res. **2**, 45–66 (2001)
28. Tong, S., Chang, E.: Support vector machine active learning for image retrieval. In: Ninth ACM International Conference on Multimedia, pp. 107–118 (2001)
29. Smailović, J., Grčar, M., Lavrač, N., Žnidaršič, M.: Stream-based active learning for sentiment analysis in the financial domain. Inf. Sci. **285**, 181–203 (2014)
30. Sculley, D.: Online active learning methods for fast label-efficient spam filtering. In: Fourth CEAS Conference on Email and Anti-Spam, pp. 143–150 (2007)
31. Nissim, N., Moskowitch, R., Rokach, L., Elovici, I.: Novel active learning methods for enhanced PC malware detection in windows OS. Expert Syst. Appl. **41**, 5843–5857 (2014)
32. Golub, T., Slonim, D., Tamaya, P., Huard, C., Gasenbeek, M., Mesirov, J., Coller, H., Loh, M., Downing, J., Calligiuri, M., Bloomfield, C., Lander, E.: Molecular classification of cancer: class discovery and class prediction by gene expression monitoring. Science **286**, 531–537 (1999)
33. Rossow, C., Dietrich, C., Grier, C., Kreibich, C., Pohlmann, N., van Steen, M.: Prudent practices for designing malware experiments: Status quo and outlook. In: IEEE Symposium on Security and Privacy, pp. 65–79 (2012)
34. Nissim, N., Cohewn, A., Elovici, I.: ALDOCX: Detection of unknown malicious microsoft office documents using designated active learning methods based on new structural feature extraction methodology. IEEE Trans. Inf. Forensics Secur. **12**, 631–646 (2017)
35. Zhao, M., Yhang, T., Ge, F., Yuan, Z.: RobotDroid: a lightweight malware detection framework on smartphones. J. Netw. **7**, 715–722 (2012)
36. Rashidi, B., Fung, C., Bertino, E.: Android malicious application detection using support vector machine and active learning. In: 13th IFIP International Conference on Network and Service Management, pp. 1–9 (2018)
37. Arp, D., Spreityenbarth, M., Hubner, M., Gascon, H., Rieck, K.: DREBIN: Effective and explainable detection of Android malware in your pocket. In: 21st Annual Network and Distributed System Security Symposium, p. 12 pages (2014)
38. Kumari, V., Varma, P.: A semi-supervised intrusion detection system using active learningSVM and fuzzy c-means clustering. In: I-SMAC, pp. 481–485 (2017)
39. Bezdek, J.: Pattern Recognition with Fuzzy Objective Function Algorithms. Plenum Press, New York (1981)
40. Tavallaee, M., Bagheri, E., Lu, W., Ghorbani, A.: A detailed analysis of the KDD cup 99 data set. In: IEEE Symposium on Computational Intelligence for Security and Defense Applications, pp. 288–293 (2009)

Chapter 5
Aiming at Comprehensibility

Although the ultimate objective of classification is to correctly classify new, unseen inputs, accuracy and other measures of the correctness of classification are nearly never the only criteria for classifier choice. The reason is that classification is usually only a support for decision making by humans, who are then responsible for the final decision made. The client of an internet shop finally decides which product to buy, not the recommender system. The administrator of a computer finally decides which software to install, not the classifier discriminating between benign software and malware. The network administrator finally decides whether the network state has to be dealt with as an attack, not the intrusion detection system. And in that supporting role, the classifier is much more useful if the human decision maker understands why a particular input has been assigned to a particular class. In many applications, the comprehensibility is, therefore, a key property taken into account when choosing a classifier, similarly important as the predictive accuracy. Moreover, some classifiers are constructed primarily with the aim of comprehensibility. They will be the topic of this chapter.

In all the kinds of classifiers presented in Chaps. 3 and 4, the knowledge explaining why a particular class is assigned to a particular feature vector has a purely numeric representation: by means of probability distributions, normal vectors of hyperplanes, parameters of non-linear surfaces or functions. The comprehensibility of such numeric representations for a human user is very low, such a representation is viewed as having a high "data fit", but a low "mental fit" [1]. On the other hand, knowledge representations considered comprehensible for a human are the following:

(i) *Visualization.* In input spaces up to 3 dimensions, which correspond to human visual perception, visualization is very intuitive and easily comprehensible. In higher dimensions, however, its comprehensibility is much lower and rapidly decreases with increasing dimension [2].

(ii) *Statements of some formal logic.* The language of the simplest formal logic, the Boolean logic, has been for more than a century used to define mathematical objects and to describe their properties. And languages of formal logics

© Springer Nature Switzerland AG 2020
M. Holeňa et al., *Classification Methods for Internet Applications*,
Studies in Big Data 69, https://doi.org/10.1007/978-3-030-36962-0_5

in general are considered more comprehensible for humans in this role than numerical representations. Also in their case, the comprehensibility decreases with the increasing dimension of the input space, but the decrease is less apparent than in the case of visualisation.

Taking into account this situation, the chapter will deal with methods using knowledge representation based on the language of some formal logic. The most popular among them are classification trees, and their popularity is to a great extent due to an easy visualization of the obtained results. Hence, classification trees actually combine both above mentioned kinds of comprehensible knowledge representation.

5.1   Classifier Comprehensibility and Formal Logic

The principle of constructing classifiers that use knowledge representation based on the language of some formal logic is most easily explained in the case of a binary classification by means of a 0/1-valued classifier on a Cartesian product,

$$\phi : \mathscr{X} \to \{0, 1\}, \text{ with } \mathscr{X} = V_1 \times \cdots \times V_n, \tag{5.1}$$

where $V_j, j = 1, \ldots, n$, is the set of feasible values of the feature $[x]_j$. Consider now a formula φ of a Boolean predicate logic such that all its atomic predicates are modelled by elements of some of the value sets $V_1, \ldots, V_n$. Then the evaluation $\|\varphi\|$ of φ is a 0/1-valued function on the Cartesian product of those value sets, and it can be easily extended to a 0/1-valued function on $\mathscr{X}$ by making it constant with respect to variables not modelling the atomic predicates of φ. Consequently, the evaluations of Boolean formulas are the same kind of functions as (5.1), which suggests to use them as 2-class classifiers. A key advantage of a classifier defined in such a way is that the insight into classification of an x from the input space $V_1 \times \cdots \times V_n$ to the class 1 or 0 is equivalent to understanding why φ is or is not valid if its atomic predicates are modelled by the corresponding components of x. Hence, instead of the function (5.1), we now deal with a Boolean formula, which is more likely to be comprehensible for a human. Moreover, this method actually uses only two kinds of formulas with a particularly simple semantics: *implications*, $\varphi \to \psi$, and *equivalences*, $\varphi \equiv \psi$. Implications and equivalences belong to important kinds of formulas not only in the Boolean logic, but also in various fuzzy logics, which allows this method to be easily generalized to them. Implications and equivalences, if used for the construction of classifiers, are more generally called *classification rules*.

5.1.1 Classification Rules in Boolean and Fuzzy Logic

In both Boolean and fuzzy logic, the evaluation $\|\varphi \to \psi\|$ of an implication $\varphi \to \psi$, as well as the evaluation $\|\varphi \equiv \psi\|$ of an equivalence $\varphi \equiv \psi$ are functions of the evaluations $\|\varphi\|$ and $\|\psi\|$. In the *Boolean logic*, these are Boolean functions on the 4-element set $\{0, 1\}^2$, which are defined by Tables 5.1 and 5.2.

In *fuzzy logics*, a general definition of the evaluation of $\varphi \to \psi$ is

$$\|\varphi \to \psi\| = \sup\{\xi \in [0, 1] | \|\varphi\| \star_t \xi \le \|\psi\|\}, \tag{5.2}$$

where $\star_t$ is the t-norm modelling conjunction in the considered fuzzy logic. Similarly, the evaluation of $\varphi \equiv \psi$ is in general defined by

$$\|\varphi \equiv \psi\| = \|\varphi \to \psi\| \star_t \|\psi \to \varphi\|. \tag{5.3}$$

For practical applications, it is useful to know which particular form the definitions (5.2) and (5.3) take in the three fundamental fuzzy logics:

(i) In the *Gödel logic*, the t-norm is the minimum, which put into (5.2)–(5.3) yields

$$\|\varphi \to \psi\| = \begin{cases} 1 & \text{if } \|\psi\| \ge \|\varphi\|, \\ \|\psi\| & \text{if } \|\psi\| < \|\varphi\|. \end{cases} \tag{5.4}$$

$$\|\varphi \equiv \psi\| = \begin{cases} 1 & \text{if } \|\psi\| = \|\varphi\|, \\ \min(\|\varphi\|, \|\psi\|) & \text{if } \|\psi\| \ne \|\varphi\|. \end{cases} \tag{5.5}$$

(ii) In the *product logic* with the t-norm product, (5.2)–(5.3) turn into

$$\|\varphi \to \psi\| = \begin{cases} 1 & \text{if } \|\psi\| \ge \|\varphi\|, \\ \frac{\|\psi\|}{\|\varphi\|} & \text{if } \|\psi\| < \|\varphi\|. \end{cases} \tag{5.6}$$

$$\|\varphi \equiv \psi\| = \begin{cases} 1 & \text{if } \|\psi\| = \|\varphi\|, \\ \frac{\min(\|\varphi\|, \|\psi\|)}{\max(\|\varphi\|, \|\psi\|)} & \text{if } \|\psi\| \ne \|\varphi\|. \end{cases} \tag{5.7}$$

(iii) In the *Łukasiewicz logic* with the t-norm $\star_L$, defined as

$$\star_L (\xi, \eta) = \max(0, \xi + \eta - 1), \tag{5.8}$$

Table 5.1 Dependence of the evaluation of $\varphi \to \psi$ on the evaluations of φ and ψ

$\|\varphi \to \psi\|$	$\|\psi\| = 0$	$\|\psi\| = 1$
$\|\varphi\| = 0$	1	1
$\|\varphi\| = 1$	0	1

Table 5.2 Dependence of the evaluation of $\varphi \equiv \psi$ on the evaluations of φ and ψ

$\|\varphi \equiv \psi\|$	$\|\psi\| = 0$	$\|\psi\| = 1$
$\|\varphi\| = 0$	1	0
$\|\varphi\| = 1$	0	1

they turn into

$$\|\varphi \to \psi\| = \begin{cases} 1 & \text{if } \|\psi\| \geq \|\varphi\|, \\ 1 + \|\psi\| - \|\varphi\| & \text{if } \|\psi\| < \|\varphi\|. \end{cases} \tag{5.9}$$

$$\|\varphi \equiv \psi\| = \begin{cases} 1 & \text{if } \|\psi\| = \|\varphi\|, \\ 1 + \min(\|\varphi\|, \|\psi\|) - \max(\|\varphi\|, \|\psi\|) & \text{if } \|\psi\| \neq \|\varphi\|. \end{cases} \tag{5.10}$$

In general, the set of classification rules is constructed by a heuristic method that joins literals describing available features (such as: "age $\leq$ 30&income > 25000 $\to$ ¬Married") that are subsequently simplified by a set of pruning operators.

An early approach, called reduced error pruning (REP) [3] allowed the deletion of the last literal from a clause or dropping a whole clause from the finalized (over-trained) ruleset. Therefore, this approach is sometimes denoted as global pruning. The original algorithm applied the pruning operations independently to all clauses in the overtrained ruleset and executed only one pruning operation at a time that resulted in the highest increase of accuracy on the pruning set. The process was repeated until the accuracy on pruning set increased. The recalculation of accuracy makes REP inefficient, resulting in complexity of up to $\mathcal{O}(n^4)$ on random data.

One of the critical optimisations to reduced error pruning, proposed in [4], is the integration of a pre-pruning approach (also called incremental pruning). This approach, in general, deliberately ignores some of the training samples before the addition of the clause to the rule set and thus aims at more general clauses, rather than best coverage of input instances. Pre-pruning without the use of pruning set was introduced separately as an alternative approach in FOIL [5] and further improved. A combination of pre-pruning and reduced error pruning, however, allowed to prune each rule right after its training. The proposed approach stops the training when the new clause has an accuracy lower than an empty clause.

Further improvements to the Incremental Reduced Error Pruning by [6] change the metric used for the pruning of individual clauses and introduce a limit on clause length. The clauses obtained by this approach, called IREP*, are then confronted with two other clauses trained and pruned to minimise the error of the newly proposed rule set; one newly trained from an empty clause, second based from the original clause. This results in a method called Repeated Incremental Pruning to Produce Error Reduction, in short RIPPER.

5.1.2 Observational Calculus

An implication $\varphi \rightarrow \psi$ or equivalence $\varphi \equiv \psi$ has to be evaluated, in general, for a particular combination $x = ([x]_1, \ldots, [x]_n)$ of feature values, a combination typically corresponding to some particular real object. However, classification rules that are valid only for particular objects, are not very useful to predict the correct class of new, unseen objects. Indeed, they then have to be evaluated for combinations $([x]_1, \ldots, [x]_n)$ different from all those used as they have been obtained. To make predictions for new objects, we would need rather statements of formal logic that are valid for the complete set of combinations $x = ([x]_1, \ldots, [x]_n)$ of features recorded in the data, i.e., that characterize those features in general.

Unfortunately, Boolean logic allows obtaining only one specific kind of such statements: statements quantified by the universal quantifier $\forall$. The validity of such a statement corresponds to the simultaneous validity of the classification rule for all combinations of features recorded in the data, and is taken as an indication that the classification rule is valid in general. In particular, the validity of the statement $(\forall \xi_1 \ldots \xi_\ell) \, \varphi \rightarrow \psi$, where $\xi_1, \ldots, \xi_\ell$ are all free variables in $\varphi \rightarrow \psi$, corresponds to the simultaneous validity of $\varphi \rightarrow \psi$ for all combinations of features $x = ([x]_1, \ldots, [x]_n)$ recorded in the data. Needless to say, it is very rare to find a statement $(\forall \xi_1 \ldots \xi_\ell) \, \varphi \rightarrow \psi$ to be valid if no single exception is allowed from the requirement that $\varphi \rightarrow \psi$ has to hold for all such combinations, even not an exception due to noise. And even rarer is to encounter the validity of a statement $(\forall \xi_1 \ldots \xi_\ell) \, \varphi \equiv \psi$ because the validity of $\varphi \equiv \psi$ implies the simultaneous validity of both $\varphi \rightarrow \psi$ and $\psi \rightarrow \varphi$. Therefore, it is desirable to have statements that are valid if the implication $\varphi \rightarrow \psi$ or equivalence $\varphi \equiv \psi$ hold not for all, but only for a sufficiently large proportion of combinations of features recorded in the data. To define and evaluate such statements, observational calculi have been developed in the 1970s, that is why we will call those statements *observational rules*.

Observational calculi are extensions of Boolean logic, proposed in the 1970s for exploratory data analysis [7] and more recently for association rules [8]. For the purpose of classification rules, the following simple observational calculus is sufficient:

(i) The language of an observational calculus is a superset of the language of some Boolean predicate calculus with only unary predicates, aka *monadic predicate calculus*, extending that language with symbols for generalized quantifiers. In particular, let Q be a symbol not belonging to the symbols of the original language of the monadic predicate calculus, $r \in \mathbb{N}$, $\varphi_1, \ldots, \varphi_r$ be formulas of the language containing only predicates and possibly connectives, and Tf_Q be a binary function on Boolean matrices with k columns, i.e.,

$$\mathrm{Tf}_Q : \bigcup_{k \in \mathbb{N}} \{0, 1\}^{k,r} \rightarrow \{0, 1\}. \tag{5.11}$$

Then Q is called a *generalized quantifier* of arity k, and Tf_Q is called *truth function* of Q. Quantifying the formulas $\varphi_1, \ldots, \varphi_r$ with the quantifiers Q is denoted $(Q\xi)(\varphi_1(\xi), \ldots, \varphi_r(\xi))$, usually simplified to $Q(\varphi_1, \ldots, \varphi_r)$. In the case of a binary quantifier $\sim$, the notation $\sim (\varphi_1, \varphi_2)$ is used, further simplified to $\varphi_1 \sim \varphi_2$.

Apart from the symbols for generalized quantifiers, the language of observational calculus has all the symbols of the original language, i.e., symbols for the unary predicates and their single variable (usually left out due to its uniqueness), logical connectives $(\neg, \&, \vee, \ldots)$, logical constants $\bar{1}$ and $\bar{0}$, and the quantifiers $\forall$ and $\exists$.

(ii) Like in the Boolean predicate calculus, formulas of the language containing only predicates and possibly connectives, i.e., not containing quantifiers, are called *open formulas*. For example,

$$\text{age}(\xi) \leq 30 \ \& \ \text{income}(\xi) > 25,000 \ \& \ \neg\text{Married}(\xi),$$

usually simplified, due to the uniqueness of ξ, to

$$\text{age} \leq 30 \ \& \ \text{income} > 25,000 \ \& \ \neg\text{Married}.$$

Also like in the Boolean predicate calculus, formulas of the language containing quantifiers, no matter whether $\forall$, $\exists$ or the generalized quantifiers introduced in (i), are called *closed formulas*.

(iii) Consider a data matrix $D \in \mathbb{R}^{p,n}$, the rows $d_1, \ldots, d_p$ of which are combinations $x = ([x]_1, \ldots, [x]_n)$ of values of features for inputs $x_1, \ldots, x_p$ from the feature space $\mathscr{X}$. If for $d_k, k = 1, \ldots, p$, the atomic predicates modelled by elements of some of the sets $V_1, \ldots, V_n$ have been evaluated, then any open formula can be evaluated using standard rules for the evaluation of Boolean connectives. In the example above, if the predicates $\text{age} \leq 30$, $\text{income} > 25,000$ and Married have been evaluated for d_k, then the above open formula is evaluated true provided the predicates $30 \geq \text{age}$ and $25,000 < \text{income}$ are true and the predicate Married is false. Further, denoting $\|\varphi_j\|_k$ the evaluation of φ_j for d_k, the formula $\forall \varphi_j$, respectively $\exists \varphi_j$ is evaluated true based on D if $\|\varphi_j\|_k = 1$ holds for all, respectively for at least one $k = 1, \ldots, p$. Finally, the *evaluation of* $Q(\varphi_1, \ldots, \varphi_r)$ based on D is

$$\|Q(\varphi_1, \ldots, \varphi_r)\| = \text{Tf}_Q \begin{pmatrix} \|\varphi_1\|_1 & \ldots & \|\varphi_r\|_1 \\ \ldots & \ldots & \ldots \\ \|\varphi_1\|_p & \ldots & \|\varphi_k\|_p \end{pmatrix} \tag{5.12}$$

The theory underlying the observational calculus has been most extensively explained in the monographs [7, 8]. Here, only the important generalized quantifiers will be surveyed. All of them are binary and all are based either on the estimation of particular probabilities, or on simple hypotheses tests.

5.1.3 Observational Rules Based on Probability Estimation

If the evaluation $\|\varphi\|$ of the formula φ is a binary function on $V_1 \times \cdots \times V_n$, then its composition with the values $[x]_1, \ldots, [x]_n$ of features is a realization of a $0/1$-valued random variable. Similarly, for two formulas φ, ψ, the composition of $(\|\varphi\|, \|\psi\|)$ with $[x]_1, \ldots, [x]_n$ is a realization of a 2-dimensional random vector that has $0/1$-valued marginals and assumes values in the set $\{(0, 0), (0, 1), (1, 0), (1, 1)\}$. The probabilities and conditional probabilities of those values can be easily estimated from data, several such estimates will be recalled below. Various assertions about such probabilities is the most common way to define generalized quantifiers.

Although according to (5.11), the truth function of a generalized binary quantifier $\sim$, $\mathrm{Tf}_\sim$, is a general $0/1$-valued function on 2-column Boolean matrices, the truth functions of common binary quantifiers has always a specific form, depending on the contingency table of the evaluations $(\|\varphi\|_1, \|\psi\|_1), \ldots, (\|\varphi\|_p, \|\psi\|_p)$,

$$
\begin{array}{c|cc}
 & \psi & \neg\psi \\
\hline
\varphi & a & b \\
\neg\varphi & c & d
\end{array}
=
\begin{array}{c|cc}
 & \psi & \neg\psi \\
\hline
\varphi & |\{k \mid \|\varphi\|_k = \|\psi\|_k = 1\}| & |\{k \mid \|\varphi\|_k = 1, \|\psi\|_k = 0\}| \\
\neg\varphi & |\{k \mid \|\varphi\|_k = 0, \|\psi\|_k = 1\}| & |\{k \mid \|\varphi\|_k = \|\psi\|_k = 0\}|
\end{array},
$$

namely the form

$$
\mathrm{Tf}_\sim \begin{pmatrix} \|\varphi\|_1 & \|\psi\|_1 \\ \cdots & \cdots \\ \|\varphi\|_p & \|\psi\|_p \end{pmatrix} = \tau_\sim(a, b, c, d), \tag{5.13}
$$

where $\tau_\sim : \mathbb{N}_0^4 \to \{0, 1\}$. For simplicity, the function $\tau_\sim$ will also be denoted $\mathrm{Tf}_\sim$ and called truth function.

By far most frequently encountered generalized quantifier of the observational calculus is the binary quantifier $\to_\theta$. It asserts that, for the rule $\psi \to_\theta \phi$, the conditional probability of ψ being valid, conditioned on φ being valid is at least a prescribed threshold $\theta \in (0, 1)$. This is commonly called *confidence* of the rule $\varphi \to_\theta \psi$.

$$
P(\|\psi\| = 1 \mid \|\varphi\| = 1) \geq \theta. \tag{5.14}
$$

The truth function of $\to_\theta$ is based on the unbiased estimate of $P(\|\psi\| = 1 \mid \|\varphi\| = 1)$, which is $\frac{a}{a+b}$. Therefore, for the numbers a, b, c, d forming the contingency table,

$$
\mathrm{Tf}_{\to_\theta}(a, b, c, d) = \begin{cases} 1 & \text{if } \frac{a}{a+b} \geq \theta, \\ 0 & \text{else.} \end{cases} \tag{5.15}
$$

In addition, it is required that the simultaneous validity of φ and ψ has at least a prescribed *support* $s \in (0, 1)$, i.e., $P(\|\varphi\| = \|\psi\| = 1) \geq s$. Replacing in this condition $P(\|\varphi\| = \|\psi\| = 1)$ by its unbiased estimate, which is $\frac{a}{p}$, yields

$$
a \geq sp, \text{ recalling that } p = a + b + c + d. \tag{5.16}
$$

The requirement for φ and ψ to have a prescribed minimal support, and the corresponding condition (5.16), is actually nothing specific for $\rightarrow_\theta$. It is usually considered with any binary generalized quantifier.

Originally, the quantifier fulfilling (5.15)–(5.16) was in observational calculus called *founded implication* [7]. Since the seminal paper [9], however, rules that hold with this quantifier are mainly known under the name *association rules*.

Notice that, due to (5.15)–(5.16), $\rightarrow_\theta$ has the following property:

$$\text{Tf}_{\rightarrow_\theta}(a, b, c, d) = 1, a' \geq a, b' \leq b, c', d' \in \mathbb{N}_0 \rightarrow \text{Tf}_{\rightarrow_\theta}(a', b', c', d') = 1. \tag{5.17}$$

Generalized quantifiers with this property are called *implicational* in the observational calculus. In Sect. 5.1.4, we will learn another quantifier from this group.

Besides implicational quantifiers, there is another specific group of generalized quantifiers, called *associational*. Their truth functions fulfil

$$\text{Tf}_{\sim}(a, b, c, d) = 1, a' \geq a, b' \leq b, c' \leq c, d' \geq d \rightarrow \text{Tf}_{\sim}(a', b', c', d') = 1. \tag{5.18}$$

Notice that (5.17) entails (5.18), thus implicational quantifiers are a subgroup of associational, in particular, $\rightarrow_\theta$ is also associational. In addition, three commonly used generalized quantifiers based on probability estimation belong to associational, but not to implicational:

- The quantifier *double founded implication*, $\leftrightarrow_\theta$ asserts that the conditional probability of the simultaneous validity of both φ and ψ conditioned by the validity of any one of them, $P(\|\varphi\| = \|\psi\| = 1 | \|\varphi\| = 1 \vee \|\psi\| = 1)$, is at least θ. Replacing again that probability with its unbiased estimator leads to the truth function

$$\text{Tf}_{\leftrightarrow_\theta}(a, b, c, d) = \begin{cases} 1 & \text{if } \frac{a}{a+b+c} \geq \theta, \\ 0 & \text{else.} \end{cases} \tag{5.19}$$

- The quantifier *founded equivalence*, $\equiv_\theta$ asserts that the probability of φ and ψ being either both valid of both invalid, or equivalently, the probability of φ and ψ evaluating equally, $P(\|\varphi\| = \|\psi\|)$, is at least θ. Replacing that probability with its unbiased estimator yields

$$\text{Tf}_{\equiv_\theta}(a, b, c, d) = \begin{cases} 1 & \text{if } \frac{a+d}{p} \geq \theta, \\ 0 & \text{else.} \end{cases} \tag{5.20}$$

- The quantifier *simple deviation*, $\sim_\delta$ asserts that the logarithmic interaction of φ and ψ is above a prescribed threshold $\delta \in (0, 1)$, i.e.,

$$\ln \frac{P(\|\varphi\| = \|\psi\| = 1) P(\|\varphi\| = \|\psi\| = 0)}{P(\|\varphi\| = 1, \|\psi\| = 0) P(\|\varphi\| = 0, \|\psi\| = 1)} > \delta. \tag{5.21}$$

Notice that from (5.21) follows

$$\sqrt{P(\|\varphi\| = \|\psi\| = 1)P(\|\varphi\| = \|\psi\| = 0)} >$$
$$> e^{\delta/2}\sqrt{P(\|\varphi\| = 1, \|\psi\| = 0)P(\|\varphi\| = 0, \|\psi\| = 1)}, \tag{5.22}$$

where $e^{\delta/2} > 1$, thus the geometric mean of probabilities of cases when φ and ψ are either both valid or both invalid is higher than the geometric mean of probabilities of cases when one of them is valid and the other invalid. For the truth function of this quantifier, the probabilities $P(\|\varphi\| = \|\psi\| = 1)$, $P(\|\varphi\| = \|\psi\| = 0)$, $P(\|\varphi\| = 1, \|\psi\| = 0)$ and $P(\|\varphi\| = 0, \|\psi\| = 1)$ in (5.21) are replaced, respectively, by their unbiased estimates a, d, b and c, yielding the consistent estimate $\ln \frac{ad}{bc}$ of the logarithmic interaction of φ and ψ, and the truth function

$$\mathrm{Tf}_{\sim_\delta}(a, b, c, d) = \begin{cases} 1 & \text{if } \frac{ad}{bc} > \delta, \\ 0 & \text{else.} \end{cases} \tag{5.23}$$

Finally, there are also two generalized quantifiers based on probability estimation not having even the weaker property (5.18), the quantifiers *above average*, $\rightarrow_\delta^+$, and *below average*, $\rightarrow_\delta^-$, with $\delta \in (0, 1)$. Their assertions are

$$\frac{P(\|\psi\| = 1 | \|\varphi\| = 1)}{P(\|\psi\| = 1)} \geq 1 + \delta, \text{ and } \frac{P(\|\psi\| = 1 | \|\varphi\| = 1)}{P(\|\psi\| = 1)} \leq 1 - \delta, \text{ respectively.} \tag{5.24}$$

The truth functions are again obtained by replacing the probabilities in (5.24) with their unbiased estimates, therefore

$$\mathrm{Tf}_{\rightarrow_\delta^+}(a, b, c, d) = \begin{cases} 1 & \text{if } \frac{a}{a+b} \geq (1 + \delta)\frac{a+c}{p}, \\ 0 & \text{else,} \end{cases} \tag{5.25}$$

$$\mathrm{Tf}_{\rightarrow_\delta^-}(a, b, c, d) = \begin{cases} 1 & \text{if } \frac{a}{a+b} \leq (1 - \delta)\frac{a+c}{p}, \\ 0 & \text{else.} \end{cases} \tag{5.26}$$

5.1.4 Observational Rules Based on Hypotheses Testing

A number of further generalized quantifiers assert results of testing some properties of the 2-dimensional random vector resulting from the composition of $(\|\varphi\|, \|\psi\|)$ with values $[x]_1, \ldots, [x]_n$ of features. Each such statistical test has three essential ingredients:

(i) *Null hypothesis* that the distribution P belongs to some particular set $\mathcal{D}_0$ of probability distributions, $H_0 : P \in \mathcal{D}_0$. Each null hypothesis is always tested on an initial assumption that P definitely belongs to some broader set of distributions $\mathscr{D} = \mathscr{D}_0 \cup \mathscr{D}_1$, where $\mathscr{D}_1$ is a set of distributions disjoint from $\mathscr{D}_0$. The $\mathscr{D}_1$-analogy of H_0, $H_1 : P \in \mathscr{D}_1$, is called *alternative hypothesis* to H_0 with respect to $\mathscr{D}$.

(ii) *Test statistic S*, which is a function of $[x]_1, \ldots, [x]_n$. Because the values of the features are realizations of random variables, also the value of S is a realization of a random variable. All tests used in the definition of generalized quantifiers actually use test statistics computed by means of the contingency table of the evaluations $(\|\varphi\|_1, \|\psi\|_1), \ldots, (\|\varphi\|_p, \|\psi\|_p)$. Consequently, (5.13) is valid also for quantifiers based on hypotheses testing.

(iii) *Critical region c_α* for a *significance level* $\alpha \in (0, 1)$ is a subset of real numbers (usually, an interval or the union of two intervals) that has the property

$$(\forall P \in \mathscr{D}_0)\ P(S \in c_\alpha) \leq \alpha, \tag{5.27}$$

or the weaker property

$$(\forall P \in \mathscr{D}_0)\ \lim_{p \to \infty} P(S \in c_\alpha) \leq \alpha. \tag{5.28}$$

The critical region together with the test statistic determine how the test is performed, in the following way: the null hypothesis H_0 is rejected at the significance level α in favour of the alternative hypothesis H_1 in those cases when $S \in c_\alpha$. If (5.27) is fulfilled, then the risk that we commit the error of rejecting a valid hypothesis, aka *error of the 1st kind*, is at most α. If only (5.28) is fulfilled, then that error cannot be restricted, we only know that for any prescribed $\varepsilon > 0$, there exists a number of data p_ε such that for at least as many data, the error of the 1st kind will be at most $\alpha + \varepsilon$. However, we don't know how large p_ε for a particular ε is. In that case, we call α an *asymptotic significance level*, and the test an *asymptotic test*.

The way how statistical tests are performed, and its connection with the condition $S \in c_\alpha$ suggest two natural possibilities how to define the truth function of a generalized quantifier $\sim$ based on hypotheses testing:

$$\text{either Tf}_\sim(a, b, c, d) = \begin{cases} 1 & \text{if } t \in c_\alpha, \\ 0 & \text{else,} \end{cases} \tag{5.29}$$

$$\text{or Tf}_\sim(a, b, c, d) = \begin{cases} 1 & \text{if } t \notin c_\alpha, \\ 0 & \text{else.} \end{cases} \tag{5.30}$$

An advantage of the definition (5.29) is that the validity of $\varphi \sim \psi$ then coincides with the rejection of H_0 at the significance level α in favour of H_1. Hence, apart from the error of the 1st kind, it can be interpreted as the validity of H_1. That is why quantifiers with a truth function defined according to (5.29) are employed much more often than those with a truth function defined according to (5.30), and we restrict attention only to them.

Several quantifiers based on hypotheses testing are actually counterparts of quantifiers based on probability estimation. All of them assert that some particular probability is greater than some threshold $\theta \in (0, 1)$. For example, the counterpart of the most common generalized quantifier $\rightarrow_\theta$ is the quantifier $\rightarrow_{\theta,\alpha}$, called *lower critical implication*, which similarly to $\rightarrow_\theta$ asserts $P(\|\psi\| = 1 | \|\varphi\| = 1) > \theta$. However, instead on an unbiased estimate of that probability, these quantifiers derive the assertion from the test, at a significance level α, of H_0 that the probability is less or equal to θ by means of the binomial distribution, in particular $H_0 : P(\|\psi\| = 1 | \|\varphi\| = 1) \leq \theta$ for $\rightarrow_{\theta,\alpha}$. Then as was noted above, the validity of $\varphi \sim \psi$ can be interpreted as the validity of H_1, in particular $H_1 : P(\|\psi\| = 1 | \|\varphi\| = 1) > \theta$ for $\rightarrow_{\theta,\alpha}$, which is the intended assertion. Notice that for the sets $\mathscr{D}_0$ and $\mathscr{D}_1$ of probability distributions, the hypotheses H_0 and H_1 for $\rightarrow_{\theta,\alpha}$ entail

$$\mathscr{D}_0 = \{P - \text{probability on } \{0, 1\}^2 | P(1, 1) \leq \theta P(\{(1, 0), (1, 1)\})\}, \quad (5.31)$$

$$\mathscr{D}_1 = \{P - \text{probability on } \{0, 1\}^2 | P(1, 1) > \theta P(\{(1, 0), (1, 1)\})\}. \quad (5.32)$$

The test statistic for testing H_0 against H_1 is a random variable with the binomial realizations $\sum_{i=a}^{a+b} \binom{a+b}{i} \theta^i (1 - \theta)^{a+b-i}$, and the critical region for a significance level α is $(0, \alpha]$. Then according to (5.29), the truth function of the quantifier $\rightarrow_{\theta,\alpha}$ is

$$\text{Tf}_{\rightarrow_{\theta,\alpha}}(a, b, c, d) = \begin{cases} 1 & \text{if } \sum_{i=a}^{a+b} \binom{a+b}{i} \theta^i (1 - \theta)^{a+b-i} \leq \alpha, \\ 0 & \text{else.} \end{cases} \quad (5.33)$$

It can be shown that (5.33) implies the condition (5.17), thus $\rightarrow_{\theta,\alpha}$ belongs to implicational quantifiers.

Apart from $\rightarrow_{\theta,\alpha}$, two other commonly used quantifiers are based on a test by means of the binomial distribution:

- The quantifier *double lower critical implication*, $\leftrightarrow_{\theta,\alpha}$ is a counterpart of $\leftrightarrow_\theta$ because it asserts $H_1 : P(\|\varphi\| = \|\psi\| = 1 | \|\varphi\| = 1 \lor \|\psi\| = 1) > \theta$. Its test statistic has the binomial realizations $\sum_{i=a}^{a+b+c} \binom{a+b+c}{i} \theta^i (1 - \theta)^{a+b+c-i}$, its critical region for a significance level α is again $(0, \alpha]$, and its truth function is

$$\text{Tf}_{\leftrightarrow_{\theta,\alpha}}(a, b, c, d) = \begin{cases} 1 & \text{if } \sum_{i=a}^{a+b+c} \binom{a+b+c}{i} \theta^i (1 - \theta)^{a+b+c-i} \leq \alpha, \\ 0 & \text{else.} \end{cases} \quad (5.34)$$

- The quantifier *lower critical equivalence*, $\equiv_{\theta,\alpha}$, is a counterpart of $\equiv_\theta$ because it asserts $H_1 : P(\|\psi\| = \|\varphi\|) > \theta$. It is characterized by the test statistic that has realizations $\sum\limits_{i=a+d}^{n} \binom{n}{i}\theta^i (1 - \theta)^{n-i}$, the critical region $(0, \alpha]$, and the truth function

$$
\mathrm{Tf}_{\equiv_{\theta,\alpha}}(a, b, c, d) = \begin{cases} 1 & \text{if } \sum\limits_{i=a+d}^{n} \binom{n}{i}\theta^i (1 - \theta)^{n-i} \le \alpha, \\ 0 & \text{else.} \end{cases} \tag{5.35}
$$

Tests relying on the binomial distribution are, however, not the only ones used in the definitions of generalized quantifiers. Actually, the two most frequently encountered quantifiers based on hypotheses testing use tests of the null hypothesis of independence of the evaluations $\|\varphi\|$ and $\|\psi\|$) against the alternative of their positive dependence. For the sets $\mathcal{D}_0$ and $\mathcal{D}_1$ of probability distributions, these hypotheses entail:

$$
\begin{aligned}
\mathcal{D}_0 = \{P - \text{ probability on } &\{0, 1\}^2 | P(1, 1) = P(\{(1, 1), (1, 0)\})P(\{(1, 1), (0, 1)\}), \\
&P(1, 0) = P(\{(1, 1), (1, 0)\})P(\{(1, 0), (0, 0)\}), \\
&P(0, 1) = P(\{(0, 1), (0, 0)\})P(\{(1, 1), (0, 1)\}), \\
&P(0, 0) = P(\{(1, 0), (0, 0)\})P(\{(0, 1), (0, 0)\})\}
\end{aligned} \tag{5.36}
$$

$$
\begin{aligned}
\mathcal{D}_1 = \{P - \text{ probability on } &\{0, 1\}^2 | P(1, 1) > P(\{(1, 1), (1, 0)\})P(\{(1, 1), (0, 1)\}), \\
&P(1, 0) < P(\{(1, 1), (1, 0)\})P(\{(1, 0), (0, 0)\}), \\
&P(0, 1) < P(\{(0, 1), (0, 0)\})P(\{(1, 1), (0, 1)\}), \\
&P(0, 0) > P(\{(1, 0), (0, 0)\})P(\{(0, 1), (0, 0)\})\}.
\end{aligned} \tag{5.37}
$$

In statistics, also other alternatives to (5.36) are sometimes used, namely negative dependence, with inequalities opposite to those in (5.37), and general dependence, i.e., positive or negative. In generalized quantifiers, however, only positive dependence has been considered so far.

The two tests of independence underlying the frequently encountered generalized quantifiers are the one-sided Fisher exact test and the χ^2 asymptotic test. Their test statistics and critical regions take into account the fact that a consistent estimate of $\frac{P(1,1)P(0,0)}{P(1,0)P(0,1)}$ is $\frac{ad}{bc}$. Due to that, the definitions of the critical regions allow the null hypothesis of independence to be rejected in favour of the alternative hypothesis of positive dependence only if $\frac{ad}{bc} > 1$. In particular, the one-sided Fisher exact test of independence uses the test statistics with realizations

$$
t = \sum_{i=a}^{\min(a+b,a+c)} \frac{\binom{a+c}{i}\binom{b+d}{a+b-i}}{\binom{a+b+c+d}{a+b}} \tag{5.38}
$$

and the critical region for a significance level α

$$c_\alpha = \begin{cases} (0, \alpha] & \text{if } ad > bc, \\ \emptyset & \text{else.} \end{cases} \tag{5.39}$$

whereas the χ^2 asymptotic test of independence uses the test statistics with realizations

$$t = \frac{(a+b+c+d)(ad-bc)^2}{(a+b)(c+d)(a+c)(b+d)}, \tag{5.40}$$

and the critical region for a significance level α

$$c_\alpha = \begin{cases} [\chi_1^2(1-\alpha), \infty) & \text{if } ad > bc, \\ \emptyset & \text{else,} \end{cases} \tag{5.41}$$

where $\chi_1^2(1-\alpha)$ denotes the $1-\alpha$ quantile of the χ^2 distribution with one degree of freedom.

Based on those tests are the following two generalized quantifiers, parametrized by the considered significance level α: the *Fisher quantifier* $\sim_\alpha^F$, with the truth function

$$\text{Tf}_{\sim_\alpha^F}(a, b, c, d) = \begin{cases} 1 & \text{if } ad > bc, \quad \sum_{i=a}^{\min(a+b,a+c)} \frac{\binom{a+c}{i}\binom{b+d}{a+b-i}}{\binom{a+b+c+d}{a+b}} \leq \alpha, \\ 0 & \text{else,} \end{cases} \tag{5.42}$$

and the χ^2 quantifier $\sim_\alpha^{\chi^2}$, with the truth function

$$\text{Tf}_{\sim_\alpha^{\chi^2}}(a, b, c, d) = \begin{cases} 1 & \text{if } ad > bc, \quad \frac{(a+b+c+d)(ad-bc)^2}{(a+b)(c+d)(a+c)(b+d)} \geq \chi_1^2(1-\alpha), \\ 0 & \text{else.} \end{cases} \tag{5.43}$$

5.1.5 Classification Rules in Spam Filtering

Classification rules in the form of implications $\phi \to C$, where C denotes a set of target message classes, is widespread in email message filtering. In many of the email user agents, users are even allowed to create their own rules based on the content of any message field. Incoming messages are then passed through the list of user-defined rules and actions associated with the target class are executed on such messages.

One of the main advantages of classification rule approaches is that categorical data can be utilised directly without a transformation to the space of numerical values. For example, IPv4 addresses of the first Mail Transfer Agent that introduced the message to the Internet can be used directly in the condition, Instead of transforming its address into almost 2^{32} one-hot encoded features (some addresses are not allowed or are reserved). Recall that one-hot encoded IPv6 address would take almost 2^{128} features. Selecting ranges of the IP addresses may decrease the dimensionality but will inevitably lose the required specificity on that feature.

The application of text matching rules is also much more flexible in the rule-based systems. If a particular sequence of words needs to be recognised in the message body, there is no need to generate all possible n-grams from a dictionary, where n is the number of words in the longest sequence, to capture such feature. In a rule-based system, only the particular sequence of words (or more generally tokens) needs to be stored.

Although certain handcrafted blacklists (trivial rule based systems) may be used in spam filters until today, we will discuss the available approaches to learning such classification rules. Therefore, we should take into account also the speed of the rule learning and more importantly, the speed of pruning. The speed requirement also favours the rule-based systems, as the state-of-the-art extensions to reduced error pruning achieve a complexity of $\mathcal{O}(n \log^2 n)$ [4], which allows us to train the rule-based classifier even on large datasets.

The rules are most commonly constructed with a *separate-and-conquer* strategy, so that each rule tries to cover only a part of the given training set (the *separate* part). This data is then removed from the training set, and recursion of the learning algorithm tries to cover some of the remaining samples (the *conquer* part). More on this approach is presented in [10].

In the case of binary classification, such as spam filtering, the rules are created only for the positive class, while the negative class is left as a default. All of the created rules are therefore in the form: $\phi \rightarrow C_{spam}$.

The creation of rule set, based on reduced error pruning introduced in [3], then consists of two primary operations: rule growth and pruning. To this end, the available data are split into a growing and pruning set.

To illustrate the rules generated by RIPPER, we provide, in Fig. 5.1, a ruleset trained on the count of words in messages from the SpamAssassin dataset with an implementation called JRip from the Weka (Waikato Environment for Knowledge Acquisition) Workbench [11]. Contrary to the previous examples, we also introduced the list of non-local IPv4 addresses from Received fields. Therefore, we should be able to capture the address of a server that introduced the spam message to the internet in the feature space.

Although the accuracy on test data (newer messages from the SpamAssassin dataset) is lower than in many of the previously mentioned classification algorithms (our experiments state the accuracy of 90.81%), the derived rules are comprehensible to the user of the system. For example, messages incoming from server 193.120.211.219 that are not a reply, do not mention Perl and contain the word "you" more than twice are considered spam. Some rules then seem apparent: word "your" six or more times, poorly formatted HTML message with word "nbsp" in plaintext or messages containing the word "Nigeria" twice or more will lead to classification as spam.

our $\geq 1 \wedge$ please $\geq 1 \wedge$ it $\leq 3 \wedge$ writes $\leq 0 \rightarrow C_{spam}$

`193.120.211.219` $\wedge$ re $\leq 0 \wedge$ you $\geq 2 \wedge$ perl $\leq 0 \rightarrow C_{spam}$

nbsp $\geq 1 \wedge$ charset $\leq 0 \rightarrow C_{spam}$

your $\geq 6 \rightarrow C_{spam}$

it $\leq 0 \wedge$ date $\leq 0 \wedge$ re $\leq 0 \wedge$ your $\geq 1 \wedge$ you $\geq 1 \rightarrow C_{spam}$

`193.120.211.219` $\wedge$ visit $\geq 1 \rightarrow C_{spam}$

re $\leq 0 \wedge$ url $\leq 0 \wedge$ the $\leq 0 \wedge$ from $\leq 0 \wedge$ list $\leq 0 \wedge \neg$`64.161.22.236` $\wedge$ spambayes ≤ 0 $\rightarrow C_{spam}$

offer $\geq 1 \wedge$ but $\leq 0 \wedge$ you $\geq 1 \rightarrow C_{spam}$

visit $\geq 1 \wedge$ in $\leq 1 \rightarrow C_{spam}$

here $\geq 1 \wedge$ click $\geq 1 \wedge$ but $\leq 0 \rightarrow C_{spam}$

`216.136.204.119` $\rightarrow C_{spam}$

nigeria $\geq 2 \rightarrow C_{spam}$

the $\leq 0 \wedge$ wm $\geq 1 \rightarrow C_{spam}$

Fig. 5.1 ruleset for spam classification trained on SpamAssassin dataset with JRip. Rules are based on count of words in the message and presence of IP address in any `Received` header, the later denoted in typewriter font

5.1.6 Classification Rules in Recommender Systems

Opposed to spam filtering, where the right-hand side of implication (consequent, succedent) is chosen from the available set of classes, recommender systems use a more general approach where both sides of the implication are taken from the sets of available features.

Usually, implications used in recommender systems take a form of $A \rightarrow_\theta S$, here both A (antecedent) and S (succedent) are constructed as a conjunction of literals in which each predicate (or its negation in some of the systems) occurs at most once.

The approach employed in the system called GUHA (General Unary Hypotheses Automaton), as presented in [7], are one of the oldest data mining methods. They aim at a systematic extraction of all interesting hypotheses from provided data (given the theoretical background of observational calculus in Sect. 5.1.2) and are most suitable for an exploratory analysis with dozens of features. Many recommender systems, however, usually operate on thousands of features.

One of the possible association rule applications to recommender systems is known as market basket analysis, where individual rows of the input data represent transactions of anonymous users and features (columns) correspond to the purchase of a given product in the respective transaction. This allows discovering groups of products that are commonly bought together and recommend the currently shopping user a list of products that may complement the current purchase. In an offline scenario, market basket analysis may be used for supermarket floor planning to maximise the number of different supermarket areas visited by the clients in the shop and, therefore, to increase exposure to as many products as possible, and consequently profit. If the identity of a user is known and some sociodemographic attributes are available, these attributes might be used in the antecedent of the association rule. Such recommendations are then based on the user profile, not on the current content of the basket, which eliminates the problem of a cold start (no recommendation can be given if no input data is available).

The major caveat of this task is the volume of available data and simultaneously its sparseness. Take, for example, the UCI dataset on Online Retail presented in [12]. The dataset contains over 22 thousand transactions on more than four thousand distinct items, where only 0.59% of the transaction-item matrix has non-zero values. Exhaustive exploratory analysis is unfeasible.

A straightforward remedy is to select *large itemsets*, i.e. sets of items with support over a threshold θ, prior to the rule mining. The pre-selection of large itemsets limits the set of items considered later in the association rule mining and, therefore, speeds up the rule mining significantly. For example, if only the itemsets $\{i_1, i_2\}$, $\{i_1, i_4\}$ and $\{i_2, i_4\}$ of four items with more than two items have sufficient support, only the association rules $i_1 \rightarrow_\theta i_2, i_2 \rightarrow_\theta i_1, i_1 \rightarrow_\theta i_4, i_4 \rightarrow_\theta i_1, i_2 \rightarrow_\theta i_4$ and $i_4 \rightarrow_\theta i_2$ need to be checked for sufficient confidence.

The original implementation of the well-known algorithm Apriori [9] starts with itemsets containing individual items (called 1-itemsets), counts transactions that contain the considered item and keeps only the itemsets that have sufficient support. In further iterations, k-itemsets are constructed by joining $(k-1)$-itemsets, and support for each k-itemset is counted by another full pass over a list of transactions.

While this approach uses a minimal amount of memory for the support counters, repeated passes over the transaction data (that did not fit into memory) were required. Therefore, different approaches that tried to count several itemset sizes at once and pruned the itemset lattice even during one pass were designed. One of the algorithms called Dynamic Itemset Counting [13] checks the itemset counters in regular intervals (of M transactions), and if some of the counters surpass the support threshold, the joined itemset starts counting. If an itemset does not achieve the threshold when presented with all available transactions, it is marked as *small itemset* and no successors in the itemset lattice are considered. This approach allows gathering all large itemsets in less than two passes over transaction list with small M and randomly distributed data. For an illustration of the process, see Fig. 5.2.

In Table 5.3, we present an excerpt of rules extracted from the UCI Online Retail dataset with R implementation of Apriori ordered by the lift–rule confidence divided by the support of antecedent. The preprocessing of the data included a selection of only purchase transactions (`InvoiceNo` does not start with "C") and filtering-out items with dubious descriptions (containing words such as "wrong", "lost" or "crushed"). Filtered `StockCode` fields were then grouped by the transactions (`InvoiceNo`) and rules with support of at least 0.005 and confidence above 0.75 were detected.

The rules extracted by Apriori (or other association rule miner) are, however, too much focused on the task of market basket analysis. Rules required in recommender systems are different in the sense that the antecedent is already available for the

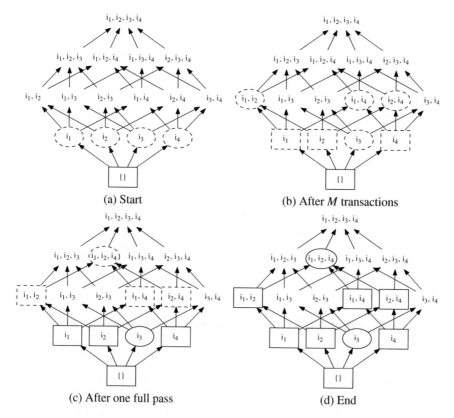

Fig. 5.2 States of itemsets of four items in the itemset lattice. Boxes denote large itemsets (support over θ), ellipses small itemsets (support under θ); dashed contour denotes that counting for given itemset is not finished, full contour denotes that all transactions were counted

current user and only a search for appropriate succedents (recommendations) in the given data is required. Another issue with earlier mentioned approaches is that the required level of support has to be specified before the rule mining and that all rules exceeding such support are mined. In recommender systems, only a limited number of recommendations can be given to the user, and minimal required rule support is not the same for all users and items.

Efficient Adaptive-Support Association Rule Mining for Recommender Systems [14] first tries to gather a required number of association rules from the perspective of the given user. Support threshold is adapted to available data, and if an insufficient number of rules is extracted, mining is repeated with a lower threshold. If the number of rules is still too low, associations over items are considered. These associations are mined offline and separately for each item so that new items with insufficient interactions are not disadvantaged due to the low overall support. The mentioned paper also deals with transformation of the rating matrix into the set of *like* (and *dislike*) transactions.

Table 5.3 Ten rules with highest lift extracted from the UCI Online Retail Dataset with R library arules. Item descriptions are shortened

Rule	Supp.	Conf.	Lift
{spaceboy bowl, dolly cup, spaceboy cup} → {dolly bowl}	0.0053	0.981	103.2
{spaceboy bowl, dolly cup} → {dolly bowl}	0.0058	0.959	100.9
{dolly bowl, spaceboy bowl, spaceboy cup} → {dolly cup}	0.0053	0.955	98.4
{xmas decoration bell} → {xmas decoration heart}	0.0054	0.880	98.2
{dolly bowl, spaceboy cup} → {dolly cup}	0.0057	0.942	97.1
{pink beaker, red beaker} → {green beaker}	0.0050	0.842	91.5
{blue beaker, green beaker} → {pink beaker}	0.0057	0.851	89.6
{paper cups, paper napkins} → {paper plates}	0.0050	0.857	89.2
{thyme, rosemary, parsley, mint, basil} → {chives}	0.0078	0.912	87.7
{dolly bowl, dolly cup, spaceboy cup} → {spaceboy bowl}	0.0053	0.930	87.3

5.1.7 Classification Rules in Malware and Network Intrusion Detection

Also in malware detection and network intrusion detection, the most often encountered is association rule learning, i.e., learning of rules that hold with the quantifier founded implication. Not surprisingly, rule learning was first employed in Windows malware detection, whereas in recent years, applications to Android malware detection prevail.

The authors of [15] proposed to use a modification of the well known algorithm for association rules learning, Apriori [9], for malware detection in collections of Windows portable executables incrementally updated with new encountered malware. That modification, called *Efficient incremental associative classification algorithm (EIAC)*, makes an explicit use of the fact that the consequents of the learned rules state the membership of the input in one of the classes malware and benign software. Because of the incremental updating of the collection of executables, EIAC repeats the following three steps:

(i) update the original malware rules,
(ii) update the original benign rules,
(iii) generate new malware rules and add them to the original ones.

The EIAC algorithm has been validated on a comprehensive collection of portable executables provided by the anti-virus laboratory of the KingSoft corporation. From that collection, 20,000 executables have been used for rules learning, and additional 10,000 executables for testing the learned rules. The accuracy of the algorithm on those test data was 93.5%.

In [16, 17], an approach to malware detection in Windows portable executables is presented, based on a modification of the Apriori algorithm proposed in [18] and called *objective-oriented association mining (OOA)*. Its specificity is that it complements the requirements for a minimal confidence of association rules (5.15) and their minimal support (5.16) with a requirement for a minimal utility, where utility is a function with values determined by domain experts, defined on the set of considered transactions. In terms of the Apriori algorithm, a transaction is a set of items of any considered kind. In the application of OOA to malware detection presented in [16, 17], an item is a system API call and a transaction is the set of all API calls in a portable executable. To this end, the binary code of the executable is disassembled, thus the presented approach does not rely on the information in the executable's import table. As to obfuscated executables, the learned association rules are based only on those that can be unpacked. The approach has been validated on more than 9,000 portable executables, including nearly 2,000 benign ones and more than 7,500 malware instances. Among the malware instances, 1,662 were packed, and only 21 of them could not be unpacked and had to be excluded from rules learning. In the code of the available executables, altogether 6,181 different APIs have been called. For the purpose of rules learning, only two groups of them have been considered:

- 1,000 APIs called in the highest number od executables.
- 1,000 APIs that lead to the highest decrease of entropy if the set of available executables is divided according to whether they call the respective API or not. This is a specific case of impurity decrease, which will be addressed in general in Sect. 5.3.

The union of both above mentioned groups included 1,351 APIs. Rule learning has then been restricted to those APIs. The utility function is based on expert labelling of executables and takes into account different number of benign and malware instances among the available portable executables, it is described in detail in [16]. Moreover, from all the rules fulfilling the requirements for confidence, support and utility, pairs of rules with identical antecedents and consequent in one of them "malware" and in the other "benign software" are finally removed, provided both supports exceed a given threshold. If a new executable is to be classified, all learned rules with the antecedent matched by that executable are found and the confidences of all among them with the consequent "malware" are summed up. separately from confidences of those with consequent "benign software". Then the executable is classified according to the consequent to which corresponds the higher of both sums of confidences. If the executable does not match any of the learned rules, then it is classified as malware for safety reasons. Apart from using the original OOA, [16, 17] presented also its modification retaining a similar accuracy, but learning approximately a third of rules, which increases the detection speed for new executables.

The method presented in [19] differs from the above mentioned applications of association rules to malware detection by aiming specifically at *ransomware*. It performs a static analysis of disassembled Windows portable executables, from which it uses the following three kinds of information:

(i) used assembly instructions, with their frequency;
(ii) called library functions;
(iii) loaded dynamic-link library (DLL) files.

The assembly instructions are used to compute cosine similarity to selected representatives of ransomware and of benign software. Selected function calls are used
to built a signature database for known ransomware types. Finally, the information
about loaded DLLs is used for learning association rules of the kind

$$\mathrm{DLL}_1 \& \dots \mathrm{DLL}_\ell \rightarrow_\theta \mathrm{DLL}_{\ell+1}, \tag{5.44}$$

where $\ell \in \mathbb{N}$ and $\rightarrow_\theta$ is the quantifier founded implication (5.15) with confidence θ
and some support s (5.16). To this end, an algorithm for learning a set of association
rules called frequent pattern growth (FP-growth) [20] is used. The importance of such
rules consists in the fact that rules learned for ransomware are substantially different
from those learned for benign software. In particular, the authors of [19] have found
that if a test set of executables is classified solely by means of DLL association rules
and a threshold of fulfilling at least 60 learned for a given training set of ransomware
executables, then this classification has approximately 70% precision.

As to the detection of Android malware, on rules learning relies the approach *Droid
Detective*, presented in [21]. The Droid Detective first decompresses the Android
application package (apk file) into the executable file and several files associated with
it, among which their approach uses the AndoridManifest.xml file, stating the permissions requested by the application. These permissions, as well as combinations of
2–6 of them, are then used as features in the learned rules (Fig. 5.3). Differently to the
above mentioned applications of rules to malware detection, the rules learned in the
Droid Detective are not association rules, but rules with the quantifier above average,
simplified for the specific case that there are only two possible consequents (in the
Droid Detective again "malware" and "benign software"). The experiments reported
in [21] used malware applications from the Malware Genom project [22], which
comprises 1,260 Android malware samples found between August 2010 and October 2011, and randomly selected 741 benign applications from the official Android
market, Google Play. The malware applications, which are sometimes repacked versions of benign applications with malicious content, primarily sign up for premium

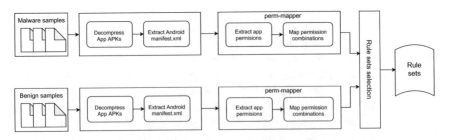

Fig. 5.3 The process of rules extraction in the Droid Detective approach (inspired by [21])

rate service by sending SMS, block incoming SMS confirmations, steal personal information via SMS, and leak contacts via internet connections. Therefore, it is not surprising that such applications request permissions for sending and receiving SMS, accessing contacts or personal information much more frequently than benign applications.

Similarly to the Droid Detective, also the algorithm *Android Risk Patterns Miner (ARP-Miner)* [23] uses rules learned from permission requests in the AndoridMani-fest.xml file for Android malware detection. However, the ARP-Miner differs from the Droid Detective in two important respects:

1. The association rules in ARP-Miner are learned using the Apriori algorithm, but instead of the requirement (5.16), the support is required to fulfil the condition

$$a \geq s(a + c). \tag{5.45}$$

2. Instead of the consequent "malware" in Droid Detective rules, the consequents of ARP-Miner rules correspond to specific malware families.

Also in the experiments with the ARP-Miner reported in [23], the 1260 malware applications from the Malware Genom project [22] and benign applications from the Google Play were used, though the number of the latter was much higher then in the experiments with the ARP-Miner reported in [21]—27,274. Most of the learned malware association rules concern the following malware families:

- *YZHC family* are SMS trojans that sends SMS to premium rate numbers.
- *Kmin family* are also trojans, with the capability of stealing and leaking sensitive information from SMS messages and uploading them to a remote server.
- *GoldDream family* spies on SMS messages and phones calls. It may also turn into a bot controlled by a remote server.
- *ADRD family*. This threat, often injected into a repacked application, tries to steal private information concerning the mobile phone and upload them to a remote server. If no network is available, it tries to open a network connection by itself. It may also send search requests to increase the ranking of the searched site.
- *DroidKungFu family*. It has various variants that may try to get root access, to send private information concerning the phone to a remote server, and to install an app called Legacy, which pretends to be a benign Google Search application, but is actually a backdoor.

Whereas those malware families are the consequents of the association rules learned in the reported experiments, their antecedents reveal several common behaviour patterns of Android malware:

(i) accessing SMS, which may result in privacy leaking and signing up premium rate service;
(ii) collecting private information;
(iii) checking the network setting or automatically changing the network setting without user awareness;

(iv) trying to kill other apps, especially antivirus applications;
(v) installing packages for subsequent attacks.

Finally, also in [24], association rules obtained with the Apriori algorithm were a part of Android malware detection. This time, however, only in such a way that the antecedents of those association rules were used, together with the particle swarm optimization method, for feature extraction in a classifier that performed the malware detection itself. To this end, a naïve Bayes classifier was employed.

In the area of network intrusion detection, association rules were used to help the *detection of coordinated botnet attacks* [25]. Such attacks are particularly difficult to detect, mainly for the following reasons:

- It is necessary to investigate a huge amount of captured data and to look in them for common sequential patterns.
- The number of infections depends on the IP address of the honeypot.
- The number of infections is not stable: there are periods when many infections happen, interleaved with periods when no malware is observed at all.
- The duration of a coordinated attack is not long.
- The pattern of coordinated attacks is constantly renewed as new malware is developed.

In the project reported in [25], association rules concerning coordinated attacks were learned with the Apriori algorithm from the logs of 94 independent honeypots periodically rebooting every 20 min and observing, under the coordination of a Cyber Clean Center, the malware traffic at a Japanese backbone from May 2008 till April 2009.

Also in [26], association rules were used for a rather specific kind of intrusion detection: this time for discovering behaviour patterns of possible attackers in data from darknet, i.e., the part of allocated IP space that contains no advertised services. Because of the absence of legitimate hosts, any darknet traffic is already by its presence aberrant: it is caused either by a malicious intent or by a misconfiguration. In particular, many kinds of malware scan also darknet during their search for the next potential victims.

The data used in [26] were based on a long-term observation over a group of darknet sensors hosted in the NICTER project [27, 28] and monitoring different IP ranges belonging to the darknet. Using the FP-growth algorithm, two kinds of association rules have been learned from that data:

(i) rules describing relationships among destination ports probed by the same scan, i.e., rules ascertaining

$$\text{port}_1 \& \ldots \& \text{port}_k \rightarrow_\theta \text{port}_{k+1}, \tag{5.46}$$

(ii) rules describing relationships among sensors that detected the same scan, ascertaining

$$\text{sensor}_1 \& \ldots \text{sensor}_\ell \rightarrow_\theta \text{sensor}_{\ell+1}, \tag{5.47}$$

The reason why only ports and sensors were analysed using association rules is that most of the observed hosts sent only a couple of connections initializing packets to the darknet, containing solely basic communication information, such as time stamp, source and destination IP address, source and destination ports, and used IP protocols. In spite of that, the resulting rules indicate specificity for certain threats. In particular, a ruleset of the kind (5.46) corresponding to the known Carna botnet [29] have been found.

In [30], three classifiers based on classification rules have been compared on a large set of real-world network intrusion data. In addition, the comparison has included also classification by means of random forests, which will be explained in Sect. 6.2. The classifiers rely on the following methods for rules extraction from data:

- The traditional methods proposed in [31], known as *frequent itemset mining (FISM)*.
- The method called *logical itemset mining (LISM)* [32].
- The method called *Explainer* [33], which actually extracts rules from a specific kind of random forests, thus building a bridge to the compared random forest, consisting of classification trees also representing rules (cf. Sect. 5.3).

The data on which the comparison in [30] has been performed consists of 8 consecutive weeks of network traffic collected by the *Cognitive Threat Analytics (CTA)* intrusion detection system [34]. Among them, the data from the first three weeks have been used for training, and the data from the subsequent five weeks for testing. The data contains about 9 million samples, 3.75 millions recorded during the first three weeks and 5.23 millions recorded during the subsequent five weeks. However, the proportion of positive samples is very small: only 4801 in the training data and 6463 in the testing data. Each sample is a collection of all network events observed on a particular network host within a 24 h window. CTA distinguishes about 300 events, which are divided into four categories:

(i) *signature based*—produced by matching of software behaviour with behavioural signatures handmade by a domain expert;
(ii) *classifier based*—created by classifiers trained on historical data;
(iii) *anomaly based*—created by an anomaly detection engine, which consists of 70 various detectors, based on statistical, volumetric and proximity properties of the input data, as well as on domain knowledge;
(iv) *contextual*—describe various network behaviours providing additional context, e.g., file download, direct access on raw IP, software update.

The rules produced by the compared methods were considered only if they achieved, each separately, a precision at least 80% on the training data. Due to this condition, the overall precision of all methods was quite high also on the test data, at least 94% during all testing weeks. The FISM and Explainer provided stable precision, reaching roughly 99% during all five weeks. On the other hand, the precision of the other two methods varied up to 4% during that time. In particular, the LISM achieved the precision 100% in 4 of the 5 testing weeks, but only around 96%

in the remaining week. Complementarily to precision, the methods were compared also with respect to recall. Here, the differences among them are much larger. The Explainer and random forests achieved recall above 80% during the first four testing weeks and around 60% in the last week. On the other hand, the recall of the FISM was below 60% during the first four weeks and below 20% in the last week, and for the LISM, the mostly highest precision was always combined with an extremely low recall below 10%.

In addition to the precision and recall results, [30] presented several interesting observations concerning the rules produced by the compared methods. Most importantly:

- The random forest obtained using training data consisted of up to 25 levels deep decision trees, entailing a huge number of very long rules.
- Another drawback of the random forest is that many of the rules represented by the path from the root to a leaf of a classification tree correspond to *negative evidence*, i.e., to not fulfilling any of the conditions in the path nodes.
- The Explainer produced only 14 considered rules, three of them of length two, and all the remaining of length one. Whereas they indicate the real cause of a threat, they typically don't provide a context sufficient for a data analyst investigating that threat.
- The LISM produced 34 considered rules of length 3–5. They provide very logical explanations of the respective threat, but samples matching them are very rare.
- The FISM produced 34 considered rules, also of length 3–5. They provide the root cause together with a reasonable amount of additional context, which makes the method suitable for using by security analysts.

5.2 Classification Rules from Genetic Algorithms

Genetic algorithms (GA) are a very important kind of *evolutionary algorithms*, i.e., algorithms that attempt to somehow mimic the biological evolution of species. More precisely, genetic algorithms mimic the evolution of genome. Similarly to other kinds of evolutionary algorithms, they work with *populations*—sets of individuals evolving through a sequence of generations. Each generation is created from the previous one by applying the *selection of individuals* and genetic operators *recombination (crossover)*, which basically consists in swapping the values of some attribute(s) between two individuals and *mutation*, which consists in modifying the values of some attribute(s). The selection of individuals is performed according to what is considered the overall objective of evolution in the considered algorithm, and is quantified by a function called *fitness*. It is defined in some problem-specific way, but always so that individuals with higher fitness are more likely to be selected into the next generation and to contribute to the evolution by means of recombination and mutation.

The schema of a basic genetic algorithm can be summarized as follows:

Algorithm 5 (A general genetic algorithm)
Input:
- fitness function f;
- number of variables n;
- sets $V_1, \ldots, V_n$ of feasible values of the features $[x]_1, \ldots, [x]_n$;
- population size N;
- initial distribution P_0 on $V_1 \times \cdots \times V_n$;
- algorithm for the selection of individuals from the evaluated current generation;
- crossover distribution $P_c(\cdot|x, y)$ on the set $\mathscr{P}(V_1 \times \cdots \times V_n)$ of subsets of $V_1 \times \cdots \times V_n$, conditioned by $x, y \in V_1 \times \cdots \times V_n$ such that $x \neq y$;
- mutation distribution $P_m(\cdot|x)$ on $\mathscr{P}(V_1 \times \cdots \times V_n)$ conditioned by $x \in V_1 \times \cdots \times V_n$;
- algorithm for the selection of the new generation from the union of the old one and the results of crossover and mutation;
- termination criterion.

Step 1. Generate the population of an initial generation G_0 according to the distribution P_0.

Step 2. In every $x \in G_k$, where G_k is the k-th generation, evaluate the fitness $f(x)$.

Step 3. If the termination criterion is fulfilled, return G_k and its evaluation.

Step 4. According to corresponding selection algorithm, select a subset $\Lambda_s \subset G_k$ for the application of crossover and mutation.

Step 5. For each pair $(x, y) \in \Lambda_s^2$ such that $x \neq y$, obtain a set $\Lambda_c(x, y)$ (possibly empty) according to $P_c(\cdot|x, y)$.

Step 6. For each $x \in \Lambda_c$, obtain a set $\Lambda_m(x)$ (possibly empty) according to $P_m(\cdot|x)$.

Step 7. Select G_{k+1} as a subset of $G_k \cup \bigcup_{(x,y)\in\Lambda_s^2, x\neq y} \Lambda_c(x, y) \cup \bigcup_{x\in\Lambda_s} \Lambda_m(x)$, according to corresponding selection algorithm.

Step 8. Increment k and return to Step 2.

Further interesting information about genetic and other evolutionary algorithms, as well as about their applications in classification, can be found in various specialized monographs, e.g. [35–39].

5.2.1 Main GA Approaches to Classification Rules Learning

Genetic algorithms can be successfully used to learn classification rules from data. However, differently to the most common kind of GA described in Algorithm 5, GA for rules learning use features $[x]_1, \ldots, [x]_n$ in connection with particular conditions on those features, such as $[x]_j \in S$ with some set S of values if $[x]_j$ is a discrete feature, and $[x]_j < v$ or $[x]_j > v$ with $v \in \mathbb{R}$ if $[x]_j$ is a continuous feature. This entails substantially more complicated ways of representing the individuals and defining

the fitness, as well as of specifying the initial distribution and distributions for mutation and crossover. In particular, in an implication $\varphi \to \psi$, only the φ-part (called
antecedent in the context of rules learning) is explicitly represented, whereas the
ψ-part (called *consequent*) is typically represented only implicitly, through grouping rules with the same consequent together. As to the fitness, it typically takes into
account, apart from some measures of classification performance introduced in Sect.
2.2, also the complexity of the individual rules or of the resulting set of learned rules,
which is measured, e.g., by the number of variables occurring in the antecedent,
by lengths of the included conjunctions, or by the number of rules in the resulting
ruleset. Finally, the initial population of rules is obtained as a population of sufficiently general rules that describe a sufficiently large proportion of vectors randomly
sampled from $V_1 \times \cdots \times V_n$.

For the learning of rules from data, two main evolutionary approaches have been
developed, differing with respect to what are the individuals evolved by the algorithm.

In the more straightforward one, called *Michigan approach*, the individuals are
single rules. That approach is easy to implement and fast to run, however, it has two
serious drawbacks:

- The areas in the input space to which different rules apply can very much overlap.
  Consequently, even if a large number of rules is selected, they still can cover the
  input space insufficiently.
- Different rules can be correlated, which as a consequence leads to the redundancy
  of some rules in the resulting ruleset.

The first drawback can be to a large extent supressed using the technique of *sequential
covering*: whenever a rule is selected, the input data covered by that rule are removed
from the input data set. On the other hand, the second drawback is inherent to the
Michigan approach and cannot be supressed in it because correlation and redundancy
are not properties of individual rules, but of pairs or groups of them, and consequently,
they cannot be taken into account in the fitness of individual rules.

Both above mentioned drawbacks are easily supressed if the individuals are whole
data sets because then the fitness function can take into account the covering of the
input space, as well as the correlation and redundancy of their members. This is
adopted in the other evolutionary approach to rules learning from data, which is
called *Pittsburgh approach*. It has, however, much higher computational demands
than the Michigan approach.

5.2.2 Genetic Operations for Classification Rules

Differently to the GA described in Algorithm 5, GA for rules learning from data use
three different kinds of crossover:

(i) *Standard crossover*, which consists in swapping the values of an attribute or
    several attributes between two individuals.

(ii) *Generalization crossover* consists in producing a child for which each feature $[x]_j$ fulfils either the condition fulfilled by the first parent or the one fulfilled by the second parent, i.e., it fulfils the disjunction of both conditions. For example,

$$\text{if in rule 1: } 18 \leq \text{ age } < 26, \text{ and in rule 2: } 20 < \text{ age } \leq 30,$$
$$\text{then in their child rule: } 18 \leq \text{ age } \leq 30. \quad (5.48)$$

In some cases, the disjunction of the parent conditions actually covers the whole value set V_j of the feature $[x]_j$. Then that disjunction can be omitted from the child rule, thus reducing its complexity.

(iii) *Specialization crossover* consists in producing a child for which each feature $[x]_j$ fulfils both conditions fulfilled by the parents, i.e., it fulfils their conjunction. In the context of the previous example (5.48),

$$\text{if in rule 1: } 18 \leq \text{ age } < 26, \text{ and in rule 2: } 20 < \text{ age } \leq 30,$$
$$\text{then in their child rule: } 20 < \text{ age } < 26. \quad (5.49)$$

In some cases, the set covered by the conjunction of the parent conditions is empty, then the rule has to be removed from the ruleset.

Generalization crossover and specialization crossover are typically applied more frequently than standard crossover. Moreover, the frequency of their use increases with increasing fitness of the recombined rules. The reason is that these two kinds of crossover tend to produce a child with high fitness if the fitness of both parents was high, which is not the case of standard crossover.

Generalization and specialization can be achieved also through mutation. For rules involving only one variable, no specific kinds of mutations are needed to this end, differently to the case of crossover. In particular, a generalization of a rule $[x]_j < v$ is any rule

$$[x]_j < v_g \text{ with } v_g > v, \quad (5.50)$$

and its specialization is any rule

$$[x]_j < v_s \text{ with } v_s < v. \quad (5.51)$$

For a rule $[x]_j > v$, the inequalities in (5.50) and (5.51) are opposite.

On the other hand, specific kinds of mutations can be needed for rules involving more variables. For example, a conjunctive rule

$$\bigwedge_{i=1}^{k} R_k \text{ with } R_j \in \{[x]_j < v_j, [x]_j > v_j\}, v_j \in \mathbb{R} \text{ for } j = 1, \ldots, k \quad (5.52)$$

can be generalized through dropping out any of the conjuncts $R_1, \ldots, R_k$ and specialized through adding another conjunct $R_{k+1} \in \{V_{k+1} < v_{k+1}, V_{k+1} > v_{k+1}\}$, $v_{k+1} \in \mathbb{R}$.

Further information about the learning of rules from data by means of evolutionary algorithms can be found in the monograph [40].

5.3   Rules Obtained from Classification Trees

A classifier $\phi : \mathcal{X} \to C = \{c_1, \ldots, c_m\}$ is called *classification tree*, more precisely binary classification tree, if there is a binary tree $T_\phi = (V_\phi, E_\phi)$ with vertices V_ϕ and edges E_ϕ such that:

1. $V_\phi = \{v_1, \ldots, v_L, \ldots, v_{2L-1}\}$, where $L \geq 2$, v_1 is the root of T_ϕ, $v_1, \ldots, v_{L-1}$ are its forks and $v_L, \ldots, v_{2L-1}$ are its leaves.
2. If the children of a fork $v \in \{v_1, \ldots, v_{L-1}\}$ are $v^L \in V_\phi$ (left child) and $v^R \in V_\phi$ (right child) and if $v = v_i$, $v^L = v_j$, $v^R = v_k$, then $i < j < k$.
3. To each fork $v \in \{v_1, \ldots, v_{L-1}\}$, a predicate φ_v of some formal logic is assigned, evaluated on features of the input vectors $x \in \mathcal{X}$.
4. To each leaf $v \in \{v_L, \ldots, v_{2L-1}\}$, a class $c_v \in C$ is assigned.
5. For each input $x \in \mathcal{X}$, the predicate φ_{v_1} assigned to the root is evaluated.
6. If for a fork $v \in \{v_1, \ldots, v_{L-1}\}$, the predicate φ_v evaluates true, then $\phi(x) = c_{v^L}$ in case v^L is already a leaf, and the predicate φ_{v^L} is evaluated in case v^L is still a fork.
7. If for a fork $v \in \{v_1, \ldots, v_{L-1}\}$, the predicate φ_v evaluates false, then $\phi(x) = c_{v^R}$ in case v^R is already a leaf, and the predicate φ_{v^R} is evaluated in case v^R is still a fork.

Also a fuzzy classifier $\phi : \mathcal{X} \to [0, 1]^m, i = 1, \ldots, r$ can be constructed using a classification tree (called *fuzzy classification tree*), the only difference is that in 4., a fuzzy set on C is assigned to each leaf $v \in \{v_L, \ldots, v_{2L-1}\}$. In the following, however, we will restrict attention only to the above described classification trees that assign particular $c_v \in C$ to leafs. Similarly, we will also stick to the above restriction to binary trees although also more general classification trees have been proposed [41].

The predicates assigned to forks usually concern only a single feature, corresponding to a single dimension of the input $x \in \mathcal{X}$. If the j-th component $[x]_j$ of x is a realization of a discrete random variable with a finite value set V_j, then such a predicate typically has one the forms

$$[x]_j = \xi \text{ or } [x]_j \in \Xi, \tag{5.53}$$

for some particular $\xi \in V_j$ or $\Xi \subset V_j$. If $[x]_j$ is a realization of a continuous random variable, then the predicate typically has one of the forms

$$[x]_j < \theta, [x]_j \leq \theta, [x]_j > \theta, \text{ or } [x]_j \geq \theta, \tag{5.54}$$

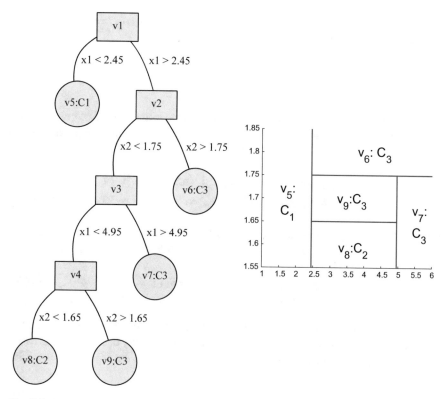

Fig. 5.4 A simple example classification tree with a 2-dimensional input space of continuous features $[x]_1$, $[x]_2$ (left), and the division of that space into quadratic regions induced by the predicates assigned to its forks (right)

for some particular $\theta \in \mathbb{R}$. Notice that each predicate of the kind (5.54) corresponds to dividing the space of continuous input dimensions by an axis-perpendicular hyperplane. Consequently, the combination of all predicates assigned to all forks induces in the space of continuous input dimensions a division into quadratic regions. A simple example of such a division for a classification tree with 2-dimensional input space and both features continuous is given in Fig. 5.4. However, there exist also classification trees in which the predicates assigned to forks concern more features simultaneously, they correspond to dividing the space of continuous input dimensions by a general hyperplane, or even by a more general surface, cf. [42].

The set of leaves $\{v_L, \ldots, v_{2L-1}\}$ can be decomposed according to the classes assigned to them into subsets $V^{(1)}, \ldots, V^{(m)}$, i.e.

$$V^{(k)} = \{v = v_i | L \leq i \leq 2L - 1, c_v = c_k\}, k = 1, \ldots, m. \qquad (5.55)$$

Then for $i = 1, \ldots, m$ and for all $x \in \mathcal{X}$, the classification $\phi(x) = c_i$ is implied by the validity of any of the conjunctions

$$\bigwedge_{v' \in \text{path}(v_1, v)} \varphi_v, v \in V^{(k)}, \qquad (5.56)$$

where $\text{path}(v_1, v)$ denotes the path from the root v_1 to the leaf v, more precisely, the set of forks belonging to that path. For example, the paths to the leafs of the classification tree in Fig. 5.4 are as follows:

$$\text{path }(v_1, v_5) = \{v_1\}, \text{path }(v_1, v_6) = \{v_1, v_2\}, \text{path }(v_1, v_7) = \{v_1, v_2, v_3\}, \quad (5.57)$$
$$\text{path }(v_1, v_8) = \text{path }(v_1, v_9) = \{v_1, v_2, v_3, v_4\}. \qquad (5.58)$$

Consequently, the classification tree ϕ can be completely described by m rules in *disjunctive normal form (DNF)* , i.e., rules expressible as disjunctions of conjunctions:

$$\bigvee_{v \in V^{(k)}} \bigwedge_{v' \in \text{path }(v_1, v)} \varphi_v(x) \rightarrow x \in c_k, x \in \mathcal{X}. \qquad (5.59)$$

However, it is not only the possibility to describe classification tree by means of DNF rules of Boolean logic that contributes to the comprehensibility of classification trees. A great contribution is also due to easy visualization of the tree T_ϕ, especially if its size is small, or at least not too large (cf. Fig. 5.4).

5.3.1   Classification Tree Learning

Like classifier learning in general, also classification tree learning consists in choosing the classifier ϕ from some set of feasible classifiers Φ according to (2.49), by means of a sequence $(x_1, c_1), \ldots, (x_p, c_p)$ of training data. In this section, we will explain what decision trees are in the set of feasible classifiers and according to what criterion the decision tree ϕ is chosen from it.

According to Chap. 2, feasible classifiers from Φ differ from each other through different values of parameters used when the classifier is constructed. Recalling the above described construction of classification trees, it is clear that parameters can be incorporated into them only through predicates p_v assigned to the forks $v \in \{v_1, \ldots, v_{L-1}\}$. In particular, in the typically used kinds of predicates (5.53) and (5.54), two kinds of parameters are encountered:

(i) A particular subset $\Xi \subset V_j$ of a finite value set V_j of a discrete variable x_j. This covers not only the predicate $x_j \in \Xi$, but also the predicate $x_j = \xi$ for some particular $\xi \in V_j$, through choosing $\Xi = \{\xi\}$.
(ii) A particular constant $\theta \in \mathbb{R}$. This covers all four encountered forms of typical predicates for continuous features (5.54).

As to the criterion for choosing ϕ from $\boldsymbol{\Phi}$, classification trees—similarly to support vector machines—do not use any of the standard classification performance measures from Chap. 2, but a specific criterion of their own. In their case, that specific criterion is based on the notion of impurity: a vertex of the tree T_ϕ is pure if it contains only data of one class. Otherwise, it is impure. Therefore, the criterion consists in maximizing the decrease of the impurity of probability distribution on C due to splitting the data according to whether they fulfil the predicate assigned to a fork. This criterion is applied hierarchically: First, it is applied to all training data, split according to the predicate p_{v_1} assigned to the root v_1. If the left child v_1^L of v_1 is a fork, then the criterion is applied to the training data fulfilling p_{v_1}, which are further split according to $p_{v_1^L}$, similarly if the right child v_1^R of v_1 is a fork, then the criterion is applied to the training data not fulfilling p_{v_1}, which are further split according to $p_{v_1^R}$. This procedure is then repeated recursively till all predicates assigned to the forks $v \in \{v_1, \ldots, v_{L-1}\}$ have been taken into consideration.

To be able to explain what exactly this criterion means, we need to introduce the concept of an impurity measure. Recall that a probability distribution on C is a vector $(p_{c_1}, \ldots, p_{c_m})$ with $0 \leq p_{c_1}, \ldots, p_{c_m} \leq 1$ and $p_{c_1} + \cdots + p_{c_m} = 1$. Then an *impurity function* is any real function i on the set of probability distributions on C that fulfils the following 3 conditions:

1. invariance with respect to permutations of the components of the probability distribution;
2. achieving its maximum at the uniform distribution on C;
3. achieving its minima at the degenerate distributions $(1, \ldots, 0)$, $(0, 1, \ldots)$, $\ldots$, $\ldots, (0, \ldots, 1)$.

Two particular impurity functions are commonly used:

(a) *entropy*, more precisely entropy with respect to an alphabet of size $a > 1$,

$$i_a(p_{c_1}, \ldots, p_{c_m}) = -\sum_{i=1}^{m} p_{c_j} \log_a(p_{c_j}), \tag{5.60}$$

where problems with the undefined expression $0 \log_a(0)$ are eliminated through defining $0 \log_a(0) = 0$,

(b) *Gini diversity index*, or simply Gini index,

$$i_{\text{Gini}}(p_{c_1}, \ldots, p_{c_m}) = \sum_{i=1}^{m} \sum_{j \neq i} p_{c_i} p_{c_j} = 1 - \sum_{i=1}^{m} p_{c_i}^2. \tag{5.61}$$

The specific criterion for choosing ϕ from $\boldsymbol{\Phi}$ in the case of classification trees can be, however, formulated in terms of an impurity function in general. For a subset of indexes $S \subset \{1, \ldots, q\}$, determining a subsequence $(x_k, c_k)_{k \in S}$ of learning data, denote π_S the empirical probability distribution on C induced by the sequence of values $(c_k)_{k \in S}$. Then for any impurity function i, the *impurity decrease* of probability distribution on C due to splitting $(x_k, c_k)_{k \in S}$ according to the predicate assigned to a fork v is defined as

$$\Delta i(S, v) = i(\pi_S) - \frac{|S^L|}{|S|} i(\pi_{S^L}) - \frac{|S^R|}{|S|} i(\pi_{S^R}), \text{ where} \tag{5.62}$$

$$S^L = \{k \in S | x_k \text{ fulfils } \varphi_v\}, \tag{5.63}$$

$$S^R = \{k \in S | x_k \text{ does not fulfil } \varphi_v\}. \tag{5.64}$$

If in particular entropy is used as the impurity function, then the impurity decrease is often called *information gain*. The criterion now consists in maximizing (5.62) with respect to the parameters of φ_v. Those parameters enter (5.62) through the sets S^L and S^R because for different values of such a parameter, different $x_k, k \in S$, can fulfil φ_v (to emphasize the dependence of S^L and S^R on the value of the parameter, we can write that value in parentheses behind the symbols of those sets). Let us have a look what this means for the two above recalled kinds of parameters encountered in the typically used kinds of predicates (5.53) and (5.54).

(i) If the parameter is a particular subset of a finite value set V_j of a discrete variable x_j, then the criterion leads to searching

$$\hat{\Xi} = \arg \max_{\Xi \subset V_j} \Delta i(S, v) =$$

$$= \arg \max_{\Xi \subset V_j} \left[i(\pi_S) - \frac{|S^L(\Xi)|}{|S|} i(\pi_{S^L(\Xi)}) - \frac{|S^R(\Xi)|}{|S|} i(\pi_{S^R(\Xi)}) \right] =$$

$$= i(\pi_S) - \arg \min_{\Xi \subset V_j} \left[\frac{|S^L(\Xi)|}{|S|} i(\pi_{S^L(\Xi)}) + \frac{|S^R(\Xi)|}{|S|} i(\pi_{S^R(\Xi)}) \right]. \tag{5.65}$$

Needless to say, computing $\frac{|S^L(\Xi)|}{|S|} i(\pi_{S^L(\Xi)}) + \frac{|S^R(\Xi)|}{|S|} i(\pi_{S^R(\Xi)})$ for all $\Xi \subset V_j$ has an exponential complexity with respect to the size of V_j, and is therefore computationally prohibitive already from V_j sizes of several dozens. In such situations, some heuristic has to be used, e.g., maximizing (5.65) stepwise in such a way that in the first step, the maximization is performed on condition $|\Xi| = 1$, whereas in the sth step, it is performed on conditions $|\Xi| = s$ and the result of the $(s - 1)$th step is a subset of Ξ.

(ii) If the parameter is a particular constant $\theta \in \mathbb{R}$, then the criterion leads to searching

$$\hat{\theta} = \arg\max_{\theta \in \mathbb{R}} \Delta i(S, v) =$$

$$= \arg\max_{\theta \in \mathbb{R}} \left[i(\pi_S) - \frac{|S^L(\theta)|}{|S|} i(\pi_{S^L(\theta)}) - \frac{|S^R(\theta)|}{|S|} i(\pi_{S^R(\theta)}) \right] =$$

$$= i(\pi_S) - \arg\min_{\theta \in \mathbb{R}} \left[\frac{|S^L(\theta)|}{|S|} i(\pi_{S^L(\theta)}) + \frac{|S^R(\theta)|}{|S|} i(\pi_{S^R(\theta)}) \right].$$

$$(5.66)$$

Although there are infinitely many $\theta \in \mathbb{R}$, the learning data $(x_k, c_k)_{k \in S}$ can have at most $|S|$ different values of the continuous feature x_j concerned by φ_v. If those values are ordered as an increasing sequence, $v_j^{\min}, \ldots, v_j^{\max}$, then the subsets S^L and S^R stay unchanged within any open interval between two subsequent values in that sequence, as well as below $v_j^{\min}$ and above $v_j^{\max}$. Consequently, computing (5.66) has a linear complexity with respect to the cardinality $|S|$ of S.

5.3.2   Classification Tree Pruning

Whereas splitting the training data during classification tree learning due to the predicates assigned to the forks attempts to maximize impurity decrease, it can simultaneously introduce 2 undesirable properties:

1. Overfitting with the learning data, hence impaired generalization.
2. High number of leaves (in the worst case as high as the number of data), which for most of them means long paths from the root to the leaves. For the DNF rules (5.59), this entails a high number of disjuncts and high numbers of conjuncts in most of them, both decreasing the comprehensibility of the rules.

A countermeasure against those disadvantageous properties is to consider not the whole tree T_ϕ, but only some subtree of it. Such a subtree can be easily obtained through *pruning* T_ϕ, i.e. replacing some of its subtrees with leaves (Fig. 5.5).

Whereas the size of the subtree is immediately seen, it is not clear how to estimate its generalization. Various heuristics have been proposed to approximate the expected error $\mathbb{E}\,\mathrm{ER}_\phi$ of ϕ. Probably best known is the one proposed in [43], which consists in using the estimate of $\mathbb{E}\,\mathrm{ER}_\phi$ based on the training data, i.e., the resubstitution error $\mathrm{ER}_{\mathrm{RE}}$ (cf. Sect. 2.2), and assuming that this estimate introduces an optimistic bias linearly increasing with the number of leaves L,

$$\mathrm{ER}_{\mathrm{RE}} = \mathbb{E}\,\mathrm{ER}_\phi - \alpha L, \tag{5.67}$$

Original classification tree

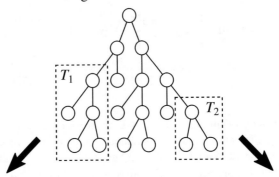

Tree pruned through replacing T_1 Tree pruned through replacing T_2

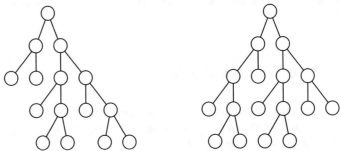

Fig. 5.5 Example of different possibilities of pruning classification trees through replacing different subtrees with leaves

where $\alpha \geq 0$. Consequently, the subtree yielding for a given choice of α the smallest generalization error is

$$T_\alpha = \arg \min_{T \in \mathscr{S}(T_\phi)} \mathbb{E}\,\mathrm{ER}_\phi = \arg \min_{T \in \mathscr{S}(T_\phi)} (\mathrm{ER_{RE}} + \alpha L). \tag{5.68}$$

This heuristic has a very interesting behaviour, which was proven in [43]. Namely, there exist constants $\alpha_0, \ldots, \alpha_s \geq 0$ and subtrees $T_0, \ldots, T_s \in \mathscr{S}(T_\phi)$ such that:

(i) $0 = \alpha_0 < \cdots < \alpha_s$,
(ii) $(\forall \alpha \in [\alpha_{k-1}, \alpha_k))\ T_\alpha = T_{k-1}$ for $k = 1, \ldots, s$, $(\forall \alpha \geq \alpha_s)\ T_\alpha = T_s$,
(iii) T_k is a subtree of T_{k-1} for $k = 1, \ldots, s$,
(iv) $T_0 = T_\phi$,
(v) T_s is the subtree of T_ϕ induced by its root v_1, i.e. $T_s = (v_1, \emptyset)$.

5.3.3 Classification Trees in Spam Filtering

In the case of spam filtering, classification trees seem to have even higher mental fit than the classification rules. Not because they contain more information (recall that classification trees can be easily transformed to a set of classification rules), but because the tree visualisation allows the user to easily extract not only the information why a particular message was labelled as spam but also what change in the attributes would lead to classification as ham and vice versa.

Another major advantage of classification trees is the method of their construction. While rules are usually created based on the coverage of remaining positive samples, forking in the tree is consistently guided by the highest decrease of node impurity. The root predicate of the tree should be, therefore, based on the most common data separation feature and subsequent forks dive into more specific distinctions in the data. This feature should again lead to better comprehensibility.

One of the earliest and most straightforward algorithms of decision tree learning is Iterative Dichotomiser 3 (ID3) [44]. It starts with the full data set in the root node, calculates entropy for all attributes and splits the data on the attribute with the lowest entropy (highest information gain). The same procedure is repeated iteratively on resulting subsets of data with remaining attributes until the node is pure or all attributes were used up. ID3 uses a greedy strategy, as the best local attribute is used, and the selection is never reconsidered; ID3, therefore, does not guarantee an optimal solution. Moreover, tree pruning is not implemented, which makes ID3 susceptible to overtraining.

On top of that, the Id3 implementation of the ID3 algorithm in Weka accepts only nominal attributes. The use of this algorithm is, therefore, limited. On the other hand, ID3 represents one of the fastest decision tree learning algorithms and is suitable for tasks with a large number of attributes.

A rather generic framework of Classification and Regression Trees (CART) by Breiman [43] discusses approaches to globally optimal tree learning and minimal cost-complexity pruning. The SimpleCart implementation in Weka, however, reveals that the approach with k-fold pruning validation is both time and memory exhaustive on a large number of attributes and, therefore, not very suitable for spam filtering.

A successor of ID3 called C4.5 [41] adds support for undefined or continuous attribute values. More importantly, C4.5 includes several methods of tree pruning. The default method uses a *confidence threshold* that calls for the pruning of nodes with a higher error rate than specified. The lower the threshold, the more aggressive is the pruning. However, C4.5 and its implementations (such as J48 in Weka) allow also the *reduced error pruning* that we discussed in Sect. 5.1.1, which may produce smaller trees with better predictive power.

For a visual comparison of the trees mined by the Id3 and J48 implementations in Weka, see Fig. 5.6. Both of these trees were trained and tested on the same SpamAssassin dataset we used for classification rule mining. Only for the ID3 algorithm, the data were binarised.

While the ID3 tree has a higher number of nodes, the tree is more balanced and has a lower height. This is caused by the greedy strategy, which always tries to find the currently best attribute. C4.5, on the other hand, tries to separate the data with a global optimum in mind and, therefore, may lead to very long branches for classification of not very distinctive data. The commonly used branches are, however, usually shorter.

The testing accuracy of both models with default settings is very similar—82.4% for ID3 and 82.6% for C4.5. Accuracy of 85.7% was achieved by C4.5 with reduced error pruning. Note, that the accuracy is still lower than in classification rules mined with reduced error pruning. One of the possible remedies may be, therefore, to grow unpruned classification trees, transform them to classification rules, which are then pruned.

Higher accuracy of other spam classifiers (such as Linear SVM that has an accuracy of 94.7%) seems to be also caused by the ability to base a single decision boundary on multiple attributes simultaneously. Although such extensions are available also for classification trees, the user comprehensibility of the resulting tree is, however, questionable.

5.3.4 Classification Trees in Recommender Systems

Direct use of classification trees in recommender systems is rather impractical. Based on the history of a given user (such as the purchase of other items), we want to deduce the next action from a rather large set of possibilities (list of all available items in the shop). Moreover, although the leaves of the classification tree are technically not limited in terms of possible values, training such a tree from sparse high-dimensional data is computationally demanding and inevitably leads to strong overfitting and unsatisfactory generalisation.

However, classification trees may be used in several stages of the recommendation process to help towards higher accuracy of the recommendation or to improve the recommender performance and scalability.

One of possible uses of a classification tree mentioned in [46] is the pre-selection of users the system will give recommendations to. Based on the user information, only the users that match a particular profile will be selected and targeted with a campaign, for example. Adding such a step to the recommender system should reduce false positives. However, this approach is not helpful at all in online scenarios, where we want to pass recommendations to a user currently visiting our site.

A method called RecTree [47] also builds a classification tree on users based on their attributes. This time, however, the construction of the tree resembles a method of hierarchical clustering (although much faster and less precise due to less flexible feature space separation), where the tree leaves correspond to a group of users. The current user is then classified as a member of one of such groups and recommendation is based on the ratings to items given by the users in that group. The process of recommendation itself is similar to a k-nn recommender. However, distances to all users do not have to be calculated or even approximated; in RecTree, the user is

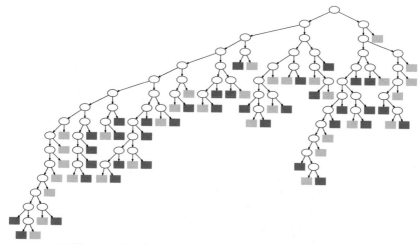

(a) The Id3 implementation in Weka [414] on binarised features

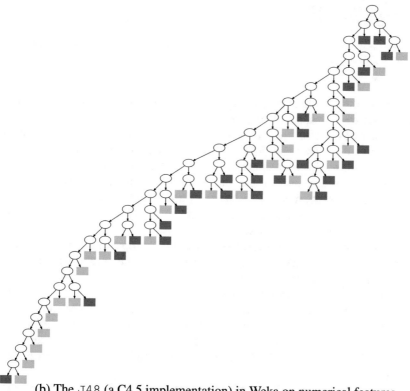

(b) The J48 (a C4.5 implementation) in Weka on numerical features

Fig. 5.6 Visual comparison of classification trees trained by two different algorithms

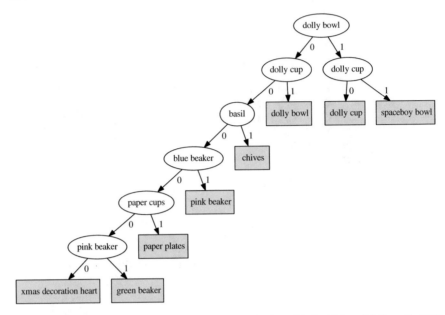

Fig. 5.7 A classification tree based on the ten Apriori rules with the highest lift from the UCI Online Retail Dataset

classified by the tree to obtain directly the neighbourhood required for score predictions of unseen items, underlying the recommendations.

Another possible use of a classification tree is connected with the estimation of item preference from navigational and behavioural patterns. The approach presented in [48] uses such preference data to eliminate the issue of sparseness in explicit data matrices (such as item purchase records) by extending the binary (not purchased/purchased) matrix to a matrix of user preferences (on a scale: will not purchase—already purchased) which is more suitable for a recommendation systems based on item rating matrix.

As exploited in [49], classification trees can be transposed to a ruleset and vice versa. This property allows to use an approach of large itemset detection and rule extraction, as mentioned in Sect. 5.1.6 and transforming such rules into a structure of a classification tree. To this end, binary information concerning the presence of an item in the antecedent is used as attributes, and the item in the succedent is used as a target class. Considering the example of the ten rules with the highest lift extracted from the UCI Online Retail Dataset, Table 5.4 represents the input matrix for tree learning. An example of a classification tree trained by ID3 on this matrix is shown in Fig. 5.7.

The transformation of the association rules to a classification tree has two major benefits. Upon a request for a recommendation, only the binary tree needs to be traversed; instead of checking possibly all rules in the system. The second major benefit is the ability of further generalisation to the classifier through tree pruning.

Table 5.4 Matrix based on ten Apriori rules with the highest lift (rule confidence divided by the support of the antecedent) from the UCI Online Retail Dataset. The column Class contains short descriptions of recommended products that the clients subsequently bought

Basil	Blue beaker	Dolly bowl	Dolly cup	Green beaker	Mint	Paper cups	Paper napkins	Parsley	Pink beaker	Red beaker	Rosemary	Spaceboy bowl	Spaceboy cup	Thyme	Xmas decoration bell	Class
0	0	0	1	0	0	0	0	0	0	0	0	1	1	0	0	Dolly bowl
0	0	0	1	0	0	0	0	0	0	0	0	1	0	0	0	Dolly bowl
0	0	1	0	0	0	0	0	0	0	0	0	1	1	0	0	Dolly cup
0	0	0	0	0	0	0	0	0	0	0	0	0	0	0	1	Xmas decoration heart
0	0	1	0	0	0	0	0	0	0	0	0	0	1	0	0	Dolly cup
0	0	0	0	0	0	0	0	0	1	1	0	0	0	0	0	Green beaker
0	1	0	0	1	0	0	0	0	0	0	0	0	0	0	0	Pink beaker
0	0	0	0	0	0	1	1	1	0	0	0	0	0	0	0	Paper plates
1	0	0	0	0	1	0	0	1	0	0	1	0	0	1	0	Chives
0	0	1	1	0	0	0	0	0	0	0	0	0	1	0	0	Spaceboy bowl

Fig. 5.8 A PLWAP tree
constructed from a set of
sequences {abac, abcac,
baba, abacc, ab}

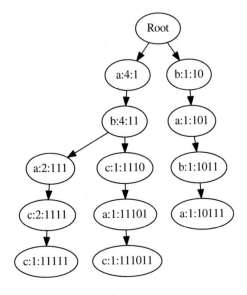

In our example, 7266 classification rules with support over 0.005 and confidence over 0.75 were extracted by Apriori on the UCI Online Retail Dataset. For manual evaluation, only a couple of rules sorted by the lift are usually selected; although our example with only ten rules is actually too small. Recommender systems, on the other hand, tend to use the whole ruleset, which makes the recommender system too large to be easily comprehensible. Training an ID3 tree (which does not use pruning) reduces the number of nodes and potentially speeds up the classification, yet still results in a classification tree with 5423 nodes. Such reduction is not much worth the additional computation cost, considering that such a classification tree covers only about 64% of the rules proposed by Apriori.

A much more aggressive reduction is obtained by the C4.5 algorithm, where the tree is reduced to 481 nodes with rule coverage of 58%. Moreover, the `SimpleCART` implementation in Weka [45] returns a classification tree with just 277 nodes and 54% rule coverage.

Just a slightly different approach can be used for analysis and recommendation based on sequential data, such as sequences of visited web pages. A set of algorithms called WAP-tree, PLWAP-tree and their extensions for online updates, is presented in [50]. These algorithms take a list of item sequences on input, filter the sequences in such a way that they consist of only frequent items, and construct a tree by appending each sequence item-by-item from a root of the tree. Each node of the tree keeps not only the identifier of the item but also a count of sequences that share this node and a unique binary identifier where "1" is appended to the first child and "0" to each new sibling. To demonstrate the construction of such a tree, see Fig. 5.8. The resulting tree structure is not used directly for classification but for a recommendation of the next item based on a set of frequent patterns extracted from such a tree.

5.3.5 Classification Trees in Malware and Network Intrusion Detection

Classification trees play an important role both in malware detection and in network intrusion detection. As to the former, they are used in the context of both operating systems most attacked by malware—Windows and Android.

In [51], classification trees were applied to static malware detection in the context of Windows portable executable files. The approach explictely takes into account that the authors of malicious Windows portable executables frequently try to avoid detection through packing. It uses mainly the following features:

- numbers of sections of different kinds (standard, non-standard, readable, writable, executable);
- number of entries in the import address table;
- content of the header fields;
- counts of various types of resources (e.g., icons, dialog boxes);
- Shanon entropy of the entire file, of the header, of the code sections and of the data sections.

From the point of view of detecting also packed files, these features are important primarily for the following two reasons:

(i) In non-packed files, a section is normally not flagged as both writable and executable, whereas in packed files, this is necessary for the unpacking to work.
(ii) In non-packed files, the entropy of various parts of the file is usually low, indicating redundancy and repetitiveness; in packed files, it is much higher.

The approach used in [51] had been originally proposed already in [52], but the classifier used there was a multilayer perceptron. The authors of [51] adapted it for classification trees to make the reasons for resulting classification more comprehensible for analysts.

An application of classification trees to dynamic malware detection in the Windows context has been proposed in [53]. The information on which the detection is based has been extracted from API logs. However, differently to usual dynamic malware detection approaches, the logged API calls do not serve directly as features. Instead, they are employed to find a small number of latent features, and only those serve as input to the classification tree.

An extensive application of classification trees to static malware detection for Android apps has been described in [54]. The construction of such trees used depends on three parameters:

(i) a minimal number of training samples in a node needed for a split;
(ii) a minimal information gain of an attribute needed for a split to be performed according to that attibute;
(iii) a parameter determining the relative importance of the information gain of an attribute and the cost of its evaluation.

The values of those parameters are set during training by means of biobjective optimization. Classification trees have been employed in [54] in two different ways.

1. Classification tree based on several simple kinds of static features, most importantly:

   - requested permissions;
   - invocation of encryption APIs;
   - invocation of dynamic loading.

2. Classification tree based on a variety of different kinds of features obtainable within static malware analysis, some of them simple, other very complex and sophisticated. More precisely, this tree consideres the following categories of features:

   (i) The simple features used by the first classification tree.
   (ii) Requesting any of 7 predefined combinations of permissions that are considered particularly dangerous.
   (iii) Several scores indicating the risk involved in installing and using the app. They are computed using generative probabilistic models with inputs formed by the requested permissions and optionally also the kind of the classified app or prior knowledge that some pemissions are more dangerous than others.
   (iv) Assessment how well the app fits signatures of known malware families. Those signatures are based on API calls obtained through static analysis.
   (v) Another application package embedded in the current one. That is a way how to avoid detection based on commonly inspected features of the current application package.
   (vi) Requesting no permissions, which with high probability indicates benign software.
   (vii) Features output by another Android malware detection system—Adagio [55]. Adagio is basically an SVM classifier, which employs the call graph of the classified app (cf. Sect. 1.5).
   (viii) Result of classifying the app to known malware families by yet another malware detection system. API1-NN [56] is a nearest neighbour classifier based on API calls obtained from static malware analysis.

   Observe that the categories (vii) and (viii) are actually outputs from other classifiers. These two classifiers, SVM and nearest neighbour, can be viewed as forming together with the classification tree considered in 2. a team. Such teams of classifiers will be the topic of the last chapter of this book.

An application of classification trees to dynamic malware detection for Android has been presented in [57]. Its dynamic features are based solely on the network communication of the device and include, for example:

- the average number of packets transferred during a session;
- the average numbers of packets sent and received per flow on a particular client-server connection;
- the average numbers of bytes sent and received per flow on a particular client-server connection;
- the ratio of incoming to outgoing bytes during a session;
- the avarage number of bytes received per second during a session.

The classification tree in [57] was trained on several publicly available datasets of Android malware with labelled families, including the Drebin dataset [58], which is the largest public dataset of that kind, and on benign software from the Google Play.

A rather specific approach to detecting Android malware by means of classification trees is used by the system *SMMDS (service-oriented mobile malware detection system)* [59]: it performs malware detection solely at the mobile operator gateway and relies solely on the packets sent by the app undergoing detection. At a gateway, a much more complex detection system can be constructed than on a mobile device, especially if that process runs in a cloud. The system SMMDS performs the following kinds of tasks (Fig. 5.9):

1. Real-time tasks, i.e., monitorig the raw packets passing through the operator gateway, preprocessing them, and running a timely initial mining of the preprocessed data.
2. Scheduled tasks, i.e., regular incremental data mining of detection results to adapt the classification tree to changes of malware. The authors include among such tasks also feature extraction from the sent packets, although it is actually not scheduled, but triggered by the real-time preprocessing.
3. Knowledge management consisting in appropriately updating 4 specific knowledge bases:

    (i) database of features of packets passing through the gateway;
    (ii) database of prior knowledge about known malware and benign software, as well as of knowledge needed for the data mining process;
    (iii) database of data mining results;
    (iv) database of detection results.

Needless to say, this approach brings several great advantages:

- Compared with detection on the device, it avoids the consumption of local resources.
- No malware database needs to be updated at the device.
- Malware detection may be based also on data mining instead of on traditional signature matching, which increases the chance that also new malware variants are detected.

On the other hand, the fact that the SMMDS system relies solely on the sent packets and their features entails a very serious disadvantage of the approach: it cannot detect any malware app that does not sent packets or whose packets do not reflect any malicious behaviour or intentions.

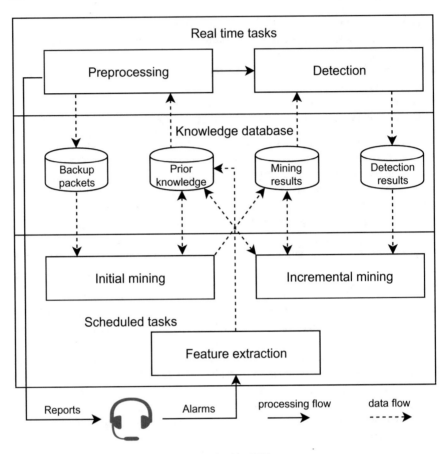

Fig. 5.9 Scheme of the SMMDS system (inspired by [59])

Three different kinds of classification trees for intrusion detection have been presented in [60–62]. The tree in [60] is a variant of the popular C4.5 tree [41]. It is called *neurotree* due to the fact that the labels in its training set are actually predictions by a trained neural network. The features used as inputs for the trained network, and consequently for the classification tree, are obtained through bi-objective evolutionary optimization with the sensitivity and specificity objectives (Fig. 5.10).

In [61], a 2-level classification system is used, which does not perform a direct intrusion detection, but the detection of peer-to-peer (P2P) traffic, very important for intrusion detection. The classification tree is applied at the flow level of network traffic, after classification with classification rules was performed at the lower packet level, and is actually applied only to packets that are classified by the classification rules as unkonwn traffic. The classification rules rely on a signature table and on heuristics interconnecting an input table and a listening table. The classification tree, in turn, uses the following groups of features:

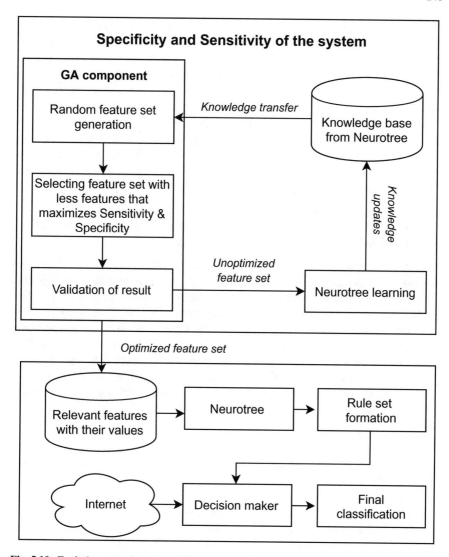

Fig. 5.10 Evolutionary optimization of the neurotree (inspired by [60])

- features related to arrival times of packets;
- features related to flow volume, both in terms of number of packets and in terms of number of bytes;
- features related to the size of packets.

The classification rules at the low level can differentiate also between different kinds of P2P traffic. The classification tree at the higher level, however, classifies only into P2P traffic and non-P2P traffic.

Like the above recalled SMMDS system, [59], represents a specific approach to Android malware detection, another specific approach that uses classification trees for network intrusion detection has been presented in [62]. It aims exclusively at detecting covert channels, i.e., secret communication not detectable through standard network monitoring. Examples of such kinds of secret communication are: file locking, ringtone volume of a device, state of the socket, size of a transferred files.

For each considered covert channel, a separate classification tree is trained. The detection of covert channel in [62] relies on recognizing an unexpectedly high energy consumption, primarily from the current and voltage values recorded in files in `/sys/class/power_supply/battery` folder. More precisely, the classification tree uses the following three features, related to the energy consumption:

(i)   average power consumption within the previous time window;
(ii) total variation of the power consumption within the previous time window;
(iii) instantaneous power consumption at the current time.

Apart from classification by a classification tree, the authors of [62] considered also regression for the detection of covert channels: a secret communication is indicated if the power consumption at the current time exceeds the prediction by a regression model predicting some upper bound of power consumtion based on the past.

Finally, [63] presents a system on the overlap of intrusion detection and anomaly detection. It is a host-based intrusion detection system, but most of its activities consist in malware detection because it aims at attacks generally known as *advanced persistent threats (APT)* [64, 65] to leak a massive quantity of important information. More precisely, the authors consider an attack gradually developing through the following stages (Fig. 5.11):

- *Advance preparation*—information about the target system is gathered and analyzed, an infected PC inside the target network is selected, allowing to alter important parts of the target's web site, and a command and control channel is established to that PC.
- *Internal network intrusion*—the attacker infiltrates the target's IT infrastructure, mainly through sending mailcious emails, connecting to the altered parts of the web site, and accessing the software update server.
- *Persistent internal activity*—collecting the data of the target's IT infrastructure, collecting additional weak points, downloading additional malware, privilege excalation to obtain access to assets accessible only with elevated privileges.
- *Goal achievement*—discovering confidental assets, leaking important data, moving to other parts of the network or to other networks.

The APT attack detection in [63] is intended only for the persistent internal activity and goal achievement stages because a chance to detect an APT attack in one of the earlier stages is negligible. The detection is host-based and relies on dynamic malware analysis. To this end, 39 features characterizing the behaviour of the analysed software have been used, derived from the process running the software and concerning primarily the use of threads, file system, registry, network and system services.

Fig. 5.11 Scheme of an advanced persistent threats attack

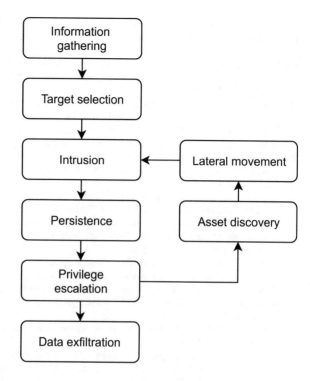

Apart from the parent process, those features have been also cumulatively gathered for all child processes. Hence, the classifier performing the APT attack detection uses altogether 78 features. That classifier is a C4.5 classification tree, implemented in the system Weka [45].

References

1. Berthold, M., Hand, D. (eds.): Intelligent Data Analysis. An Introduction. Springer (1999)
2. Chen, C., Härdle, W., Unwin, A. (eds.): Handbook of Data Visualization. Springer (2008)
3. Brunk, C., Pazzani, M.: An investigation of noise-tolerant relational concept learning algorithms. In: Machine Learning Proceedings, pp. 389–393. Elsevier Science Publishers, Amsterdam (1991)
4. Fürnkranz, J., Widmer, G.: Incremental reduced error pruning. In: Machine Learning Proceedings, pp. 70–77. Elsevier Science Publishers, Amsterdam (1994)
5. Quinlan, J.: Learning logical definitions from relations. Mach. Learn. **5**, 239–266 (1990)
6. Cohen, W.: Fast effective rule induction. In: Machine Learning Proceedings, pp. 115–123. Elsevier Science Publishers, Amsterdam (1995)
7. Hájek, P., Havránek, T.: Mechanizing Hypothesis Formation. Springer (1978)
8. Rauch, J.: Observational Calculi and Association Rules. Springer (2013)
9. Agrawal, R., Srikant, R.: Fast algorithms for mining of association rules. In: 20th International Conference on Very Large Data Bases, pp. 487–499. AAAI Press, Menlo Park (1994)

10. Fürnkranz, J.: Separate-and-conquer rule learning. Artif. Intell. Rev. **13**, 3–54 (1999)
11. Eibe, F., Hall, M., Witten, I.: The WEKA Workbench. Practical Machine Learning Tools and Techniques. Morgan Kaufmann Publishers, Online Appendix for Data Mining (2016)
12. Chen, D., Sain, S., Guo, K.: Data mining for the online retail industry: a case study of RFM model-based customer segmentation using data mining. J. Database Market. Customer Strateg. Manag. **19**, 197–208 (2012)
13. Brin, S., Motwani, R., Ullman, J., Tsur, S.: Dynamic itemset counting and implication rules for market basket data. ACM Sigmod Record **26**, 255–264 (1997)
14. Lin, W., Alvarez, S., Ruiz, C.: Efficient adaptive-support association rule mining for recommender systems. Data Min. Knowl. Discov. **6**, 83–105 (2002)
15. Feng, S., Han, Z.: An incremental associative classification algorithm used for malware detection. In: 2nd International Conference on Future Computer and Communication, pp. V1–757–V1–760 (2010)
16. Ding, Y., Yuan, X., Tang, K., Xiao, X., Zhang, Y.: A fast malware detection algorithm based on objective-oriented association mining. Comput. Secur. **39**, 315–324 (2013)
17. Xiao, X., Ding, Y., Zhang, Y.a nd Tang, K., Dai, W.: Malware detection based on objective-oriented association mining. In: IEEE International Conference on Machine Learning and Cybernetics, pp. 375–380 (2013)
18. Y.D., S., Z., Z.: Objective-oriented utility-based association mining. In: ICDM'02: IEEE International Conference on Data Mining, pp. 426–433 (2002)
19. Subedi, K., Budhathoki, D., Dasgupta, D.: Forensic analysis of ransomware families using static and dynamic analysis. In: SPW: IEEE Symposium on Security and Privacy Workshops, pp. 180–185 (2018)
20. Han, J., Pei, J., Yin, Y.: Mining frequent patterns without candidate generation. ACM SIGMOD Record **29**, 1–12 (2000)
21. Liang, S., Du, X.: Permission-combination-based scheme for android mobile malware detection. In: IEEE ICC Mobile and Wireless Networking Symposium, pp. 2301–2306 (2014)
22. Zhou, Y., Jiang, X.: Dissecting Android malware: Characterization and evolution. In: IEEE Symposium on Security and Privacy, pp. 95–109 (2012)
23. Wang, Y., Watson, B., Zheng, J., Mukkamala, S.: ARP-miner: mining risk patterns of android malware. In: MIWAI: International Workshop on Multi-disciplinary Trends in Artificial Intelligence, pp. 363–375 (2015)
24. Adebayo, O., Abdulaziz, N.: Android malware classification using static code analysis and Apriori algorithm improved with particle swarm optimization. In: WICT: Fourth World Congress on Information and Communication Technologies, pp. 123–128 (2014)
25. Ohrui, M., Kikuchi, H., Terada, M.: Mining association rules consisting of download servers from distributed honeypot observation. In: 13th IEEE International Conference on Network-Based Information Systems, pp. 541–545 (2010)
26. Ban, T., Eto M.and Guo, S., Inoue, D., Nakao, K., Huang, R.: A study on association rule mining of darknet big data. In: IEEE International Joint Conference on Neural Networks, p. article no. 7280818 (2015)
27. Inoue, D., Eto, M., Yoshioka, K., Baba, S., Suzuki, K., Nakayato, J., Ohtka, K., Nakao, K.: NICTER: an incident analysis system toward binding network monitoring with malware analysis. In: WOMBAT Workshop on Information Security Threats Data Collection and Sharing, pp. 58–66 (2008)
28. Nakao, K., Yoshioka, K., Inoue, D., Eto, M.: A novel concept of network incident analysis based on multi-layer observation of malware activities. In: JWIS'07: Second Joint Workshop on Information Security, pp. 267–279 (2007)
29. Le Malécot, E., Inoue, D.: The carna botnet through the lens of a network telescope. In: Foundations and Practice of Security, pp. 426–441. Springer (2014)
30. Kopp, M., Bajer, L., Jílek, M., Holeňa, M.: Comparing rule mining approaches for classification with reasoning. In: ITAT 2018: Information Technologies—Applications and Theory, pp. 52–58 (2018)

31. Agrawal, R., Imieliński, T., Swami, A.: Mining association rules between sets of items in large databases. In: International Conference on Management of Data, pp. 207–216 (1993)
32. Kumar, S., Chnadrashekar, V., Jawhar, C.: Logical itemset mining. In: IEEE 12th International Conference on Data Mining, Workshops, pp. 603–610 (2012)
33. Kopp, M., Holeňa, M.: Evaluation of association rules extracted during anomaly explanation. In: ITAT 2015: Information Technologies—Applications and Theory, pp. 143–149 (2015)
34. Valeros, V., Somol, P., Rehak, M., Grill, M.: Cognitive threat analytics: Turn your proxy into security device (2016). Cisco Blog Security, https://blogs.cisco.com/security/cognitive-threat-analytics-turn-your-proxy-into-security-device
35. Bandyopadhyay, S., Pal, S.: Classification and Learning Using Genetic Algorithms. Springer (2007)
36. Drugowitsch, J.: Design and Analysis of Learning Classifier Systems: A Probabilistic Approach. Springer (2008)
37. Reeves, C., Rowe, J.: Genetic Algorithms: Principles and Perspectives. Kluwer Academic Publishers, Boston (2003)
38. Schaefer, R.: Foundation of Global Genetic Optimization. Springer (2007)
39. Vose, M.: The Simple Genetic Algorithm. Foundations and Theory. MIT Press, Cambridge (1999)
40. Freitas, A.: Data Mining and Knowledge Discovery with Evolutionary Algorithms. Springer (2002)
41. Quinlan, J.: C4.5: Programs for Machine Learning. Morgan Kaufmann Publishers (1993)
42. Criminisi, A., Shotton, J., Konokoglu, E.: Decision forests: a unified framework for classification, regression, density estimation, manifold learning and semi-supervised learning. Found. Trends Comput. Graphics Vis. **10**, 81–227 (2012)
43. Breiman, L., Friedman, J., Olshen, R., Stone, C.: Classification and Regression Trees. Wadsworth, Belmont (1984)
44. Quinlan, J.: Induction of decision trees. Mach. Learn. **1**, 81–106 (1986)
45. Witten, I., Frank, E., Hall, M.: Data Mining: Practical Machine Learning Tools and Techniques, 4th Edition. Morgan Kaufmann Publishers (2016)
46. Cho, Y., Kim, J., Kim, S.: A personalized recommender system based on web usage mining and decision tree induction. Expert Syst, Appl. **23**, 329–342 (2002)
47. Chee, S., Han, J., Wang, K.: Rectree: an efficient collaborative filtering method. In: International Conference on Data Warehousing and Knowledge Discovery, pp. 141–151 (2001)
48. Kim, Y., Yum, B.J., Song, J., Kim, S.: Development of a recommender system based on navigational and behavioral patterns of customers in e-commerce sites. Expert Syst. Appl. **28**, 381–393 (2005)
49. Nikovski, D., Kulev, V.: Induction of compact decision trees for personalized recommendation. In: ACM symposium on Applied computing, pp. 575–581 (2006)
50. Ezeife, C., Liu, Y.: Fast incremental mining of web sequential patterns with PLWAP tree. Data Min. Knowl. Discov. **19**, 376–416 (2009)
51. Teh, A., Stewart, A.: Human-readable real-time classifications of malicious executables. In: 10th Australian Information Security Management Conference, pp. 9–18 (2012)
52. Shafiq, M., Tabish, S., Faroq, M.: PE-probe: Leveraging packer detection and structural information to detect malicious portable executables. In: Virus Bulletin Conference, p. 10 pages (2009)
53. Sundarkumar, G., Ravi, V., Nwogu, I., Govinaraju, V.: Malware detection via API calls, topic models and machine learning. In: IEEE International Conference on Automation Science and Engineering, pp. 2012–2017 (2015)
54. Lille, D., Coppens, B., Raman, D., De Sutter, B.: Automatically combining static malware detection techniques. In: MALWARE 2015: 10th International Conference on Malicious and Unwanted Software, pp. 48–55 (2016)
55. Gascon, H., Yamaguchi, F., Arp, D., Rieck, K.: Structural detection of Android malware using embedded call graphs. In: AISec: ACM WOrkshop on Artificial Intelligence and Security, pp. 45–54 (2013)

56. Aafer, Y., Du, W., Yin, H.: DroidAPIMiner: Mining API-level features for robust malware detection in Android. In: Security and Privacy in Communication Networks, pp. 86–103. Springer (2013)
57. Zulkifli, A., Hamid, I., Shah, W., Abdullah, Z.: Android malware detection based on network traffic using decision tree algorithm. In: Recent Advances on Soft Computing and Data Mining, pp. 485–494. Springer (2018)
58. Arp, D., Spreityenbarth, M., Hubner, M., Gascon, H., Rieck, K.: DREBIN: Effective and explainable detection of Android malware in your pocket. In: 21st Annual Network and Distributed System Security Symposium, p. 12 pages (2014)
59. Cui, B., Jin, H., Carullo, G., Liu, Z.: Service-oriented mobile malware detection system based on mining strategies. Pervasive Mobile Comput. **24**, 101–116 (2015)
60. Sindhu, S., Geetha, S., Kannan, A.: Decision tree based light weight intrusion detection using a wrapper approach. Expert Syst. Appl. **39**, 129–141 (2012)
61. Ye, W., Cho, K.: Hybrid P2P traffic classification with heuristic rules and machine learning. Soft Comput. **18**, 1815–1827 (2014)
62. Caviglione, L., Gaggero, M., Lalande, J., Mazurczyk, W., Urbański, M.: Seeing the unseen: revealing mobile malware hidden communications via energy consumption and artificial intelligence. IEEE Trans. Inf. Forensics Secur. **11**, 799–810 (2016)
63. Moon, D., Pan, S., Kim, I.: Host-based intrusion detection system for secure human-centric computing. J. Supercomput. **72**, 2520–2536 (2016)
64. Corporation, S.: Advanced persistent threats: a Symantec perspective. Tech. rep, Symantec (2011)
65. Tankard, C.: Persistent threats and how to monitor and deter them. Network Secur. **8**, 16–19 (2011)

Chapter 6
A Team Is Superior to an Individual

In the previous chapters, we have learned a number of different kinds of classification methods. Their classifiers frequently depend on parameters, e.g., a k-nn classifier on the number of neighbours, a multilayer perceptron on the number of hidden neurons, a SVM with polynomial kernel on the degree of polynomial, or a SVM with Gaussian kernel on the center and variance of the Gaussian. Consequently, we can always choose from a very large pool of classifiers, and if sufficient computational resources are available, we can easily use a whole team of them simultaneously. This has the clear advantage that a team member can specialize in a part of the input space. The team as a whole, however, is still able to correctly classify every input, provided each part of the input space is dealt with by the specialization of at least one member of the team and it is known how much to trust each member for the classification of a given input.

In this chapter, we show various ways and possibilities how a team of classifiers can be constructed. Then we concentrate on the most frequently encountered kind of classifier teams, teams of classification trees, also known as random forests.

6.1 Connecting Classifiers into a Team

At a first glance, a team is simply a set of classifiers together with a function aggregating their results:

$$\mathcal{T} = (\{\phi_k\}_{k=1}^r, A), \tag{6.1}$$

where

$$\phi_k : \mathcal{X} \to C = \{c_1, \ldots, c_m\}, k = 1, \ldots, r, \tag{6.2}$$

are the individual classifiers, and

$$A : C^r \to C \tag{6.3}$$

© Springer Nature Switzerland AG 2020
M. Holeňa et al., *Classification Methods for Internet Applications*,
Studies in Big Data 69, https://doi.org/10.1007/978-3-030-36962-0_6

is an *aggregation function*, which serves to compute the final classification $\phi : \mathcal{X} \to C$ by the team

$$\phi(x) = A(\phi_1(x), \ldots, \phi_r(x)), x \in \mathcal{X}. \tag{6.4}$$

If the team is formed by a single classifier, then the aggregation function is required to respect its classification,

$$r = 1 \Rightarrow A(\phi_1(x)) = \phi_1(x), x \in \mathcal{X}. \tag{6.5}$$

Looking at it again, one realizes that the classifiers must be treated not as a set $\{\phi_k\}_{k=1}^r$, but as a sequence $(\phi_1, \ldots, \phi_r)$:

(i) The aggregating function A may depend on the order in which the classifiers enter the aggregation. Such a function cannot be applied to a set.
(ii) A particular classifier can be used repeatedly during the aggregation.

Needless to say, the requirement that a team formed by a single classifier must respect its classification remains valid even if that classifier is used repeatedly,

$$\phi_1 = \cdots = \phi_r \Rightarrow A(\phi_1(x), \ldots, \phi_r(x)) = \phi_1(x) = \cdots = \phi_r(x), x \in \mathcal{X}. \tag{6.6}$$

In the most general case, however, a classifier team actually has a more complicated structure:

$$\mathcal{T} = ((\phi_1, \ldots, \phi_r), (\tau_{\phi_1}, \ldots, \tau_{\phi_r}), A), \tag{6.7}$$

where

- $\phi_k : \mathcal{X} \to \mathcal{F}(C) = [0, 1]^m, k = 1, \ldots, r$, hence, fuzzy classifiers need to be considered in the general case;
- $\tau_\phi : \mathcal{X} \to [0, 1]$ for $\phi = \phi_1, \ldots, \phi_r$, the function τ_ϕ is called *trust* or *confidence* in the classifier ϕ;
- $A : [0, 1]^{(m+1)r} \to [0, 1]^m$, and A is continuous.

Similarly to (6.4), the final fuzzy classification $\phi : \mathcal{X} \to [0, 1]^m$ is given by

$$\phi(x) = A(\phi_1(x), \ldots, \phi_r(x), \tau_{\phi_1}(x), \ldots, \tau_{\phi_r}(x)) \in [0, 1]^m, x \in \mathcal{X}. \tag{6.8}$$

Hence, the result of aggregation for a particular $x \in \mathcal{X}$ depends not only on the results of its classification by the individual classifiers, but also on the confidences that its classification by each of the classifiers was correct.

Finally, the requirements (6.5) and (6.6) generalize to

$$\phi_1 = \cdots = \phi_r, \tau_{\phi_1} = \cdots = \tau_{\phi_r} = \tau : \mathcal{X} \to [0, 1] \Rightarrow$$
$$\Rightarrow A(\phi_1(x), \ldots, \phi_r(x), \tau(x), \ldots, \tau(x)) = \phi_1(x) = \cdots = \phi_r(x, \in \mathcal{X}), \tag{6.9}$$

irrespectively of the confidence τ.

Recall from Chap. 2 that a crisp classifier $\phi_k : \mathcal{X} \to C$ can be viewed at the same time as a fuzzy classifier $\phi_k : \mathcal{X} \to [0, 1]^m$ mapping the feature vector into the set into the set $\{u \in \{0, 1\}^m \mid \sum_{i=1}^m u_i = 1\}$. In this way, also the aggregation of crisp classifiers can incorporate confidences, according to (6.8).

6.1.1 Aggregation Functions

In the case of crisp classifiers, the most frequently used way of aggregating the predictions of team members is simply their *majority voting*—the winning prediction is the one represented most frequently among its members (or one of them if there are several such predictions),

$$|\{k \mid \phi_k(x) = A(\phi_1(x), \ldots, \phi_r(x))\}| = \max_{i=1,\ldots,m} |\{k \mid \phi_k(x) = c_i\}|, x \in \mathcal{X}. \quad (6.10)$$

A generalization of (6.10) involving confidences is weighting the vote of each team member by its confidence,

$$\sum_{\phi_k(x)=A(\phi_1(x),\ldots,\phi_r(x),\tau_{\phi_1}(x),\ldots,\tau_{\phi_r}(x))} \tau_{\phi_k}(x) = \max_{i=1,\ldots,m} \sum_{\phi_k(x)=c_i} \tau_{\phi_k}(x), x \in \mathcal{X}, \quad (6.11)$$

or more generally, by a non-decreasing function $\nu : [0, 1] \to \mathbb{R}_{\geq 0}$ of the confidences,

$$\sum_{\phi_k(x)=A(\phi_1(x),\ldots,\phi_r(x),\tau_{\phi_1}(x),\ldots,\tau_{\phi_r}(x))} \nu(\tau_{\phi_k}(x)) = \max_{i=1,\ldots,m} \sum_{\phi_k(x)=c_i} \nu(\tau_{\phi_k}(x)), x \in \mathcal{X}. \quad (6.12)$$

The weighting by confidence (6.11) is a specific case of (6.12) with ν being identity. In the case of fuzzy classifiers, some analogy of (6.10) is

$$(A(\phi_1(x), \ldots, \phi_r(x)))_i = \frac{1}{r} \sum_{k=1}^r (\phi_k(x))_i, i = 1, \ldots, m, x \in \mathcal{X}, \quad (6.13)$$

and the corresponding analogy of (6.12) is

$$(A(\phi_1(x), \ldots, \phi_r(x), \tau_{\phi_1}(x), \ldots, \tau_{\phi_r}(x)))_i =$$

$$= \frac{1}{\sum_{j=1}^r (\nu(\tau_{\phi_j}(x)))} \sum_{k=1}^r \nu(\tau_{\phi_k}(x))(\phi_k(x))_i, i = 1, \ldots, m, x \in \mathcal{X}. \quad (6.14)$$

Notice that $(\phi(x))_i, i = 1, \ldots, m$ in (6.13) is the average of $(\phi_1(x))_i, \ldots, (\phi_r(x))_i$. And an average, in turn, is nothing else than the *mean value*, with respect to the uniform distribution on the team, of a random variable $\varphi_x(k) = \phi_k(x)$. Similarly,

$(\phi(x))_i, i = 1, \ldots, m$ in (6.14) is the mean of that random variable with respect to the distribution $\mu = \left(\frac{v(\tau_{\phi_1}(x))}{\sum_{k=1}^{r} v(\tau_{\phi_k}(x))}, \ldots, \frac{v(\tau_{\phi_r}(x))}{\sum_{k=1}^{r} v(\tau_{\phi_k}(x))} \right)$. Alternative aggregation functions are the *median* or *trimmed mean* of $\varphi_x(i) =$ with respect to μ, as well as $\max_{k=1,\ldots,r} \phi_k(x)$, respectively $\min_{k=1,\ldots,r} \phi_k(x)$ if we are particularly interested in the best, respectively worst possibilities.

A generalization of (6.14), thus a further generalization of majority voting, is aggregating the results of team members for $x \in \mathcal{X}$ by a fuzzy integral of φ_x with respect to an arbitrary distribution μ. Though various kinds of fuzzy integrals exist [1], only the two most traditional ones—Choquet integral and Sugeno integral—have been employed often enough as aggregation functions [2]. Both make use of an increasing ordering $\phi^{(i,1)}(x), \ldots, \phi^{(i,r)}(x)$ of $\phi_1(x), \ldots, \phi_r(x)$ with respect to the membership in the i-th class, and of the sets $S^{(i,1)}, \ldots, S^{(i,r)}$ of classifiers with highest membership in the i-th class,

$$(\phi^{(i,1)}(x))_i \leq \cdots \leq (\phi^{(i,r)}(x))_i, i = 1, \ldots, m, x \in \mathcal{X}, \tag{6.15}$$

$$S^{(i,k)} = \{\phi^{(i,k)} : \mathcal{X} \to [0,1]^m | k = i, \ldots, r\}, i = 1, \ldots, m. \tag{6.16}$$

With an additional notation $\phi^{(i,0)} = 0, i = 1, \ldots, m$, the *Choquet integral* of φ_x with respect to μ is defined

$$(C) \int \varphi_x d\mu = \sum_{k=1}^{r} (\phi^{(i,k)}(x) - \phi^{(i,k-1)(x)}) \mu(S^{(i,k)}), \tag{6.17}$$

whereas the *Sugeno integral* of φ_x with respect to μ is defined

$$(S) \int \varphi_x d\mu = \max_{k=1,\ldots,r} \min(\phi^{(i,k)}(x), \mu(S^{(i,k)})). \tag{6.18}$$

Among aggregation methods that are not generalizations of majority voting, *mixtures of experts* definitely deserve to be recalled [3, 4]. In this method, the input space $\mathcal{X}$ is divided into r subsets $\mathcal{X}_1, \ldots, \mathcal{X}_r$ corresponding to individual members of the team, and each of them, the aggregation function takes over the classification by the corresponding expert $k = 1, \ldots, r$:

$$A(\phi_1(x), \ldots, \phi_r(x)) = \phi_k(x), x \in \mathcal{X}_k, \tag{6.19}$$

in the case of crisp classification without confidences, and

$$(A(\phi_1(x), \ldots, \phi_r(x), \tau_{\phi_1}(x), \ldots, \tau_{\phi_r}(x)))_i = v(\tau_{\phi_k}(x))(\phi_k(x))_i,$$
$$i = 1, \ldots, m, x \in \mathcal{X}_k, \tag{6.20}$$

in the general case (6.8).

Finally, the aggregation function can also be obtained through the application of an additional classifier to the results of the members of the team, as well as to their confidences if they are available. The second level classifiers can, in principle, be of any kind, most frequently the logit method is encountered in this role. Such a hierarchical organization of classifiers is called *stacking* [5, 6]. Notice its similarity to the architecture of multilayer perceptrons, presented in Sect. 3.5. In fact, if the members of the team are multilayer perceptrons, and if the second level classifier is a perceptron or also a multilayer perceptron, then the resulting classification by the team is equivalent to the classification by a single, large multilayer perceptron, with more hidden layers than has any member of the team and with incomplete connections between subsequent layers.

6.1.2 Classifier Confidence

There are many possibilities how to define a classifier confidence. If we restrict attention to the usual crisp classifiers (6.2), then probably most often encountered is the *Euclidean Local Accuracy (ELA)* [7]. It employs a given validation set V of data with known classification, and a given number k of nearest neighbours such that the cardinality $|V|$ of V is at least k. Using notation from earlier chapters of the book, in particular c_v for the correct classification of an input $v \in V$, and $\|\varphi\|$ for the truth value of a statement φ, ELA is defined as follows:

$$\tau_\phi^{ELA}(x) = \frac{1}{k} \sum_{i=1}^{k} \|\phi(v^{(x,i)}) = c_{v^{(x,i)}}\|, x \in \mathcal{X},\tag{6.21}$$

where $v^{(x,1)}, \ldots, v^{(x,|V|)}$ is an ordering of elements of V according to their Euclidean distance d_E from x, i.e., $d_E(v^{(x,1)}, x) \leq \cdots \leq d_E(v^{(x,|V|)}, x)$.

Another often employed confidence is the overall correct classification of the classifier ϕ on a validation set V,

$$\tau_\phi^V(x) = \frac{1}{|V|} \sum_{v \in V} \|\phi(v) = c_v\|, x \in \mathcal{X}.\tag{6.22}$$

Notice that this confidence, differently to (6.21), does not take into account the particular input to which it is applied. For a given classifier, it is constant. Such confidences are called *static*, whereas confidences that take into account their particular input, such as (6.21), are called *dynamic*.

6.1.3   Teams of Classifiers of Different Kinds and Ensembles of Classifiers of the Same Kind

Due to the continuity of the aggregation function and due to (6.9), a team of similar classifiers brings nothing new because it classifies similarly to each of its members. Hence, for a team to be useful, it needs to be diverse. As was mentioned already in the introduction to this chapter, the diversity of the team of classifiers can be achieved in two ways:

1. Individual members of the team are classifiers of different kinds, e.g., one of them is a support vector machine, another a k-nn classifier, still another a classification tree, etc. Typically, such a team has only several members and each member uses only a part of the available input features. Consequently, the individual members of the team specialize in lower-dimensional projections of the input space X corresponding to the features used by each of them.
2. All members of the team are classifiers of the same kind, but they employ different combination of values of tunable parameters. Typically, such teams have dozens to thousands of members, which differ from each other through the used input features, subsets of training data, or both.

Needless to say, both ways can be combined. This leads to teams with many classifiers of several different kinds. However, teams with many members that are all of the same kind are more common. Such teams are called *ensembles of classifiers*.

6.1.4   Teams as an Approach to Semi-supervised Learning

The fact that different members of the team can be trained with different subsets of features can be utilized in situations when semi-supervised learning is needed. Recall from Sect. 2.4.2 that these are situations when we have only a small number p of training data, $(x_1, c_1), \ldots, (x_p, c_p)$, but in addition a set $\mathscr{U}$ of unclassified feature vectors. For semi-supervised learning, we need to train each classifier ϕ_k, $k = 1, \ldots, r$ with training pairs containing only a specific subset of features of the feature vectors from training data, namely

$$(([x_1]_j)_{j \in \mathscr{F}_k}, \ldots, ([x_p]_j)_{j \in \mathscr{F}_k}), \tag{6.23}$$

$$\text{where } \mathscr{F}_k \cap \mathscr{F}_{k'} = \emptyset \text{ for } k \neq k', \bigcup_{k=1}^{r} \mathscr{F}_k = \{1, \ldots, n\}. \tag{6.24}$$

Then, the trained classifier ϕ_k predicts the class for $x \in \mathscr{U}$, and the resulting pairs $(x, \phi_k(x))$ complete the training data for the remaining classifiers ϕ_ℓ, $\ell \neq k$. Typically, only the feature vectors $x \in \mathscr{U}$ with highest confidence $\tau_k(x)$ are used to this end.

This kind of semi-supervised learning is called *co-training*. By far most frequently, it is used with teams consisting of only two classifiers. The reason is that for $r \geq 3$ classifiers, problems occur if classifiers τ_k, $\tau_{k'}$, $k \neq k'$ suggest to complete the training data for the remaining classifiers ϕ_ℓ, $\ell \neq k, k'$ with the pairs $(x, \phi_k(x))$ and $(x, \phi_{k'}(x))$, where $x \in \mathscr{U}$ is the same, but $\phi_{k'}(x) \neq \phi_k(x)$.

The success of co-training, i.e., the improvement of the accuracy of the resulting classification compared with classification based only on the original training data, depends on the properties of the sets of features $([x]_j)_{j \in \mathscr{F}_k}$ for $k = 1, \ldots, r$. The expected improvement is highest if the following two conditions are fulfilled, by the random vectors the realizations of which are those sets of features:

(i) For classification into the considered classes, each such random vector alone is with a high probability sufficient.

(ii) Those random vectors are conditionally independent conditioned on the classes $c \in C$.

On the other hand, if these conditions are not fulfilled, and especially if the random vectors corresponding to $([x]_j)_{j \in \mathscr{F}_k}$, $k = 1, \ldots, r$ are dependent, then the accuracy of the resulting classifier can be even lower than if it were trained only with the original training data $(x_1, c_1), \ldots, (x_p, c_p)$.

6.1.5 Main Methods for Team Construction

The most natural way to construct a team of classifiers is to train each individual classifier with all available data using a state-of-the-art training algorithm, and then to use some aggregation function to combine the results of their classification. However, this works only if all classifiers in the team are of different types, which already assures a sufficient diversity among them. In an ensemble, training all its members with the same data using the same training algorithm leads to classifiers that are identical or nearly identical, making the obtained ensemble useless. A principal remedy is to train each member of the ensemble on a specific subset of the available data and/or using a specific modification of the training algorithm. This principle underlies several methods for ensemble construction, by far most important among which are bagging and boosting.

Bagging is an abbreviation of *bootstrap aggregating*, which originates from the fact that learning pairs $(x_1^{(k)}, c_1^{(k)}), \ldots, (x_t^{(k)}, c_t^{(k)})$ for training the k-the ensemble member, are obtained by *bootstrap sampling*. This is a sampling in which each training pair $(x_j^{(k)}, c_j^{(k)})$, $j1, \ldots, t$ is sampled from the original data $(x_1, c_1), \ldots, (x_p, c_p)$ independently, thus the same pair can occur repeatedly in $(x_1^{(k)}, c_1^{(k)}), \ldots, (x_t^{(k)}, c_t^{(k)})$ (therefore, bootstrap sampling is sometimes called *sampling with replacement*). Bagging can be, in principle, combined with any aggregation function, which can in addition incorporate dynamic or static confidence. The probability that a particular (x_ℓ, c_ℓ), $\ell = 1, \ldots, p$ occurs among $(x_1^{(k)}, c_1^{(k)}), \ldots, (x_t^{(k)}, c_t^{(k)})$ is

$$P((\exists \ell')\ (x_{\ell'}^{(k)}, c_{\ell'}^{(k)}) = (x_\ell, c_\ell)) = 1 - \left(1 - \frac{1}{p}\right)^t. \tag{6.25}$$

Thus the expected number of the original data in the k-th bootstrap sample is

$$\mathbb{E}(|\{\ell|(\exists \ell')\ (x_{\ell'}^{(k)}, c_{\ell'}^{(k)}) = (x_\ell, c_\ell)\}|) =$$

$$= \sum_{\ell=1}^{p} P((\exists \ell')\ (x_{\ell'}^{(k)}, c_{\ell'}^{(k)}) = (x_\ell, c_\ell)) = p\left(1 - \left(1 - \frac{1}{p}\right)^t\right). \tag{6.26}$$

Notice that this expected number is the same for all ensemble members, independent of k. The training pairs not included into the k-th bootstrap sample,

$$\{(x_\ell, c_\ell)|(\forall \ell')\ (x_{\ell'}^{(k)}, c_{\ell'}^{(k)}) \neq (x_\ell, c_\ell)\}, \tag{6.27}$$

are called *out-of-bag data* for the k-th ensemble member. They are a bagging counterpart of the validation folds in cross-validation, described in Sect. 2.4. In particular, the average error of all the ensemble members $\phi_1, \ldots, \phi_r$ on their respective out-of-bag data, called *out-of-bag error*, is a counterpart of the cross-validation error (the average error of all the classifiers trained during cross-validation on their respective validation folds).

Most frequently, the bootstrap size is taken to be the size of the original data, i.e., $t = p$. In that case, applying to (6.25) the well-known calculus result

$$\lim_{p \to \infty} \left(1 - \frac{1}{p}\right)^p = e^{-1} \tag{6.28}$$

yields for (x_ℓ, c_ℓ), $\ell = 1, \ldots, p$

$$\lim_{p \to \infty} P((\exists \ell')\ (x_{\ell'}^{(k)}, c_{\ell'}^{(k)}) = (x_\ell, c_\ell)) = 1 - \lim_{p \to \infty} \left(1 - \frac{1}{p}\right)^p = 1 - e^{-1} \doteq 0.632. \tag{6.29}$$

For a sufficiently large p, the expected number of the original data in a bootstrap sample is therefore

$$\mathbb{E}(|\{\ell|(\exists \ell')\ (x_{\ell'}^{(k)}, c_{\ell'}^{(k)}) = (x_\ell, c_\ell)\}|) = \sum_{j=1}^{p} P((\exists \ell')\ (x_{\ell'}^{(k)}, c_{\ell'}^{(k)}) = (x_\ell, c_\ell)) \doteq 0.632p. \tag{6.30}$$

Boosting is a method of ensemble construction in which new ensemble members focus on data misclassified by previous ensemble members. In this way, it is possible to achieve a high accuracy of the aggregated classification by the ensemble even in situations when the accuracy of the classification by individual ensemble members was only slightly better than 50%, i.e., only slightly better than a random guess. From the point of view of classifier learning, this is usually characterized as the ability of the method to create a *strong learner* from *weak learners* (and this ability accounts for the name of the method).

So far, more than a dozen boosting algorithms have been proposed. Here, we recall only the most popular among them (which is, at the same time, one of the most traditional), the algorithm *AdaBoost*, in its variant for binary classification [8]. Its aggregation function uses the weighted average of its members, the weights being logarithms of their confidences,

$$
A(\phi_1, \ldots, \phi_r, \tau_1, \ldots, \tau_r) = \begin{cases} 1 | \sum_{k=1}^{r} \ln(\tau_k)(\phi_k - \frac{1}{2}) \geq 0 \\ 0 | \text{else.} \end{cases}
\tag{6.31}
$$

The confidences are in this case static, i.e., $\tau_k, k = 1, \ldots, r$ is constant for each ensemble member (cf. Sect. 6.1.2). On the other hand, τ_k is specific for given k, taking into account the success of classification by the ensemble members $\phi_1, \ldots, \phi_r$, as follows:

$$
\tau_k = \sqrt{\frac{\sum_{\ell=1}^{p} w_\ell^k \|\phi_k(x_\ell) = c_\ell\|}{\sum_{\ell=1}^{p} w_\ell^k \|\phi_k(x_\ell) \neq c_\ell\|}} = \exp\left(\frac{1}{2} \ln \frac{\sum_{\ell=1}^{p} w_\ell^k \|\phi_k(x_\ell) = c_\ell\|}{\sum_{\ell=1}^{p} w_\ell^k \|\phi_k(x_\ell) \neq c_\ell\|}\right).
\tag{6.32}
$$

Here, $w_\ell^k, \ell = 1, \ldots, p$, depend on the confidence τ_k of ϕ_k, in such a way that $w_\ell^{k+1} = w_\ell^k \tau_k^{-2\|\phi_k(x_\ell)=c_\ell\|}$ for $j = 1, \ldots, r - 1$, and $w_1^1 = \cdots = w_p^1$.

In implementations of boosting, conventionally, $w_j^1 = \frac{1}{p}$ is used, which turns $(w_1^1, \ldots, w_p^1)$ in a probability distribution. But this is actually immaterial because (6.32) does not depend on the value of $w_1^1 = \cdots = w_p^1$. Observe that (6.32) can be calculated only if neither $\phi_k(x_\ell) = c_\ell$ for all $\ell = 1, \ldots, p$, nor $\phi_k(x_\ell) \neq c_\ell$ for all $\ell = 1, \ldots, p$. The case $\phi_k(x_\ell) = c_\ell$ for all $\ell = 1, \ldots, p$ suggests that instead of the whole ensemble, the individual classifier ϕ_k should be used. The very unlikely case that $\phi_k(x_\ell) \neq c_\ell$ for all $\ell = 1, \ldots, p$ suggests that the classifier ϕ_ℓ should be removed from the ensemble.

Algorithm 6 (AdaBoost)
Input:
- classifiers $\phi_1, \ldots, \phi_r$;
- training pairs $(x_1, c_1), \ldots, (x_p, c_p)$, $w > 0$.
Step 1. Set $k = 1$, $w_1^1 = \ldots w_p^1 = w$.
Step 2. Calculate τ_k according to (6.32).
Step 3. If $\phi_k(x_\ell) \neq c_\ell$ for all $\ell = 1, \ldots, p$, then remove ϕ_k from the ensemble.
Step 4. If $\phi_k(x_\ell) = c_\ell$ for all $\ell = 1, \ldots, p$, then stop with $A(\phi_1, \ldots, \phi_r, \tau_1, \ldots, \tau_r) = \phi_k$.
Step 5. If $k = r$, then stop with $A(\phi_1, \ldots, \phi_r, \tau_1, \ldots, \tau_r)$ calculated according to (6.31).
Step 6. Else increment k, set $w_\ell^k = w_\ell^{k-1} \tau_{k-1}^{-2\|\phi_k(x_\ell)=c_\ell\|}$ for $\ell = 1, \ldots, p$, and return to Step 2.
Output: Definition of the aggregation function $A(\phi_1, \ldots, \phi_r, \tau_1, \ldots, \tau_r)$, and possibly the information that some ϕ_k has to be removed from the ensemble.

6.1.6 Using Teams of Classifiers with Multimedia Data

Multimedia is defined as content that combines several media forms, aka modalities. Television broadcast, movies, video content published online and streamed video, for example, consist of a motion picture, one or more audio channels and possibly several tracks of closed captions. Web pages and social media posts may contain (aside from the text data) animations or still images. Each content form may contain parts of the transmitted message, some of which may be duplicated (such as spoken text and closed captions), complementary (commentary over a sporting event), to some extent independent (announcement over illustrative video), irrelevant, or even perceived as contradictory (image depicting sarcasm attached to a post on social media).

The overall goal of the multimedia classification is to gather the required information from the content while using the most of the available data to achieve the highest possible confidence. However, the individual modalities of multimedia content are very different from each other, raw data is usually too large for direct use in a classifier and individual modalities need to utilize different methods for extraction of a high-level (semantic) information.

The size of raw data is commonly reduced by the use of feature extractors. These take the raw data of a specific modality and extract a significantly smaller numerical descriptor of a certain aspect of the original content while neglecting other aspects. Pictures or individual frames of the video may be, for example, reduced to a colour histogram, set of areas sharing similar texture with their shape characteristics or a list containing points of interest (visual corners) and their local texture descriptions (e.g. SIFT, SURF, ORB). The motion picture can be reduced to a set of motion vectors. The audio signal can be transformed to a frequency domain (spectrum), for example, by a Fourier transform, and reduced to information on power in certain frequency bands.

All of these features are mere projections of the original data and are, therefore, still considered as low-level features. Moreover, as the individual feature extractors are tightly coupled with the modality they operate on, feature extractors do not directly solve the issue of processing all modalities simultaneously to obtain a single set of high-level (semantic) features.

Two principal approaches to fusion of the modalities are available, called early and late fusion. Early fusion takes extracted features from individual modalities and concatenates them to a single vector. Such a vector is then classified with an off-the-shelf classifier as a whole. We should be, however, aware of the fact that the extracted features might have a very different semantics and value ranges. Therefore, a careful selection of a classifier is recommended. For classifiers that are sensitive to feature mean and variance, the second issue can be solved with a value normalisation if the ranges are known.

Late fusion is just a different name for a team of classifiers in the multimodal scenario. Individual modalities are classified with the best performing classifier (or a team of classifiers) with the same set of target classes, and the resulting labels are combined by one of the aggregation functions mentioned in Sect. 6.1.1. This approach allows a much higher flexibility in the selection of the classifier for given modality. However, this approach is to no avail if the required information can be gathered only from a combination of modalities and not from any individual modality.

To reduce the shortcomings of the fusion strategies, they can be combined in a hybrid (double) fusion approach. The resulting team of classifiers then includes also at least one member that uses concatenated features from different modalities as an input.

The most commonly considered multimedia content, video, has yet another issue: the different sampling rate of each modality. Images are usually captured and distributed in sampling rates ranging from 23.976 to 60 frames per second. The audio signal is captured at 44.1 to 126 kHz. Closed captions may cover variable periods ranging from several seconds to tens of seconds, based on the speed of the dialogue. The fusion system, therefore, needs to select which time section of each modality will be used for the classification and which method is used for aggregation of the samples in the selected period.

An extensive survey of multimodal fusion for multimedia analysis is available in [9].

As an example of a multimedia classifier, SESAME [10] is a system for event detection from video content. This system uses a late fusion of two SVMs based on different coding schemes of visual features, three SVMs with different encodings of dense motion features, one SVM based on bag-of-audio-words, an SVM and Random forest on detected visual concepts and one SVM on features from automated speech recognition.

6.1.7 Teams of Classifiers in Malware and Network Intrusion Detection

For malware and network intrusion detection, teams of classifiers of different kinds have been used more often than ensembles of classifiers of the same kind. Most efforts have been invested to teams of classifiers for malware detection in the context of Android, which is nowadays the most common operating system in end-user devices and attracts most attention of malware creators through its openness.

The authors of [11] used teams of different kinds of classifiers for static malware analysis. They considered various kinds of classifiers, including naïve Bayes, SVM and classification trees, and features obtained through feature selection from the following two groups:

- Permissions requested by the classified app, extracted by analysing the Android manifest. The importance of permissions for malware detection follows from the fact that for various types of malware, particular permissions are needed: For example, apps stealing user credentials and selling user information typically request the permissions to access internet, coarse location and to read phone state, for apps aimed at obtaining a financial gain through premium SMS, the permission to send sms is essential and also the permission to read phone state is frequently requested, whereas for apps involved in distributed denial of service attacks, the permissions to access internet and network state are typically requested. Altogether 135 permissions were made available for feature selection.

- Operation system API calls. Their importance follows from the fact that malicious behaviour typically requires some way of communicating to the outside world. Hence, API calls in the category "urlopenconnection" are needed, in particular the calls "connect" and "openconnection". In addition, apps stealing user credentials and selling information about user behavior need the calls "getnetworkinfo" and "getwifistate" to know the whereabouts of the user, and the calls "getdeviceid" and "getcelllocation" to get the identification information of the device. Finally, apps monitoring smartphone activity typically need the calls "startservice" and "startactivity". Altogether 210 API calls were considered.

A team of different classifiers has been more extensively applied to static analysis of Android malware in [12]. That team consists of 5 classifiers—a k-nearest neighbour and a naïve Bayes classifier, a support vector machine, a classification tree and a random forest, which is an ensemble of classification trees and will be addressed below in Sect. 6.2 (Fig. 6.1). Its application to the detection of Android malware is more extensive compared to the team from [11] in two aspects:

(i) Although similarly to the above approach, only only features obtained through static malware analysis were used, their number was much larger compared to the 135 permissions and 210 API calls in [11]. In [12], altogether, 2,374,340 features, in an overwhelming majority API calls, were extracted from the training data, formed by 8,701 malware apps and 107,327 benign apps. All the extracted features were used for a preliminary classification by a support vector machine. Due to the very high dimension of the feature space, malware and benign apps were linearly separable in it. The components of the normal vector were then used for feature selection—features corresponding to components with a very small absolute value were discarded, leaving 34,630 features that were used for the final classification by all members of the team.

(ii) In addition to the classification into malware and benign software, the team performs an additional classification of benign apps into game apps and non-game apps, and finally a classification of each of those categories into subcategories, such as "brain cards and casual" or "sports and racing" for games, and "book readers and magazines" or "themes and wallpaper" for non-games.

In [13], a stacked team for Android malware detection has been described. It has two layers, the lower one combines four classifiers, the upper one three classifiers. The employed kinds of classifiers include k nearest neighbours, artificial neural networks, the logit method, and variants of classification trees.

In this context, we would like to recall the sophisticated decision-tree-based system for Android malware detection presented in [14], which we introduced in Sect. 5.3.5. As was mentioned there, the second, more complex from the two classification trees employed in that system uses 8 different categories of features, among

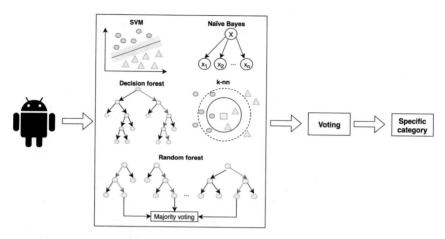

Fig. 6.1 Team of 5 classifiers for Android malware detection considered in [12]

which the categories 7 and 8 are outputs of an SVM classifier and a nearest neighbour classifier, respectively. These two classifiers can be viewed as forming a team with the classification tree, more precisely a stacked team with the SVM and nearest neighbour classifier in the lower layer, and the classification tree in the upper layer.

Teams of different classifiers have not been used only for Android malware detection. In [15], their application to intrusion detection has been reported, more precisely, to the detection of botnets used for DDoS attacks (cf. Sect. 1.5). The available team members include a Bayesian network and variants of k nearest neighbours and decision trees. Because an overwhelming majority of internet traffic is TCP-based, the employed features are primarily the details of transferred TCP packets.

Malware detection using an ensemble of classifiers of the same kind can be performed within the framework *FENOC (Framework based on ENsemble One-Class learning)* [16]. It has been developed for malware detection in the context of the Windows operating system and the employed ensemble has the following properties:

- Its members are a particular kind of support vector machines, called one-class SVM, their learning is denoted as one-class learning.
- As aggregation function, majority voting is used.
- For each member SVM, a random subset of features is used, which implies that the SVM is constructed on a random subspace of the original feature space.
- Before the member SVMs are constructed, the training data are clustered in such a way that each cluster corresponds either to malware or to benign software. Each SVM is then trained for one particular cluster.

FENOC performs both static and dynamic malware analysis (Fig. 6.2). The used features can be divided into three groups:

(i) *n*-grams of Op-codes, i.e., of CPU instructions, which are obtained through static malware analysis if the original binary file is disassembled;

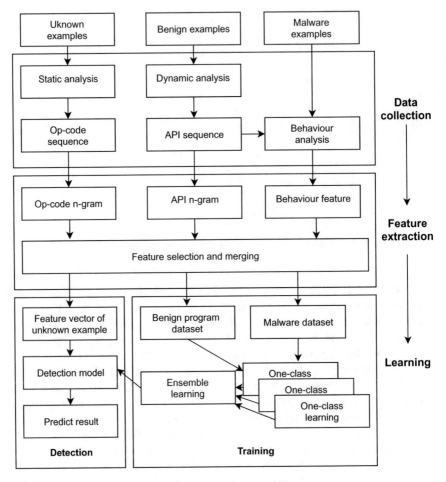

Fig. 6.2 Scheme of the Framework FENOC (inspired by [16])

(ii) *n*-grams of Windows API calls and their arguments for the main process and its child processes, which are obtained through dynamic malware analysis;

(iii) behavioural features with higher level semantics represented using behavioural graph, also obtained through dynamic malware analysis.

The dynamic malware analysis is performed in the sandbox Osiris [17]. The random subspace method in ensemble construction, which is robust to noise, partially alleviates the main issue of dynamic analysis, namely loss of information due to incomplete execution of a malware that is caused by some trigger condition being not satisfied (in extreme case meaning sandbox evasion).

Another malware detection system using an ensemble of SVMs is the SBMDS, presented in [18]. This acronym stands for "string based malware detection system", due to the fact that the system performs static analysis of portable executables (PE files), which is based on two kinds of interpretable strings:

- API calls;
- strings contained in the PE file that can indicate intents of attackers.

The SBMDS trains the SVMs ensemble by bagging and uses the aggregation function majority voting. Malware detection by this system is carried out in four major steps (Fig. 6.3):

(i) Obtaining interpretable strings with a feature parser.
(ii) Selection of features related to different kinds of malware.
(iii) Training an ensemble of SVMs.
(iv) Using the trained ensemble as malware detector.

An important property of the SBMDS is that the used ensemble not only predicts whether new software is malware but also to which particular family of malware included in the training data it belongs.

Also, the authors of [19] performed static and dynamic Windows malware analysis using an ensemble of classifiers of the same kind. For that ensemble, a particular variant of perceptrons for binary classification has been used, called one side class perceptrons. It does not require linear separability of the two classes but consequently cannot guarantee zero classification error on the training data, only zero false positives or zero false negatives. Their ensemble is constructed in the following way:

1. An initial set of 500 features is made available for perceptron training, resulting from feature selection applied to the altogether 6,275 features obtained through static and dynamic malware analysis.
2. The available training data are clustered.
3. Two ensemble members are created with as many input neurons as is the number of features available for training and such that one perceptron achieves zero false positives and as little as possible false negatives on the clustered training data, whereas the other achieves zero false negatives and as little as possible false positives.
4. If all available training data were correctly classified by the two ensemble members created in the previous step, the construction of the ensemble is finished, otherwise the training data correctly classified by those ensemble members are discarded from the available training data.
5. The set of features available for training is extended with additional 200 features resulting from feature selection applied to the features obtained through static and dynamic malware analysis and the ensemble construction returns to step 2.

Finally, in [20], intrusion detection has been performed using an ensemble of SVMs in combination with active learning. To this end, the set $\mathcal{U}$ of unlabelled inputs used in active learning is divided into a given number of disjoint subsets $\mathcal{U}_1, \ldots, \mathcal{U}_r$. On each of them, a SVM is trained using active learning according to Algorithm 4 in Sect. 4.4, thus at the end, an ensemble of SVMs $\phi_1, \ldots, \phi_r$ is obtained.

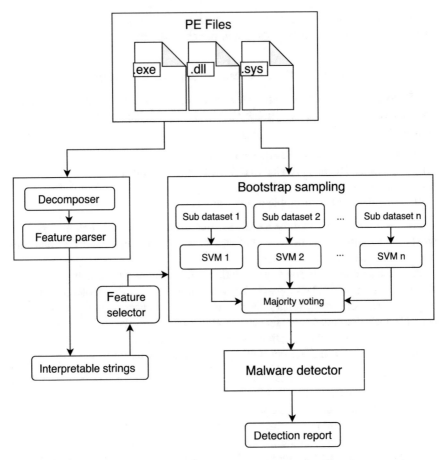

Fig. 6.3 Scheme of malware detection by the SBMDS system (inspired by [18])

6.2 Ensembles of Trees—Random Forests

The most frequently encountered kind of classification ensembles are ensembles in which the individual members are classification trees. Such ensembles have become popular under the name *random forests* [21]. They are constructed by bagging, i.e., bootstrap aggregation of individual trees (Fig. 6.4). In addition, there is the possibility that each tree is trained using only a random subset of considered input features, i.e., only a random subset of dimensions of the considered input space. Typical size of random forests encountered in applications are dozens to thousands of trees, and their usual aggregation function is majority voting, or some of its fuzzy generalizations (6.13) or (6.14). Because standard classification trees introduced in Sect. 5.3 are crisp classifiers, they need to be viewed as fuzzy classifiers mapping into the set $\{u \in \{0, 1\}^m | \sum_{i=1}^{m} u_i = 1\}$ (cf. Sect. 2.1) if (6.13) or (6.14) is to be applied. An example

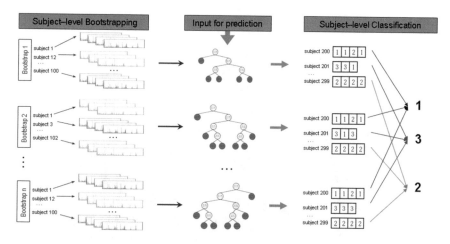

Fig. 6.4 Random forests are trained by means of bagging, i.e., bootstrap aggregation of individual trees. Subsequently, when new subjects are input to the forest, each tree classifies them separately, according to the leaves at which they end, and the final classification by the forest is obtained by means of majority voting (from [22])

fuzzy membership function of a class for a random forest trained on a 2-dimensional space of continuous variables is shown in Fig. 6.5.

This way of construction entails two main advantages of random forests:

1. Due to the large number of classification trees aggregated in a random forest, it is possible to obtain a rich variety of membership functions, in spite of the fact that classification trees themselves are crisp.
2. If individual trees are trained with only subsets of input features, the random forest tends to decrease the impact of noise in training data provided the noise is due to only some particular features.

6.2.1 Kinds of Random Forests

As was mentioned already in the introduction to this section, the randomness of random forests always concerns the data employed for training individual ensemble members, due to the fact that the forest is constructed by bagging, but in addition, it can concern also involved input features, i.e., dimensions of the input space. According to whether also randomness concerning features is present, two broad groups of random forests can be differentiated:

1. *Random forests grown in the full input space.* Each tree is trained using all considered input features. Consequently, any feature has to be taken into account when looking for the split condition assigned to an inner node of the tree. However,

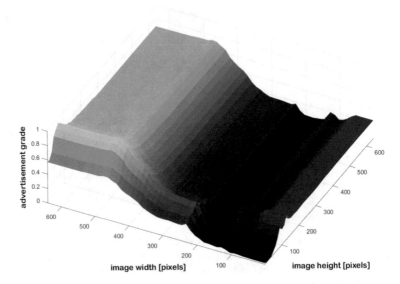

Fig. 6.5 Grade of membership in the fuzzy class "advertisement" predicted by a classifier classifying web images into advertisements and non-advertisements based on their height and width, trained with the UCI-repository data set "Internet advertisements" [23]

features actually occurring in the split conditions can be different from tree to tree, as a consequence of the principle of bagging that each tree is trained with another set of training data, sampled randomly with replacement from the original training pairs $(x_1, c_1), \ldots, (x_p, c_p)$. Due to the same principle, even if a particular feature occurs in split conditions of two different trees, those conditions can be assigned to nodes at different levels of the tree.

A great advantage of this kind of random forests is that each tree is trained using all the information available in its set of training data. Its main disadvantage is the above mentioned high complexity. In addition, if several or even only one variable are very noisy, that noise gets nonetheless incorporated into all trees in the forest. Because of those disadvantages, random forests are grown in the complete input space primarily if its dimension is not high and no input feature is substantially noisier than the remaining ones.

2. *Random forests grown in subspaces of the input space.* Each tree is trained using only a randomly chosen fraction of features, typically a small one. This means that a tree t is actually trained with projections of the training data into a low-dimensional space spanned by some randomly selected dimensions $i_{t,1} \leq \cdots \leq i_{t,d_t} \in \{1, \ldots, d\}$, where d is the dimension of the input space, and d_t is typically much smaller than d. Using only a subset of features not only makes forest training much faster, but also allows to eliminate noise originating from only several features. The price paid for both these advantages is that training makes use of only a part of the information available in the training data.

Apart from this broad basic division of random forests, specific group of them can be delimited with specific conditions on the tree structure and/or node splitting. Here, we mention only one such specific kind of random forests—ferns. A *random fern* is a random forest in which splitting in all nodes on the same level, i.e., at the same distance from the root, is performed according to the same condition [24]. More precisely, there is a specific input dimension j_ℓ, $1 \le j_\ell \le d$ for each level ℓ in the tree, and

- if the j_ℓ-th input feature is continuous, then there exists a threshold $\theta_{\ell \in \mathbb{R}}$ such that to each node on the level ℓ, the predicate $[x]_{j_\ell} < \theta_\ell$ (or $[x]_{j_\ell} \le \theta_\ell$), is assigned;
- if the j_ℓ-th input feature is categorical, then there exists a subset $S_{j_\ell} \subset V_{j_\ell}$ of its value set V_{j_ℓ} such that to each node on the level ℓ, the predicate $[x]_{j_\ell} \in S_{j_\ell}$ is assigned.

In Fig. 6.6, the levels 1 and 2 of an example fern are schematically depicted.

6.2.2 Random Forest Learning

Principally, random forest learning is very simple—it transfers to learning all the individual trees forming the forest. Individual tree learning is connected to the forest learning in two ways:

(i) Each tree is trained with a subset of the forest training data. Different trees are trained with different subsets, obtained by bagging (Fig. 6.4).
(ii) Each tree is trained using either all features included in the forest training data or some random subset of them. In the first case, the tree is grown in the full input space to which the training data belong, whereas in the second case, it is grown in some its subspace, to which the training data are projected.

Consequently, training a random forest is a two-step process:

Step 1. A subset of the forest training data is assigned to each tree, possibly also a subset of features to be used in its training (if not all features should be used).

Step 2. Each tree is trained with the assigned data and features, by means of standard classification tree learning, as was described in Sect. 5.3.

In the context of random forest learning, active learning is frequently encountered, which was introduced earlier in connection with SVM learning, in Sect. 4.4. Actually, support vector machines and random forests are the two kinds of classifiers that are combined with active learning most often. The fact that random forests are ensembles can be exploited in a straightforward active learning method, for which we will now provide a brief sketch.

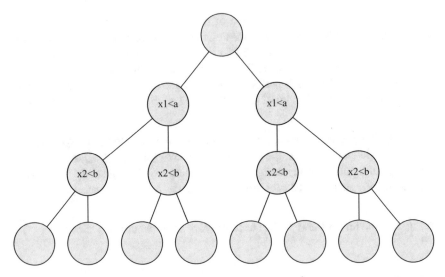

Fig. 6.6 Predicates determining the splits on the 1st and on the 2nd level of an example fern

As usually, consider a sequence of training data $(x_1, c_1), \ldots, (x_p, c_p)$ with which a random forest consisting of trees $\phi_1, \ldots, \phi_r$ has been trained, such that

$$\phi_1(x_\ell) = \cdots = \phi_r(x_\ell) = c_\ell \text{ for } \ell = 1, \ldots, p. \tag{6.33}$$

This can be achieved, e.g., through keeping all leaves of all trees pure (Sect. 5.3). In addition, let a finite set $\mathcal{U}$ of unclassified feature vectors be available. According to (6.33), the random forest is completely certain as to the classification of $x_\ell, \ell = 1, \ldots, p$, or equivalently, it has no uncertainty as to their classification. Consequently, the classification of x_ℓ has zero entropy, provided that entropy is based on the empirical probability with respect to the random forest:

$$H(x_\ell; c_1, \ldots, c_m) = \sum_{\ell'=1}^{m} P(c_\ell = c'_\ell) \ln P(c_\ell = c'_\ell) =$$

$$= \sum_{\ell'=1}^{m} \frac{|\{k|\phi_k(x_\ell) = c'_\ell\}|}{r} \ln \left(\frac{|\{k|\phi_k(x_\ell) = c'_\ell\}|}{r} \right).$$

$$\tag{6.34}$$

This suggests to use maximization of that entropy as the principle according to which, in active learning, the additional feature vectors from $\mathcal{U}$ are chosen for obtaining the correct class.

That active learning method can be summarized in the following algorithm.

Algorithm 7 (Active learning in random forests)

Input:
- training data $(x_1, c_1), \ldots, (x_p, c_p)$,
- set of unclassified inputs $\mathscr{U}$,
- classification trees $\phi_1, \ldots, \phi_r$ that have been trained with $(x_1, c_1), \ldots, (x_p, c_p)$,
- number p_a of additional feature vectors for which the correct class should be obtained.

Step 1. for $x \in \mathscr{U}$, define

$$H(x; c_1, \ldots, c_p) = \sum_{i=1}^{m} \frac{|\{k|\phi_k(x) = c_i\}|}{r} \ln\left(\frac{|\{k|\phi_k(x) = c_i\}|}{r}\right). \tag{6.35}$$

Step 2. Set $k = 1$.
Step 3. Denote $x_{p+k} = \arg\max_{x \in \mathscr{U}} H(x; C)$.
Step 4. For x_{p+k}, obtain the correct class c_{p+k}.
Step 5. Remove x_{p+k} from $\mathscr{U}$.
Step 6. If $k < p_a$, increment $k \to k + 1$ and return to Step 3.
Output: Training data extended to $(x_1, c_1), \ldots, (x_{p+p_a}, c_{p+p_a})$.

6.2.3 Random Forests in Spam Filtering

In [25], a spam filter based on random forests and preprocessing by means of clustering into a prescribed number of clusters (k-means clustering) was compared with spam filters based on a naïve Bayes classifier, a support vector machine, and a k-nn classifier, and it achieved a noticeably higher area under the ROC curve. However, the spam filter used solely textual features, and the comparison relied on a single spam dataset—the Spam Email Corpus used at the 2007 TREC conference, which contains 75,419 messages received by an email server at the University of Waterloo between April 8 and July 6, 2007. Maybe due to these two limitations, the above comparison does not seem to have noticeably influenced further development of spam filters—random forests do not belong to classifiers most commonly employed in them. On the other hand, they are encountered in less standard, more specific cases of dealing with malicious email, as in the four situations described below.

Image spam filtering based not on OCR, but only on visual features. In [26], a system RoBoTs (Robust BootsTrap based spam detector) has been proposed for such kind of image spam filtering. Its name refers to the fact that it aims at robustness against images from a common template, and that it uses bootstrap resampling. In addition, RoBoTs has several other important features:

- It performs the classification into the commonly used two classes: spam and ham.
- As classifier, random forests are employed, consisting of classification trees of the kind described in [27].

- Visual features are extracted at multiple resolution levels, and for each resolution level, a separate random forest is constructed.
- The extracted visual features include color, texture and shape, they are extracted for each subblock of each resolution level.
- The high-dimensional vector of extracted visual features is not used directly as the input to the forest, but is mapped to a lower-dimensional space first. The resulting forest is called *linear discriminant forest* in [26].
- The linear transformation of the extracted features is performed block-wise according to a random partition of the set of all visual features extracted at the considered resolution level.
- For each block of that linear transformation, a separate subset of training data is used, obtained by means of bootstrap resampling.

Phishing, a related email threat recalled in Sect. 1.1.5. In [28], random forests were used to detect phishing. The presented approach relies on the assumption that phishers often use the same infrastructure repeatedly to deliver their emails and collect the victim's information. Therefore, clustering based on the eigenvalues of the similarity matrix between emails (aka spectral clustering) was performed, and various relationships between individual emails and the resulting clusters served as features for the random forest classifier. As the impurity measure in the trees forming the forest, Gini index was used.

Spam filtering in Twitter messages. In [29], random forests were used to this end. In Twitter, spam spreading is similarly easy as in usual email messages because each user can reply to anyone on Twitter without being his/her follower or being followed by him/her. Due to the specific restriction on text length, Twitter messages are especially dangerous through inciting unaware users to malicious websites, using the construct "URL shortening". At the same time, that restriction also makes it difficult to find suitable representations for tweets and for users, as well as to extract the most important features from tweets.

Malicious emails targeted to individual users or small groups of users. In [30], random forests were used to filter this kind of malicious emails that is, in a way, complementary to spam. Such emails are much more difficult to detect than spam because they lack several properties that play a crucial role in spam filtering:

- Message content is usually very specific to the recipient, there are no words or expressions indicating targeted emails in general.
- Identical or similar targeted emails are not sent to many users simulateneously.
- Targeted emails are typically not sent from addresses with bad reputation, thus blacklisting does not work for them.

6.2.4 Random Forests in Recommender Systems

In connection with recommender systems, random forests belong to the most successful classifiers. This can be documented with the five particular examples below.

It is also confirmed by a comparison reported in [31], where four important kinds of classifiers were assessed within the same recommender system. Apart from random forests, they included classification trees of the kind proposed in [32], naïve Bayes classifier, and classification rules. The considered recommender system recommended helpful hotel reviews available in the reviewing and booking system TripAdvisor [33].

Random forests are primarily used in situations when apart from the basic information needed in the respective kind of recommender systems, i.e., information about the behaviour of users in a collaborative filtering system and about their preferences in a content-based system, also additional information concerning users is systematically collected. In that context, mainly the following two kinds of additional information are encountered:

1. Demographic features, such as gender, age, education, occupation, level of income. They can much influence whether a particular product, service, contact, etc. is interesting for the user, and consequently, whether it is worth being recommended. For example, a luxury car is more likely to be interesting for a middle-aged person with high income than for an arbitrary web surfer.
2. Relationships between users, especially friendship relationships, typically taken over from social networks. They, too, can substantially influence the impact of a recommendation. For example, a user is more likely to buy a recommended product if the recommendation is supported by references to his/her friends who have already bought that product.

That additional information allows to specify particular demographic profiles and/or to delimit specific communities in social networks. Recommender systems based on such specific demographic profiles or social communities are sometimes called *model-based*, whereas those based on non-specific information about the behaviour and/or preferences of all users are called *memory-based*. An important feature of model-based recommender systems is that they allow to avoid the *new user problem*, i.e., the problem of recommendations for a user about the behaviour and/or preferences of whom no or very little information is available.

The recommender system reported in [34] tackles, in addition, also the *new item problem*, i.e., the problem that new items entered into the system, do not have enough ratings nor other feedback of users to recommendations, as well as the combination of both problems, which is called *double cold start* in [34]. Aggregation functions implemented in the recommender system include majority voting as well as various kinds of weighted voting, most importantly voting with weights based on probability distributions of the predictions by individual classification trees, estimated using the training data. The system has been tested on the well-known *Movie-Lens dataset* [35], which contains more than one million movie ratings.

Not in every recommender system makes sense to separate basic and additional information. In particular in medical recommender systems, all kinds of patient information—his/her medical history, family history, demographic features, subjective complaints, measurements—are treated in a similar way. For example, in [36], random forests are used as the classifier underlying a system for online medical advice. Providing such advice to online patients consists in four steps:

1. The healthcare provider initiates a request for performing particular home medical tests of an online patient, i.e., measurements that can be performed by the patient himself/herself, e.g., blood pressure, blood sugar level, weight. Results of the home tests are then merged with results of laboratory tests, which are performed in a hospital, but typically in much longer time intervals.
2. Using the results of currently valid laboratory and home tests, as well as other relevant parts of the patient data, a trained random forest classifies current diagnosis and disease condition.
3. The current diagnosis and disease condition output by the random forest is sent to an advisory component of the system that keeps the physicians' medical advice. The advisory component finds advice best corresponding to the diagnosis and the patient's physiologic state.
4. Since responsibility for the advice is, finally, with the respective physician, they are first send to him/her for approval and/or correction. Only then, they are send to the patient and added to his/her medical record.

The recommender systems presented in [34, 36] belong to the standard kind of recommender systems, surveyed in Sect. 1.2. However, random forests have been successfuly used also in more sophisticated recommender systems, enriched through combining with additional approaches, such as the above recalled active learning.

In [37], the system *SPrank* is presented (the name stands for "semantic path-based ranking"), enriching a recommender system through combining it with regression trees (a regression counterpart of classification trees), semantic networks and an ontology. That system is characterized by the following properties:

- The classifier underlying the recommender system is a random forest.
- Also the regression trees are aggregated into ensembles, using a kind of boosting called gradient boosting.
- Semantic networks are employed to describe a huge decentralized knowledge base commonly known as Linked Open Data (LOD) [38].
- The employed ontology has been derived from DBpedia [39], a subset of LOD dealing with encyclopedic data.

The SPrank system has been tested on the above mentioned Movie-Lens dataset [35], as well as on a dataset collected from the *Last.fm* music system [40], which contains nearly 100,000 items of listening data for 1,892 users.

An important kind of such more sophisticated systems based on classification by means of random forests are *three-way recommender systems*, i.e., those comprising instead of the usual two classes "recommend" and "not recommend" actually three classes (Fig. 6.7)

$$\{\text{recommend, not recommend, consult the user}\}. \tag{6.36}$$

The third class, consult the user, is used if the system is uncertain whether to recommend or not. Needless to say, this clearly corresponds to the paradigm of active learning. As with any classifier, unless the classification error (2.12) is used as an

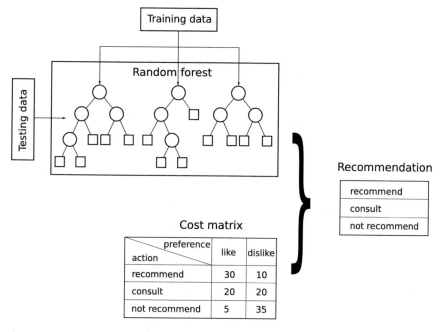

Fig. 6.7 The scheme of a 3-way recommender system based on random forests

error function, the weights in the more general cost-weighted error (2.14) need to be specified. This includes, in particular, the following costs:

1. costs of the consequences of having recommended an item although the user dislikes it;
2. cost of the consequences of not having recommended an item although the user would like it;
3. costs of consulting the user, which is normally independent of whether the user likes or dislikes the item.

A recommender system of that kind has been recently presented in [41]. The active learning in this system consists in estimating the probability that the considered user likes the considered item and comparing it with two thresholds $\alpha^\star$, $\beta^\star$, $\beta^\star < \alpha^\star$, as follows:

$$P(\text{user likes the item}) \begin{cases} \geq \alpha^\star & \rightarrow \text{ item is recommended to the user,} \\ \leq \beta^\star & \rightarrow \text{ item is not recommended to the user,} \\ \in (\beta^\star, \alpha^\star) & \rightarrow \text{ the user is consulted.} \end{cases}$$

$$(6.37)$$

Also that system has been tested on the above mentioned Movie-Lens dataset [35].

While random forests are typically employed in situations when additional, mostly demographic information is available, they can also be used to solve the problem of obtaining such an information. The most natural way of obtaining demographic information, namely inquiring of the user, has two serious disadvantages:

- Some of the demographic data (e.g., age, income level) may be perceived as sensitive and the user may feel uncomfortable giving them.
- The inquiry may be perceived as disturbing and/or boring.

Therefore, in [42], a different approach has been used: Possible demographic profiles were divided into a finite number of classes, and a random forest was used to classify users to those classes, based on their clickstream data, such as visited pages and clicked links.

6.2.5   Random Forests in Malware and Network Intrusion Detection

For malware detection, random forest is the most frequently encountered kind of classifier teams in the context of both most targeted operating systems—Windows and Android.

In [43], random forests have been used to detect malicious portable executables (PE) files. Individual trees were grown on random subspaces of the overall feature space. Recall from Sect. 6.1.7 that such a random subspace approach has been employed also to construct an ensemble of support vector machines for malware detection in the system *FENOC* [16]. The feature space in [43] is 376-dimensional and its individual dimensions correspond to the following two groups of features:

(i) 188 features describe the structure of the PE binary file, such as file header, section headers, anonymous sections, transferred control function, entry point and resource directory. They indicate malicious intentions especially if predefined sizes of headers have been changed.

(ii) 188 features record calls to API functions of the operating system. API calls are important in the context of the fact that the malicious actions can be performed only by means of certain API functions.

In [44, 45], a substantially richer set of features has been used for the detection of malicious PE files. In addition to the PE file structure and the operating system API calls, they record, in particular, also:

- information on packing, encryption and archiving;
- data concerning digital certificates;
- information whether structured exception handling, data execution prevention and address space layout randomization are used;
- information whether the file contains rigidly fixed IP addresses;
- information whether direct cookie links are present;
- information whether databases are in use.

Apart from a random forest, that feature set has been in [44, 45] used also for malware detection by a multilayer perceptron with one hidden layer.

Stacked random forests have been used for Windows malware detection in [46]. Apart from stacking, that ensemble classifier has several other specific properties:

(i) Although it does not cover all the features encountered in [43–45] (for example, from the Intel x86 instruction set, it uses only 280 instructions based on their frequency in malicious applications), its feature set is actually much more powerful because it includes all unigrams and bigrams of bytes of the classified binary file.

(ii) Feature selection is performed stepwise: the classifier starts with no features, and then gradually augments the feature set by adding them one by one.

(iii) The ensemble attempts to detect malware even from packed binary files, without unpacking them beforehand. This is based on the assumption that malware packers commonly use weak encryption and compression algorithms. In such a case, certain statistical and structural properties are preserved, allowing classification even without unpacking.

(iv) The ensemble attempts to classify not only malware and benign software, but also different malware families. However, it does not manage classification of zero-day malware that is dissimilar to the previously seen malware.

Using random forests for the detection of Android malware, more precisely, for static malware detection, has been reported in [47]. As static features, primarily permissions requested by the classified app are used: altogether 401, although only 135 of them are documented in the Android specification. The permissions are further complemented by 55 hardware and software features, all of them documented in the specification. From all those 456 features, 136 remain after feature selection. A specificity of the malware detector described in [47] is that the detection itself is performed off-device. In the device, only the monitoring whether new apps are installed or existing ones upgraded is performed, using a broadcast receiver component.

Whereas the detector in [47] is an example of a detector based on static malware analysis, the Android malware detector described in [48] uses exclusively dynamic malware analysis. The classified app is run in isolation on an Android virtual machine, together with a tool that generates pseudo-random streams of user events such as clicks or gestures. Traced system calls from a given period of interactions of the app with that tool are collected, and the frequencies of different system calls form the feature vector. Apart from random forests, the authors of [48] have used those features also for classification with classification trees, though they achieve a lower accuracy (using the implementations in the system Weka [49], the accuracy of random forests was 88%, whereas that of classification trees was only 85%).

A comprehensive approach to the detection of Android malware based on both static and dynamic analysis, denoted *Ensemble Clustering and Classification (EC2)*, has been presented in [50]. Similarly to the detector in [46], it attempts not only to detect malware, but also to classify it to families encountered in training data. Differently to [46], it is successful also in detecting malware not belonging to any of the families on which it was trained. As to the features obtainable with static malware analysis, the EC2 relies solely on the rich and easily interpretable information contained in the app header, aka manifest, primarily because it must be in plain text, hence the contained information cannot be hidden through encryption. Features obtainable from the manifest can be divided into three groups:

(i) Permissions. The EC2 uses 183 permissions, mostly from the official Android documentation, but also some customized permissions defined by app developers. The EC2 testing has shown that both the presence and the absence of certain permissions can be a strong indicator of belonging to a particular malware family.

(ii) App components. Apps sharing similar structure may be variants of the same family. The EC2 does not consider the names of the components because these are just the names of classes defined by the app developer and hence can be easily altered, but instead their number and hierarchical organization. More generally, this group of features includes also the size of the app file.

(iii) Malware of the same family may be developed by the same author. For app from the Google Play marketplace, the developer information is published there. For third-party marketplaces, however, this does not need to be the case. Therefore, the EC2 assumes that apps digitally signed with the same certificate have been developed by the same author.

To perform dynamic malware analysis, the classified app runs for 120 seconds in the sandbox DroidBox [51], without simulating any user input. The EC2 then uses unigrams, bigrams and trigrams mainly of the following kinds of events recorded by the sandbox:

- file reading and writing activities;
- sending and receiving network activities;
- using cryptographic primitives;
- dynamic loading of classes;
- starting services as background processes;
- generating system events.

The name Ensemble Clustering and Classification reflects the fact that the approach uses clustering to alleviate problems with small families. This is particularly important in the context of using random forests for malware classification. The authors of [50] compared random forests with 5 other kinds of classifiers (nearest neighbours, naïve Bayes, logistic regression, SVM and classification trees), as well as with 4 existing malware detectors in terms of a performance measure generalizing the F-measure for multiclass classification. While random forests were the best for malware families of size ≥ 10, they were inferior to some others for smaller families. Principally, the EC2 combines random forests with clustering in the following way:

1. A clustering method is used to cluster all available apps, no matter whether they belong to the training set or to the test set.
2. A random forest viewed as a fuzzy classifier is trained on the training set.

3. For each app from the test set, its membership in each family is predicted.
4. If the maximum of the predicted memberships of an app from the test set in the different families is higher than a given threshold $\delta_1 > 0$, then the app is assigned to that family. Otherwise, it is considered unlabelled.
5. If more than a given proportion $\delta_2 \in (0, 1)$ of a cluster C have been assigned to a single family, then the whole C is assigned to that family. Otherwise, a new family is created for the apps in C. This allows on the one hand discovering salient features of each family, on the other hand focusing the work of security analysts on the cluster or clusters forming new families.

The EC2 has been tested on two publicly available datasets of Android malware with labelled families:

(i) The *DREBIN dataset* [52] is the largest public dataset of that kind. It contains 156 malware families, which are fine-grained and always quite specifically delineated: For example, the family FakeInstaller contains apps that pretend to install or uninstall other apps from the device, whereas their primary purpose is to send premium SMS without user consent, or the family MobileTx contains trojans that both steal information and also try to send premium SMS. Main consequences of the specificity of DREBIN families are:

- The specific features allowing to discriminate a family from the others. Also the features allowing to discriminate a family from the others are specific. Returning to the two examples above, the former is better distinguished through static features, whereas the latter through dynamic features.
- Most of the malware families in the DREBIN dataset are small. From the 156 families, 112 contain less than 10 samples.

(ii) The *Koodous dataset* [53] is more recent than DREBIN, but smaller. It contains only 7 malware families, which are not as small as in DREBIN (the smallest one includes 31 apps) and are much more coarse-grained, e.g., SMS fraud, adware, information theft, or ransomware. The performance of the EC2 on Koodous is worse than on DREBIN, which might have the following reasons:

- Classification into the coarse-grained Koodous families is more difficult because they are less specifically delineated than the DREBIN families.
- Clustering is less helpful if the families are larger.
- Since the malware apps in the Koodous dataset are more recent, they use more sophisticated obfuscation techniques, making them more difficult to detect.

Finally, [54] and [55] deal with the application of random forests to intrusion detection. In [54], a network intrusion detection system is described, which consists of two layers – the lower *anomaly detection layer* and the upper *classification layer*. In the lower layer, an anomaly score is assigned to all the network traffic, using more than 30 anomaly detectors, based on empirical estimates of probabilities and conditional probabilities, time series analysis, and specific knowledge of network protocols. The 10 % of network traffic with the highest anomaly score are then passed to the upper layer, where classification by means of a random forest is performed. The random forest uses 357 features describing network users and network traffic. According to [54], that IDS is used in more than 500 enterprise networks, ranging from small to large companies with tens of thousands of users.

The random forest in the classification layer uses the aggregation function *Bayesian tree aggregation*, which determines the resulting class c^* as

$$c^* = \arg\max_{c \in C} P(\phi_1(x), \ldots, \phi_T(x)|c) \tag{6.38}$$

To make the computation of (6.38) tractable, the authors of [54] make an assumption inspired by (3.31), that the predictions of individual trees are independent given the true class label c. Consequently,

$$c^* = \arg\max_{c \in C} P(c) \prod_{k=1}^{T} P(\phi_k(x)|c). \tag{6.39}$$

As the probability and conditional probabilities in (6.39) are unknown, we use their estimates based on the training data $(x_1, c_1), \ldots, (x_p, c_p)$ to determine c^*:

$$c^* = \arg\max_{c \in C} \hat{P}(c) \prod_{k=1}^{T} \hat{P}(\phi_k(x)|c), \tag{6.40}$$

where

$$\hat{P}(c) = \frac{|\{j|c_j = c\}|}{p}, \quad \hat{P}(\phi_k(x)|c) \frac{|\{j|c_j = \phi_k(x) = c\}|}{|\{j|c_j = c\}|}. \tag{6.41}$$

Due to the finite number of data, it can easily happen that $\hat{P}(\phi_k(x)|c) = 0$ for some $k = 1, \ldots, T, c \in C$, which would make (6.40) unusable. In [2], several possibilities to deal with this problem have been presented, the most simple being to set $\hat{P}(\phi_k(x)|c)$ to some small value $\varepsilon > 0$ in that case.

In [54], Bayesian tree aggregation has been compared with majority voting on extensive real world data. In particular, training data consisted of proxy logs recorded during October and November 2015. Among them, the negative logs were uniformly downsampled, whereas all the 10 % of data with the highest anomaly score were kept. This resulted in 3500000 training instances, out of which 150000 were attributed to the positive class. For testing, on the other hand, data from a single busy working day in January 2016 were used and no downsampling was performed, yielding 10900000 test instances. Although the random forest classified individual proxy logs, the evaluation has been performed on the level of users, a user being considered infected if any of the proxy logs related to his or her communication was labelled as positive. Evaluation on the user level has been adopted because it reflects the efficacy of the system perceived by the customer, who is interested mainly in the number of user devices that have to be maintained, and not in specific network communications of a device. The comparison has shown that Bayesian tree aggregation has a slightly lower precision than majority voting, but a substantially higher recall. Most importantly, also the F-measure, i.e., the harmonic mean of precision and recall (2.22), is higher for Bayesian tree aggregation than for majority voting.

In [55], an approach to the detection of DDoS attacks is presented, relying on the following assumptions:

(i) Attack commands to the whole botnet are issued from a centralized command and control server through a particular communication channel.
(ii) After the bot establishes a connection to the command and control server, it primarily receives commands. Hence, there is substantially more communication from the command and control server to the bot than in the opposite direction.
(iii) Typically, the commands send from the command and control server to the same bot are similar. Therefore, the size of most packets is likely to be concentrated in an interval.

Due to these assumptions, a DDoS attack can be detected already at its early stage through detecting that some communication sessions between two computers in the network are actually sessions of communication between a command and control server and a bot. As features for that detection, characteristics of packets transferred in both directions within a communication session between the respective two computers are used. Examples of such features are for the respective session the amount of packets and the total size of packets, the average, variance, maximum, minimum of packet size, and the histogram of packet sizes.

References

1. Grabisch, M., Nguyen, H., Walker, E.: Fundamentals of Uncertainty Calculi with Applications to Fuzzy Inference. Kluwer Academic Publishers, Dordrecht (1994)
2. Kuncheva, L.: Combining Pattern Classifiers: Methods and Algorithms. Wiley, New York (2004)
3. Jacobs, R.: Methods for combining experts' probability assessments. Neural Comput. 7, 867–888 (1995)
4. Nowlan, S., Hinton, G.: Evaluation of adaptive mixtues of competing experts. In: Methods For Combining Experts' Probability Assessments, pp. 774–7780 (1991)
5. Smyth, P., Wolpert, D.: Linearly combining density estimators via stacking. Mach. Learn. 36, 59–83 (1999)
6. Wolpert, D.: Stacked generalization. Neural Netw. 5, 241–259 (1992)
7. Štefka, D., Holeňa, M.: Dynamic classifier aggregation using interaction-sensitive fuzzy measures. Fuzzy Sets Syst. 270, 25–52 (2015)
8. Freund, Y., Schapire, R.: A decision-theoretic generalization of on-line learning and an application to boosting. J. Comput. Syst. Sci. 55, 119–139 (1997)
9. Atrey, P., Hossain, M., El Saddik, A., Kankanhalli, M.: Multimodal fusion for multimedia analysis: a survey. Multimed. Syst. 16, 345–379 (2010)
10. Myers, G., Nallapati, R., van Hout, J., Pancoast, S., Nevatia, R., Sun, C., Habibian, A., Koelma, D., van de Sande, K.E., Smeulders, A., Snoek, C.: Evaluating multimedia features and fusion for example-based event detection. Mach. Vis. Appl. 25, 17–32 (2014)

11. Sheen, S., Anitha, R., Natarjan, V.: Android based malware detection using a multifeature collaborative decision fusion approach. Neurocomputing **151**, 905–912 (2015)
12. Wang, W., Li, Y., Wang, X., Liu, J., Zhang, X.: Detecting Android malicious apps and categorizing benign apps with ensemble of classifiers. Future Gener. Comput. Syst. **78**, 987–994 (2018)
13. Ouyang, L., Dong, F., Zhang, M.: Android malware detection using 3-level ensemble. In: IEEE CCIS, pp. 393–397 (2016)
14. Lille, D., Coppens, B., Raman, D., De Sutter, B.: Automatically combining static malware detection techniques. In: MALWARE 2015: 10th International Conference on Malicious and Unwanted Software, pp. 48–55 (2016)
15. Samson, F., Vaidehi, V.: Hybrid botnet detection using ensemble approach. J. Theor. Appl. Inf. Technol. **95**, 1646–1654 (2017)
16. Liu, J., Song, J., Miao, Q., Cao, Y., Quan, Y.: An ensemble cost-sensitive one-class learning framework for malware detection. Int. J. Pattern Recognit. Artif. Intell. **29**, 1550,018/1–27 (2015)
17. Cao, Y., Miao, Q., Liu, J., Li, W.: Osiris: a malware behavior capturing system implemented at virtual machine monitor layer. Math. Probl. Eng. **2013**, 402,438/1–11 (2013)
18. Ye, Y., Chen, L., Wang, D., Li, T., Jiang, Q., Zhao, M.: SBMDS: an interpretable string based malware detection system using SVM ensemble with bagging. J. Comput. Virol. Hacking Techn. **5**, 283–293 (2009)
19. Vatamanu, C., Cosovan, D., Gavriluţ, D.: Perceptron-based ensembles and binary decision trees for malware detection. In: International Conference on Artificial Neural Networks, pp. 250–259 (2017)
20. Zao, M., Zhai, J., He, Z.: Intrusion detection system based on support vector machine active learning and data fusion. In: ISICA, LNCS 6382, pp. 272–279. Springer (2010)
21. Breiman, L.: Random forests. Mach. Learn. **45**, 5–32 (2001)
22. Karpievitch, Y., E.G., H., Leclerc, A., Dabney, A., Almeida, J.: An introspective comparison of random forest-based classifiers for the analysis of cluster-correlated data by way of RF++. PLoS One **4**, e7087 (2009)
23. Kushmerick, N.: Learning to remove internet advertisments. In: ACM Conference on Autonomous Agents, pp. 175–181 (1999)
24. Özuysal, M., Calonder, M., Lepetit, V., Fua, P.: Fast keypoint recognition using random ferns. IEEE Transa. Pattern Anal. Mach. Intell. **32**, 448–461 (2010)
25. De Barr, D., Wechsler, H.: Spam detection using clustering, random forests, and active learning. In: Sixth Conference on E-Mail and Anti-Spam (2009)
26. Shen, J., Deng, R., Cheng, Z., Nie, L., Yan, S.: On robust image spam filtering via comprehensive visual modeling. Pattern Recognit. **48**, 3227–3238 (2015)
27. Breiman, L., Friedman, J., Olshen, R., Stone, C.: Classification and Regression Trees. Wadsworth, Belmont (1984)
28. De Barr, D., Ramanathan, R., Wechsler, H.: Phishing detection using traffic behavior, spectral clustering, and random forests. In: Intelligence and Security Informatics, pp. 67–72 (2013)
29. Meda, C., Bisio, F., Zunino, R.: A machine learning approach for twitter spamers detection. In: 2014 International Carnahan Conference on Security Technology, pp. 1–6 (2014)
30. Deshmukh, P., Shelar, M., Kulkarni, N.: Detecting of targeted malicious email. In: IEEE Global Computing on Wireless Computing and Networking, pp. 199–202 (2014)
31. O'Mahony, M., Cunnigham, P., Smyth, B.: An assessment for machine learning techniques for review recommendation. In: Coyle, L., Freyne, J. (eds.) Artificial Intelligence and Cognitive Science, pp. 241–250. Springer (2010)
32. Quinlan, J.: C4.5: Programs for Machine Learning. Morgan Kaufmann Publishers (1993)
33. Tripadvisor. https://www.tripadvisor.com
34. Zhang, H.R., Min, F., He, X.: Aggregated recommendation through random forests. Scientific World J. **2014**, 649,596/1–11 (2014)
35. Movie-lens. http://www.grouplens.org

36. Hussein, A., Omar, W., Li, X.: Efficient chronic disease diagnosis prediction and recommendation system. In: EMBS: IEEE International Conference on Biomedical Engineering and Sciences, pp. 2209–214 (2012)
37. Ostuni, V., Di Noia, T., Di Sciascio, E., Mirizzi, R.: Top-N recommendations from implicit feedback leveraging linked open data. In: RecSys'13—7th ACM Conference on Recommender Systems, pp. 85–92 (2013)
38. Bizer, C., Heath, T., Berners-Lee, T.: Linked data—the story so far. Int. J. Semant. Web Inf. Syst. **5**, 1–22 (2009)
39. Dbpedia. http://dbpedia.org
40. Last.fm. http://www.lastfm.com
41. Zhang, H., Min, F.: Three-way recommender systems based on random forests. Knowl. Based Syst. **91**, 275–286 (2016)
42. De Bock, K., Van den Poel, D.: Predicting website audience demographics for web advertising targeting using multi-website clickstream data. Fundam. Informaticae **98**, 49–70 (2010)
43. Krawczyk, B., Woźniak, M.: Evolutionary cost-sensitive ensemble for malware detection. Adv. Intell. Syst. Comput. **299**, 433–442 (2014)
44. Kozachok, A., Bochkovb, M., Kochetkova, E.: Heuristic malware detection mechanism based on executable files static analysis. In: 3rd International Conference Information Technology and Nanotechnology, pp. 132–139 (2017)
45. Kozachok, A., Kozachok, V.: Construction and evaluation of the new heuristic malware detection mechanism based on executable files static analysis. J. Comput. Virol. Hacking Techn. **14**, 225–231 (2018)
46. Zhang, Y., Huang, Q.M.X., Yang, Z., Jiang, J.: Using multi-features and ensemble learning method for imbalanced malware classification. IEEE TrustCom/BigData SE/ISPA **2016**, 965–973 (2017)
47. Morales-Ortega, S., Escamilla-Ambrosio, P., Rodriguez-Mota A. Coronad-De-Alba, D.: Native malware detection in smartphones with Android OS using static analysis, feature selection and ensemble classifiers. In: MALWARE 2016: 11th International Conference on Malicious and Unwanted Software, pp. 67–74 (2017)
48. Bhatia, T., Kaushal, R.: Malware detection in Android based on dynamic analysis. In: International Conference on Cyber Security And Protection Of Digital Services, p. article no. 8074847 (2017)
49. Witten, I., Frank, E., Hall, M.: Data Mining: Practical Machine Learning Tools and Techniques, 4th Edition. Morgan Kaufmann Publishers (2016)
50. Chakraborty, T., Pierazzi, F., Subrahmanian, V.: EC2: Ensemble clustering and classification for predicting Android malware families. IEEE Trans. Dependable Secure Comput. **15**, 913–928 (2018)
51. Dorneanu, V.: Android dynamic code analysis—mastering DroidBox (2014). Blog post on http://blog.dornea.nu
52. Arp, D., Spreityenbarth, M., Hubner, M., Gascon, H., Rieck, K.: DREBIN: Effective and explainable detection of Android malware in your pocket. In: 21st Annual Network and Distributed System Security Symposium, p. 12 pages (2014)
53. Mannella, L.: Heuristics and evolutionary algorithms for Android malware signature optimization. Master's thesis, Technical University of Torino (2018)
54. Brabec, J., Machlica, L.: Decision-forest voting scheme for classification of rare classes in network intrusion detection. In: IEEE International Conference on Systems, Man and Cybernetics, pp. 3325–3330 (2018)
55. Lu, L., Feng, Y., Sakurai, K.: C&C session detection using random forest. In: 11th International Conference on Ubiquitous Information Management and Communication, p. 6 pages (2017)

Chapter 7
Conclusion

This book has brought an overview of main classification methods in the context of six important kinds of internet applications: spam filtering, recommender systems, sentiment analysis, example-based web search, malware detection, and network intrusion detection. Those kinds of applications have occurred in the book in two roles: In the first chapter, a survey of all of them has been given, at the same time indicating why classification is needed for each of them. Then in Chaps. 2–6, particular internet applications of the explained classification methods have been reviewed.

Due to the fast development of internet in general and of the considered applications in particular, the parts of the book concerning them won't stay up-to-date for a longer time. However, the principles of the presented classification methods will, as well as their applicability to the considered internet applications. Moreover, we expect that more advanced and more general versions of classification methods can come into use in such applications in the future. Therefore, the book has included a number of advanced and general modifications of the presented methods, even if some of them cannot be witnessed by particular internet applications yet.

© Springer Nature Switzerland AG 2020
M. Holeňa et al., *Classification Methods for Internet Applications*,
Studies in Big Data 69, https://doi.org/10.1007/978-3-030-36962-0_7

Printed in the United States
By Bookmasters